Rail Infrastructure Resilience

Woodhead Publishing Series in Civil and Structural Engineering

Rail Infrastructure Resilience

A Best-Practices Handbook

Edited by

Rui Calçada
Sakdirat Kaewunruen

Woodhead Publishing is an imprint of Elsevier
50 Hampshire Street, 5th Floor, Cambridge, MA 02139, United States
The Boulevard, Langford Lane, Kidlington, OX5 1GB, United Kingdom

Copyright © 2022 Elsevier Inc. All rights reserved.

No part of this publication may be reproduced or transmitted in any form or by any means, electronic or mechanical, including photocopying, recording, or any information storage and retrieval system, without permission in writing from the publisher. Details on how to seek permission, further information about the Publisher's permissions policies and our arrangements with organizations such as the Copyright Clearance Center and the Copyright Licensing Agency, can be found at our website: www.elsevier.com/permissions.

This book and the individual contributions contained in it are protected under copyright by the Publisher (other than as may be noted herein).

Notices

Knowledge and best practice in this field are constantly changing. As new research and experience broaden our understanding, changes in research methods, professional practices, or medical treatment may become necessary.

Practitioners and researchers must always rely on their own experience and knowledge in evaluating and using any information, methods, compounds, or experiments described herein. In using such information or methods they should be mindful of their own safety and the safety of others, including parties for whom they have a professional responsibility.

To the fullest extent of the law, neither the Publisher nor the authors, contributors, or editors, assume any liability for any injury and/or damage to persons or property as a matter of products liability, negligence or otherwise, or from any use or operation of any methods, products, instructions, or ideas contained in the material herein.

ISBN: 978-0-12-821042-0 (print)
ISBN: 978-0-12-821043-7 (online)

For information on all Woodhead publications
visit our website at https://www.elsevier.com/books-and-journals

Publisher: Matthew Deans
Acquisitions Editor: Glyn Jones
Editorial Project Manager: Emily Thomson
Production Project Manager: Kamesh Ramajogi
Cover Designer: Greg Harris

Typeset by STRAIVE, India

 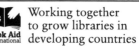

Contents

Contributors	xi
About the Editors	xv
Foreword	xvii
Acknowledgments	xix

1 Introduction   1
Rui Calçada and Sakdirat Kaewunruen
- 1.1 Background   1
- 1.2 Railway infrastructure resilience   2
- 1.3 Advanced condition monitoring   3
- 1.4 Conclusions   3
- References   4

2 Railway vulnerability and resilience   5
Qing-Chang Lu, Pengcheng Xu, Xin Cui, and Jing Li
- 2.1 Railway system vulnerability and resilience analyses   5
- 2.2 Methodologies for railway vulnerability and resilience   10
- 2.3 Railway vulnerability and resilience practices in Chinese cities   19
- 2.4 Conclusions   32
- Acknowledgments   33
- References   33

3 Rail resilience to climate change: Embedding climate adaptation within railway operations   37
Emma J.S. Ferranti, Andrew D. Quinn, and David J. Jaroszweski
- 3.1 Introduction   37
- 3.2 Best practice in climate adaptation and resilience   44
- 3.3 Rail Adapt Framework for climate change adaptation   47
- 3.4 Conclusions   57
- References   58

4 Rail transport resilience to demand shocks and COVID-19   65
Erik Jenelius
- 4.1 Introduction   65
- 4.2 Rail transport demand shocks   66
- 4.3 Impacts of COVID-19 on rail demand and supply   67
- 4.4 Rail system resilience   70

4.5	Supply losses	70
4.6	Demand spikes	72
4.7	Demand losses	74
4.8	Increasing demand loss resilience	75
4.9	Concluding remarks	77
	References	78

5 Management of railway stations exposed to a terrorist threat — 81
Sakdirat Kaewunruen and Hamad Alawad

5.1	Introduction	81
5.2	Previous studies	83
5.3	Security risk analysis	84
5.4	Managing terrorism risk	87
5.5	The technology and terrorist threat	89
5.6	Emergency and preincident management	91
5.7	Conclusion	92
	References	93

6 Rail infrastructure systems and hazards — 97
Chayut Ngamkhanong, Keiichi Goto, and Sakdirat Kaewunruen

6.1	Extreme temperature	97
6.2	Earthquakes	102
6.3	Flooding	104
6.4	Summary	107
	References	107

7 Wheel-rail dynamic interaction — 111
Zhen Yang and Zili Li

7.1	Introduction	111
7.2	Modeling of wheel-rail dynamic interaction	114
7.3	Detection and maintenance	126
	References	131

8 Wheel-rail interface under extreme conditions — 137
Milan Omasta and Hua Chen

8.1	Introduction into the wheel-rail interface	137
8.2	Basics of the wheel-rail contact	138
8.3	The wheel-rail interface under extreme conditions	141
	References	156

9 Train and track interactions — 161
Wanming Zhai, Shengyang Zhu, and Stefano Bruni

9.1	Introduction: An overview of train and track interactions	161
9.2	Models of train and track interactions	162
9.3	Vehicle-track interaction due to differential subgrade settlement	168

	9.4	Vehicle-track interaction due to polygonal wheel under traction condition	171
	9.5	Vehicle-track interaction under extreme weather conditions	175
	9.6	Conclusions	177
	Acknowledgments		177
	References		177
10	**Approaches for weigh-in-motion and wheel defect detection of railway vehicles**		**183**
	Araliya Mosleh, Pedro Aires Montenegro, Pedro Alves Costa, and Rui Calçada		
	10.1	Introduction	183
	10.2	Propose approaches to obtain WIM and wheel defect detection	184
	10.3	Numerical modeling	187
	10.4	Results and discussion	196
	10.5	Conclusion	203
	Acknowledgments		204
	References		205
11	**Railway ground-borne vibrations: Comprehensive field test development and experimental validation of prediction tools**		**209**
	Aires Colaço, Alexandre Castanheira-Pinto, Pedro Alves Costa, and Rui Calçada		
	11.1	Introduction	209
	11.2	Experimental characterization of the Carregado test site	210
	11.3	Numerical modeling of ground-borne vibrations	230
	11.4	Experimental validation	234
	11.5	Conclusions	238
	Acknowledgments		239
	References		239
12	**Lateral resistance of different sleepers for the resilience of CWR tracks**		**243**
	Guoqing Jing and Peyman Aela		
	12.1	Introduction	243
	12.2	Fastening/sleeper resistance	246
	12.3	The effect of ballast specifications on the lateral resistance	247
	12.4	Influence of the sleeper type and shape on lateral resistance	250
	12.5	Numerical assessment of sleeper lateral resistance	261
	12.6	Ballast components contribution to lateral resistance for different sleepers	262
	12.7	Conclusion	262
	References		266

13	**Diagnostics and management methods for concrete sleepers**		**271**
	Dan Li, Sakdirat Kaewunruen, Alex Remennikov, and Ruilin You		
	13.1	Sleeper design process	272
	13.2	Prestressed concrete railway sleeper subject to dynamic load	273
	13.3	Fatigue assessment for prestressed concrete sleeper	278
	13.4	Rail seat abrasion	284
	13.5	Time-dependent behavior of prestressed concrete sleepers	285
	13.6	Summary	292
	References		292
14	**Railway ballast**		**295**
	Yunlong Guo, Valeri Marikine, and Guoqing Jing		
	14.1	Introduction	295
	14.2	Ballast degradation	299
	14.3	Ballast inspection and assessment	306
	References		310
15	**Railway turnouts and inspection technologies**		**319**
	Mehmet Z. Hamarat, Mika Silvast, and Sakdirat Kaewunruen		
	15.1	Introduction	319
	15.2	Components of turnouts	320
	15.3	Inspection	325
	15.4	Maintenance	334
	15.5	Lifecycle cost	338
	15.6	Conclusions	339
	Acknowledgments		339
	References		339
16	**Risk-based maintenance of turnout systems**		**341**
	Serdar Dindar and Sakdirat Kaewunruen		
	16.1	Introduction	341
	16.2	Railway turnouts	342
	16.3	Identification of a risk analysis method	343
	16.4	Establishment of risk-based maintenance	349
	16.5	Environmental impact consideration into a maintenance chain	350
	16.6	Concluding remarks	351
	References		352
17	**Railway bridge under increased traffic demands**		**355**
	Shervan Ataei and Amin Miri		
	17.1	Introduction	355
	17.2	Monitoring program	357
	17.3	Processing data	363

	17.4	Analysis of results	364
	17.5	Conclusion	380
	References		386
18	**Structural health monitoring strategy for damage detection in railway bridges using traffic induced dynamic responses**		**389**
	Andreia Meixedo, Diogo Ribeiro, João Santos, Rui Calçada, and Michael Todd		
	18.1	Introduction	389
	18.2	Literature review on SHM for damage detection	390
	18.3	Railway bridge over the Sado River	394
	18.4	Strategy for damage detection using train induced dynamic responses	398
	18.5	Conclusions	405
	Acknowledgments		406
	References		406
19	**Improved dynamic resilience of railway bridges using external dampers**		**409**
	Sarah Tell, Andreas Andersson, Amirali Najafi, Billie F. Spencer, Jr., and Raid Karoumi		
	19.1	Dampers for structural vibration mitigation	409
	19.2	Equation of motion for a bridge with viscous damper	411
	19.3	Real-time hybrid simulation and testing	414
	19.4	Response of the bridge-damper systems	415
	19.5	Laboratory testing of an FVD	418
	19.6	Comparison of damper performance	419
	19.7	Conclusions	422
	Acknowledgments		422
	References		422
20	**Responses of mast structure and overhead line equipment (OHLE) subjected to extreme events**		**425**
	Chayut Ngamkhanong, Sakdirat Kaewunruen, Rui Calçada, and Rodolfo Martin		
	20.1	Introduction	425
	20.2	Maintenance criteria	427
	20.3	Vibration characteristics of OHLE	428
	20.4	OHLE under harsh environment	431
	20.5	Summary	437
	References		438
21	**Reliability quantification of the overhead line conductor**		**441**
	Sakdirat Kaewunruen, Chayut Ngamkhanong, and Jiabao Jiang		
	21.1	Introduction	441
	21.2	Concept of reliability analysis	443

21.3	Load calculations	**447**
21.4	Results and discussions	**453**
21.5	Conclusions	**460**
Data availability		**461**
Acknowledgments		**461**
References		**461**

Index **463**

Contributors

Peyman Aela Beijing Jiaotong University, Beijing, China

Hamad Alawad University of Birmingham, Birmingham, United Kingdom

Andreas Andersson Royal Institute of Technology (KTH), Stockholm, Sweden

Shervan Ataei School of Railway Engineering, Iran University of Science and Technology, Tehran, Iran

Stefano Bruni Department of Mechanical Engineering, Politecnico di Milano, Milano, Italy

Rui Calçada CONSTRUCT—LESE, Faculty of Engineering (FEUP), University of Porto, Porto, Portugal

Alexandre Castanheira-Pinto CONSTRUCT—LESE, Faculty of Engineering (FEUP), University of Porto, Porto, Portugal

Hua Chen Railway Technical Research Institute, Tokyo, Japan

Aires Colaço CONSTRUCT—LESE, Faculty of Engineering (FEUP), University of Porto, Porto, Portugal

Pedro Alves Costa CONSTRUCT—LESE, Faculty of Engineering (FEUP), University of Porto, Porto, Portugal

Xin Cui Chang'an University, Xi'an, China

Serdar Dindar Civil Engineering, Izmir Katip Celebi University, Izmir, Turkey

Emma J.S. Ferranti School of Engineering; Birmingham Centre for Railway Research and Education (BCRRE), University of Birmingham, Birmingham, United Kingdom

Keiichi Goto Railway Dynamic Division, Railway Technical Research Institute, Tokyo, Japan

Yunlong Guo Delft University of Technology, Delft, The Netherlands

Mehmet Z. Hamarat Department of Civil Engineering, School of Engineering, University of Birmingham, Birmingham, United Kingdom

David J. Jaroszweski Birmingham Centre for Railway Research and Education (BCRRE); School of Geography, Earth and Environmental Science, University of Birmingham, Birmingham, United Kingdom

Erik Jenelius Department of Civil and Infrastructural Engineering, KTH Royal Institute of Technology, Stockholm, Sweden

Jiabao Jiang China Construction Seventh Engineering Division Corp Ltd., Zhengzhou, China

Guoqing Jing Beijing Jiaotong University, Beijing, China

Sakdirat Kaewunruen Department of Civil Engineering; Birmingham Centre for Railway Research and Education, School of Engineering, University of Birmingham, Birmingham, United Kingdom

Raid Karoumi Royal Institute of Technology (KTH), Stockholm, Sweden

Dan Li Department of Civil Engineering; Birmingham Centre for Railway Research and Education, School of Engineering, University of Birmingham, Birmingham, United Kingdom

Jing Li Chang'an University, Xi'an, China

Zili Li Delft University of Technology, Delft, The Netherlands

Qing-Chang Lu Chang'an University, Xi'an, China

Valeri Marikine Delft University of Technology, Delft, The Netherlands

Rodolfo Martin Evoleo Technology Pty Ltd., Porto, Portugal

Andreia Meixedo CONSTRUCT-LESE, Faculty of Engineering, University of Porto, Porto, Portugal

Amin Miri School of Railway Engineering, Iran University of Science and Technology, Tehran, Iran

Pedro Aires Montenegro CONSTRUCT—LESE, Faculty of Engineering (FEUP), University of Porto, Porto, Portugal

Contributors xiii

Araliya Mosleh CONSTRUCT—LESE, Faculty of Engineering (FEUP), University of Porto, Porto, Portugal

Amirali Najafi University of Illinois at Urbana-Champaign, Urbana-Champaign, IL, United States

Chayut Ngamkhanong School of Engineering, University of Birmingham, Birmingham, United Kingdom; Department of Civil Engineering, Faculty of Engineering, Chulalongkorn University, Bangkok, Thailand

Milan Omasta Brno University of Technology, Brno, Czech Republic

Andrew D. Quinn School of Engineering; Birmingham Centre for Railway Research and Education (BCRRE), University of Birmingham, Birmingham, United Kingdom

Alex Remennikov School of Civil, Mining and Environmental Engineering, University of Wollongong, North Wollongong, NSW, Australia

Diogo Ribeiro CONSTRUCT-LESE, School of Engineering, Polytechnic of Porto, Porto, Portugal

João Santos LNEC, National Laboratory for Civil Engineering, Lisbon, Portugal

Mika Silvast Loram Finland Oy, Tampere, Finland

Billie F. Spencer, Jr. University of Illinois at Urbana-Champaign, Urbana-Champaign, IL, United States

Sarah Tell Royal Institute of Technology (KTH), Stockholm, Sweden

Michael Todd Department of Structural Engineering, University California San Diego, San Diego, CA, United States

Pengcheng Xu Chang'an University, Xi'an, China

Zhen Yang Delft University of Technology, Delft, The Netherlands

Ruilin You Railway Engineering Institute, China Academy of Railway Sciences, Beijing, China

Wanming Zhai State Key Laboratory of Traction Power, Southwest Jiaotong University, Chengdu, People's Republic of China

Shengyang Zhu State Key Laboratory of Traction Power, Southwest Jiaotong University, Chengdu, People's Republic of China

About the Editors

Rui Calçada, PhD, CEng, is Full Professor in the Faculty of Engineering at the University of Porto (FEUP) in Portugal, Head of the Civil Engineering Department, Coordinator of the CSF-Centre of Competence in Railways, and Member of the Scientific Council of FEUP. He received his PhD and Habilitation in civil engineering from the University of Porto. His research interests include advanced models for analysis of the train–infrastructure dynamic interaction, dynamic effects on bridges and transition zones, fatigue assessment of railway bridges, track–structure interaction, condition monitoring systems, and advanced algorithms for condition assessment of the railway infrastructure. He was responsible for the participation of FEUP in the European projects CAPACITY4RAIL, IN2RAIL, IN2TRACK2, and IN2TRACK3 and in the SwiTrack'EN consortia, an associated member of the Joint Undertaking SHIFT2RAIL. He is Member of the Board of Directors of the Portuguese Railway Platform (PFP) and President of Strategic Council of PFP. He is currently working on the European Commission's Rail Infrastructure Systems Engineering Network or RISEN project, which has received funding from the EU's Horizon 2020 program.

Dr. Sakdirat Kaewunruen, PhD, MBA, FIEAust, CPEng, NER, MPWI, RPEQ, is Reader in railway and civil engineering in the School of Civil Engineering at the University of Birmingham in the United Kingdom. He is also Coordinator of the RISEN project and a founding Director of the China-Europe Building Technology Association. He has extensive industry experience in the field of structural, civil, and track engineering in both industry and academia. With more than 14 years in the rail industry and regulatory environments prior to joining academia, he has an array of research interests, including rail engineering, track design, track components, structural and geotechnical engineering, and maintenance and construction. He received his PhD in civil engineering from the University of Wollongong in Australia. He has also completed an Emerging Leader Program with the John F. Kennedy School of Government at Harvard University. He has held visiting appointments at various institutions, including Massachusetts Institute of Technology (MIT), Railway Mechanics at Chalmers University of Technology in Gothenburg, Sweden, the University of Tokyo, the University of Illinois at Urbana-Champaign, and the Railway Technical Research Institute in Tokyo, Japan. He coordinates the EU-funded RISEN project and is a CI of S-CODE. He has led numerous projects, sits on various industry committees (including ISO standard committees), and has published widely in the field.

Foreword

Rail infrastructure resilience is fundamentally important to enable reliable and seamless operations through safer vehicle–track interactions, particularly under increased traffic and vulnerable situations due to natural and human-induced threats such as floods and terrorist acts, both of which seem to be increasing. Despite the fact that the operational conditions, regulatory requirements, environmental exposures, and maintenance practices of railways vary, many of the uncertainties with their consequences and thus the risks that the rail industry faces are globally quite similar for operators, regulators, and infrastructure managers. Therefore, immediate needs arise to decrypt and decentralize either specific or common solutions to the challenges posed by increased operational demands and extreme events, such as terrorism, extreme weather events, earthquakes, tsunamis, and similar factors. This book highlights the railway infrastructure resilience to additional operational demands and extreme weather conditions.

I am extremely pleased that these challenges have been emphasized within the Rail Infrastructure Systems Engineering Network (RISEN) project, which was funded by the European Commission (Grant No. 691135). The research collaboration and knowledge body in this area can be systemically used to timely address high-risk and high-impact issues in which the current state of practice is still insufficient. The book is the outcome of RISEN, aiming to gather state-of-the-art research outputs related to the most essential issues currently affecting the safety, reliability, and functionality of rail infrastructure systems worldwide.

The book is a state-of-the-art collection of a new body of knowledge focused on key research themes to improve insights into the resilience of rail infrastructure systems to climate change, extreme events from weather conditions, and future operational demands. The content of the book will significantly benefit practicing engineers, researchers, and graduate students who strive to improve the life cycle performance, response, and resilience of rail infrastructure systems.

Herbert H Einstein
Department of Civil and Environmental Engineering,
Massachusetts Institute of Technology, MA, USA.
February 2022.

Acknowledgments

The research and studies found in this book have been funded by the European Commission's RISEN project (Rail Infrastructure Systems Engineering Network), Grant Number 691135. The RISEN project aims to enhance knowledge creation and transfer of expertise using both international and intersectoral secondment mechanisms among European Advanced Rail Research Universities/SMEs and Non-EU, which are world-class rail universities. We are extremely grateful to all partners of the RISEN consortium. The European rail university group consists of the University of Birmingham (United Kingdom), University of Porto (Portugal), Politecnico di Milano (Italy), Delft University of Technology (The Netherlands), KTH Royal Institute of Technology (Sweden), Tampere University (Finland), Norwegian University of Science and Technology (Norway), Polytechnic University of Valencia (Spain), and Brno University of Technology (Czech Republic). Nonacademic and SME partners include MERMEC (France), Roadscanners (Finland), Loram (Finland), EVOLEO (Portugal and Germany), China Academy of Railway Sciences (China), and Japan Railway Technical Research Institute (Japan). International partners include the University of Illinois at Urbana-Champaign (USA), Massachusetts Institute of Technology (USA), University of California, Berkeley (USA), Tsinghua University (China), Southwest Jiaotong University (China), University of Wollongong (Australia), and Iranian University of Science and Technology (Iran). This project adds research skill mobility and innovation dimension to the existing bilateral collaborations between universities through research exchange, joint research supervision, summer courses, international training and workshops, and joint development of innovative inventions. The project spanned from 2016 to 2021. We are extremely grateful to the partner institutions for their support throughout the course of the project. We especially thank the European Commission Officer Eleftheria Lykouressi who was extremely kind and highly instrumental to the success of the project.

In the process of writing this book, Dr. Chayut Ngamkhanong (currently at Chulalongkorn University), Pasakorn Sengsri, Mohamad Ali Ridho, Hao Fu, Jessada Sressakoolchai, and other team members have contributed to a great deal of book formatting, visual and figure preparation, photographs, and editorial and data visualization. We express our heartfelt thanks to them. In addition, we thank the Elsevier editorial managers, who have been extremely kind and supportive in every stage of the book development.

Introduction

Rui Calçada[a] and Sakdirat Kaewunruen[b]
[a]CONSTRUCT—LESE, Faculty of Engineering (FEUP), University of Porto, Porto, Portugal,
[b]University of Birmingham, Birmingham, United Kingdom

1.1 Background

Recent natural disasters and man-made hazards have resulted in poor and inadequate responses from transport service providers and operators. The elevated risks and possible consequences strongly underpin an immediate need to tackle the challenges posed by extreme events from those natural and man-made disasters, such as terrorism, extreme weather events (rain, heat, snow, humidity), earthquakes, tsunamis, and so on. It is noted that the information available on the high-level effects of extreme weather on transport systems is currently improving, thanks to on-going research and development. However, the information on the actual consequences on rail infrastructure systems has not been adequately recorded and monitored. This is evident by numerous service cancelations and uncountable delays during extreme seasons. Also, the risk of climate change on the statistical deviations to extreme climate cannot be underestimated as it reinforces additional consequences onto exiting problems. Importantly, research collaboration and the knowledge body in this area are insufficient to systemically solve the high-risk and high-impact issues in the current state of practice.

This book aims to address some of the most essential issues currently affecting the safety, reliability, and functionality of European rail infrastructure systems. It is based on a collaborative project, Rail Infrastructure Systems Engineering Network (RISEN), whose emphasis is placed on the fundamental need to build a collective body of knowledge, collaboration, capacity, and capability of European researchers, governmental officers, and industry experts working on similar research themes aimed at redefining the response, resilience, and adaptation of railway and urban transport infrastructures using smart integrated systems. Thus, three critical areas of research have been identified in this book. The scope of this book can extend to organizations and staff who conduct research to improve life cycle performance, response and resilience of rail infrastructure systems to climate change and extreme events from natural and man-made hazards, and to future operational demands. The goal of this book is to establish a new body of knowledge focused on these key research themes to improve response and environmental resilience of rail infrastructure systems.

1.2 Railway infrastructure resilience

Emerging risks and their significant consequences with no sign of early warning are recently evidenced by many extreme events such as the Nepal earthquakes, the Madrid train bombing, etc. [1–4]. Much past research has emphasized the applications of technology toward solving front-line problems in the railway industry. Although practical knowledge has been developed alongside with corporate knowledge, the science and technology needed are still insufficient to innovate and revolutionize the railway industry from a fundamental principle viewpoint. Many fundamental issues, such as choice of materials, durability, capacity, engineering properties, functionality requirements, and design concepts, remain unchanged. Together, with a high turnover rate of technical staff within the rail industry worldwide, many incidents have been repeated causing high maintenance costs, service cancelations and delays, and even the loss of human lives due to catastrophic incidents. Environmental-friendly, resilient, and smart rail infrastructure will enhance future rail capacity and adaptability to climate change and extreme events due to either natural or man-made hazards.

This book compiles new findings that will help evaluate engineering requirements and performance of modern railway tracks and vehicle-track interaction to cater future demand for both passenger and freight services. The collaborative research network, RISEN, underpins the effort to build shared lessons learned in the industry as well as facilitate original and unified solutions to the practical problems associated with infrastructure resilience. The insights from the project have been collated to develop this best-practice handbook for restoring and improving railway infrastructure systems and engineering. In addition, novel work on climate change adaptation in railway and transport infrastructure using a "systems thinking approach" has been included to improve strategic planning, design, and maintenance of rail infrastructure systems to be more adaptive and resilient.

As an essential part in this best-practice handbook, the guidelines for climate change adaptation are a key part to establish appropriate preventative strategies for mitigating the effect of climate change on railway and urban infrastructures. The first part of the book will provide identification of asset types, track and operational parameters, environments, and customers. Then, the later part will highlight some of the groundbreaking research findings that will stimulate common ground and systems thinking approaches to solve infrastructure problems and enhance the monitoring of asset conditions. It is anticipated that the fundamental understanding of railway infrastructure systems in this book will be applied toward opportunities to apply and design for resilience. This can be done separately for passenger and freight transport as they have different modal parameters to consider.

This book also highlights some novel railway geotechnology research to integrate a more realistic model of ballast into train-track, train-turnout, and train-bridge analyses. This original concept will enable better maintenance, restoration, and resilience improvement methods for critical rail infrastructure. New findings to improve railway structure and wheel/rail interface, application of a systems approach to infrastructure resilience, and extreme event risk modeling are also included to share new reference resources for resilience improvement in European railway infrastructure systems as well as around the world. This part of the book is expected to (i) enhance future rail capacity and adaptability to climate change and extreme events due to either natural or man-made hazards; (ii) consider systems thinking approaches to understand trade-offs and multicriterion performance of railway and transport infrastructures; and (iii) develop original and new fundamental concepts constituting new methodology to optimize asset management frameworks without the expense of public safety.

1.3 Advanced condition monitoring

At present, aging railway infrastructure systems possess additional emerging risks due to their inability to provide early warning to maintainers so that critical components can be prioritized and managed in a timely manner. Therefore, an integrated research approach needs to be adopted to reinforce asset condition monitoring (bottom-up) and response prediction (top-down) of rail systems management, maintenance, and operation, thus providing safe and seamless railways. Novel smart sensors, wireless technologies, and on-board monitoring technology such as infrastructure-to-infrastructure and infrastructure-to-vehicle communications are critical to modernize railway infrastructure systems. Integration of synergized sensors in railway information models (RIMs) can revolutionize real-time asset maintenance, monitoring, and prioritization policy. This book will pave a pathway to enhance advanced condition monitoring for railway infrastructure systems in Europe and around the world.

The last part of this book discusses some enabling technologies aimed at (i) providing methodological approaches and practical tools for the condition-based management of the railway infrastructure and optimizing infrastructure reliability and availability while reducing whole-life cycle costs; (ii) developing methods for the automated and continuous management of diagnostic information to be made available by condition monitoring systems and allowing the best use of diagnostic information in the maintenance decision process; and (iii) exchanging best practices concerning condition monitoring and maintenance of the railway infrastructure across the partners of the RISEN network.

1.4 Conclusions

Railway infrastructure is a complex system connecting not only its own system but also linking with other transport modes and urban systems. As a result, research in this area is inter- and multidisciplinary by nature. This book has been designed to take "systems thinking" approaches into account in all aspects to improve railway infrastructure resilience and advanced condition monitoring. It is aimed at generating new paradigms and thinking approaches, assuring that the cross-disciplinary considerations will be embedded for resilience adaptation roadmaps, practical guidelines, and policy strategies. The goal of this new book is to further enhance those understandings and advance them to create and innovate new and step-change improvements in design methodologies, advanced monitoring and maintenance, and resilience of rail infrastructure systems. As a result, new state-of-the-art review is inevitable, but rather the focus will be on moving to resilience of the systems and advanced monitoring sensors. In some transport modes, such as road transport, methodologies to utilize weather-related information already exist. Similarly, the research agenda on rail transport is advanced by the EU through separate projects and programs from which experiences can be used to enrich this book without duplication of effort. This new book is primarily based on the collaborative research synergies through RISEN.

When it comes to understanding how the infrastructure system's design and operation works in real life, passenger and freight services are responding to different incentives. In passenger rail transport, the decision frameworks for infrastructure system engineering are based on public safety, reliability, performance, and resilience to minimize recovery time in case of emergencies and crises. In freight transport services, on the other hand, the engineering decision is driven by the cost and performance resulting from downtime of minerals or goods transported, including travel time and reliability, and also by the specific characteristics of goods transported (coal, iron ore, etc.). The European rail transport system consists of several service modes, such as passenger, high-speed passenger, metro, and freight services. The shared corridor provides flexibility but also adds a critical maintenance issue along the route. With the complexity of railway systems in mind, this book aims to provide new fundamental insights covering all modes of rail transport services with both dedicated and shared corridors. Evidently, the novelty of the new approach is to combine resilience, sustainability, and advanced condition monitoring with intermodal railway infrastructure systems. Moreover, from the maintainers' and operators' perspective, of great importance is the existence of robust organizational and cooperative networks (among the various academia, industry sector, transport operators, and governments) that are based on a concrete collaborative structure capable of safeguarding public safety, reliability and cost advantage from better condition monitoring for rail passengers and users, and promoting "resilience" and "recovery readiness" to exposed risks from climate change and extreme events. This book has thus served as an exciting new platform of research outcomes derived from multidimensional collaboration to advance progress toward sustainable development of railway infrastructure systems globally.

References

[1] S. Kaewunruen, J.M. Sussman, A. Matsumoto, Grand challenges in transportation and transit systems, Front. Built Environ. 2 (2016) 4, https://doi.org/10.3389/fbuil.2016.00004.

[2] A. Matsumoto, M. An, C. Van Gulijk, S. Kaewunruen, Editorial: safety, risk and uncertainties in transportation and transit systems, Front. Built Environ. 5 (2019) 25, https://doi.org/10.3389/fbuil.2019.00025.

[3] S. Bruni, S. Kaewunruen, Editorial: best practices on advanced condition monitoring of rail infrastructure systems, Front. Built Environ. 6 (2020) 592913, https://doi.org/10.3389/fbuil.2020.592913.

[4] S. Bruni, S. Dindar, S. Kaewunruen, Editorial: best practices on advanced condition monitoring of rail infrastructure systems, volume II, Front. Built Environ. 7 (2021) 748846, https://doi.org/10.3389/fbuil.2021.748846.

Railway vulnerability and resilience

Qing-Chang Lu, Pengcheng Xu, Xin Cui, and Jing Li
Chang'an University, Xi'an, China

2.1 Railway system vulnerability and resilience analyses

Railway systems, consisting of rail and urban rail transit, have been developing rapidly and play an important role in daily travel, owing to their safety, efficiency, and convenient services provided to travelers. The importance of the railway system can be observed not only from the widespread rail network all over the world but also from the large passenger flow it serves. Thus, a railway system has to be robust under regular operation and resilient under disruptive events such as natural disasters, intentional attacks, and incidents. Such incidents would pose great threats on commuters' daily travel, resulting in a vulnerable rail network as well as economic losses and even death [1,2]. Consequently, there is growing interest in the analyses of railway system vulnerability and resilience in recent decades that investigate the effect on and recovery of a railway system under disruptive incidents.

2.1.1 Railway system vulnerability analysis

Since the devastating earthquake that shocked Kobe, Japan, in 1995, transportation network vulnerability has become an important concern for transportation researchers [3–5]. Although no agreement has been reached on the exact definition of "transportation network vulnerability," the vulnerability methodology is now well established by addressing its susceptibility of incidents and consequences under disruptions. Literature on transportation network vulnerability mainly contributes to the development of methodologies that measure consequences on network performance after disruption events. These methodologies can be categorized as an exposure-importance approach [4], an accessibility measure [6,7], a game theory method [8], and so on [9]. These methods are mainly applied to road networks at the beginning based on a network scan approach [7,10,11]. Researchers then try to overcome the disadvantage of computation time of the full scan method by either identifying links for further analysis based on certain criteria [12,13] or calculating the "impact area" of the affected link to downscale the network for analysis [14].

Although major efforts have been made on road networks, railway system vulnerability has received less attention. Only a few related researches in transportation vulnerability analysis address railway networks [15,16]. Different from the road network vulnerability analysis, railway system vulnerability is initially researched based on complex network theory that addresses the physical structure of a rail network [17,18]. Cats and Jenelius [19] extended the betweenness centrality measure to a dynamic and stochastic network and applied it to the rapid public transport system in Stockholm, Sweden, to identify candidate important links. This approach of vulnerability analysis would be important for the planning and design of a railway system. A limitation of this method is that it only measures the variations of average travel time, ignoring the distributions of passenger flow, which is incapable of capturing the impacts of rail incidents on passengers.

To reveal the changes of passenger flow characteristics, there are growing research interests in railway network vulnerability analysis addressing passengers' travel time and distance changes under incidents [20,21]. Rodríguez-Núñez and García-Palomares [22] proposed a methodology considering the changes of average travel time rather than physical network characteristics, evaluating the vulnerability of rail transit network in Madrid, Spain. Lu and Lin [21] explored the vulnerability of a rail transit network within a multimodal public transport network, emphasizing the relationship of passenger flow distributions between urban rail transit and bus transit networks. As a result, the development of methods addressing passenger flow characteristics in vulnerability analysis, in general, and the study of railway system analysis, in particular, has become important directions of research that have attracted much interest recently.

In addition, unlike other transportation systems, railway systems are greatly interdependent with land use around stations resulting from either transit-oriented development (TOD) or people's preference on rail travel. Besides network topology and passenger flow, railway system vulnerability may be also affected by land use variables interacting with it. For example, passengers living around suburban stations dominated by residential land use with fewer other travel alternatives depend much more on rail transit and thus are more vulnerable to rail incidents. As argued by Li et al. [23], land use influences people's travel behavior to a certain extent and should be considered in transportation analysis as one of the most significant factors. It is critical to understand the interrelations between rail stations and different combinations of land use patterns with ever-increasing TOD applications, in which differently combined land use patterns are usually indicated by the mixed land use degree index [24]. Land use characteristics impose specific spatial constraints for most, although not all, activities, and it has been used to build different kinds of travel demand models. Jiang et al. [20] developed an accessibility approach measuring the vulnerability of urban rail transit networks addressing land use impacts and rail passenger flow characteristics. It was found that land use should be included when evaluating the vulnerability of a railway system.

As reported by the previously discussed research works, the inclusion of passenger flow and land use characteristics around stations are particularly essential for the analysis of railway system vulnerability, especially in developing countries with a spreading railway network and increasing travel. Another finding is that a majority of vulnerability studies focus on the railway network in developed countries that would have small passenger demand variation, and thus the results might not be applicable to developing countries with a growing and changing rail ridership. When disruptive events occur, people may not only want to know the vulnerability of the railway network but also which stations were affected or unaffected; however, such information is rarely provided. Methodologies of vulnerability analysis usually treat railway systems independently without considering the interdependency nature between multimodal transit networks in reality, which would overestimate the vulnerability of a railway network under disruptions. People would transfer to other nearby transport modes if a rail station failed or was closed, and exclusive of this alternative in railway systems, vulnerability analysis may reach inaccurate results and conclusions.

Current research efforts contribute to methodologies evaluating rail system vulnerability and characteristics of rail network vulnerability in topology, passenger flow, land use, and so on. However, vulnerability analysis of railway systems is still facing great challenges due to its growing important role in population and cargo transportation around the world and the complex system of systems within multimodal transportation systems.

2.1.2 Railway system resilience analysis

Another perspective of a railway system under stress is resilience analysis, which is defined as the capability of a railway system to recover rapidly from a severe shock to achieve its original state [1]. However, network resilience shares a similar concept and methodology to network vulnerability analysis. Based on recent reviews of Faturechi and Miller-Hooks [25] and Mattsson and Jenelius [15], less extensive literature on transportation system resilience than vulnerability analysis was found. Methodologies addressing the resilience of a railway network are also discussed from the topological structure and functional measures.

Reggiani [26] highlighted the role of topological connectivity in the analysis of network resilience and outlined operational measures enhancing network resilience. Derrible and Kennedy [27] interpreted robustness more specifically as alternative paths and likelihood of accidents. Based on a functional measure, De-Los-Santos et al. [28] measured passengers' resilience by comparing the combined travel time and passenger flow before and after rail failures on the Madrid rail transit network. D'Lima and Medda [29] addressed the resilience of the London Underground network from the diffusive effects of shocks on passengers. Miller-Hooks et al. [30] agreed that the resilience of a transportation network should include both topological and operational ability to cope with disruptions. Zhang et al. [31] presented a broad concept of network resilience accounting for not only the system's ability to absorb changes but also adaptive actions that can be taken to preserve or restore network performance. Cats et al. [32] measured link criticality and rapid degradation in a public transport network robustness model connecting local capacity reductions to network-wide performance changes. Based on the accessibility theory, Lu [33] contributed to the modeling of the rail network resilience measure under different operational incidents, identifying the dynamic changes of rail network resilience and critical stations with the duration of time of incidents.

Most current research works analyze the resilience of a railway system based on link failure [12] by identifying and prioritizing critical links, but rail stations are more exposed to management and operational incidents because of the complexity of interacting facility systems and passenger flow at stations. Methodologies based on station failures would provide another supplement to network resilience analysis and practical implications for railway system management under incidents.

These reviews reveal several shortcomings in railway system resilience assessment models, which are summarized as follows:

- The limitation of a time-dependent approach

The railway system resilience analysis rarely considers the system performance evolution over time. For example, the number of passengers in the network changes over time. Thus, disruptions during off-peak hours do not have the same impact as disruptions in rush hours on the railway system.

- The limitation of a network of networks model

The network of networks approach and network interdependency among correlated transportation systems are insufficiently addressed in the literature of railway system resilience researches and practices.

- The limitation of multiincident disruption scenarios

Railway network resilience is usually evaluated under one disruptive event at a time. But critical incidents could occur simultaneously at several locations of the system. Thus, the evaluation of railway system resilience should include different types of simultaneous incidents.

- The limitation of inclusion of dynamics in network evolution and passenger flow

The resilience model of a railway system often assumes that passenger flow is unchanged before and after rail incidents, but passenger flow varies with time and incidents. Also, evolution of a rail network is usually ignored in the long run.

Moreover, few railway network resilience studies are carried out in developing countries, especially those with rapid development of the railway system and large passenger volumes.

2.1.3 The relationship between railway system vulnerability and resilience

According to a recent review of researches in transportation network vulnerability and resilience [15,34], the concept of resilience should be even more comprehensive to include recovery from disasters; the definition of railway system vulnerability suggested by Berdica [3] is the susceptibility to incidents that can result in considerable reductions in railway network serviceability. It could be found that fewer researches on transportation network resilience than vulnerability analysis are carried out, which is especially obvious for railway networks. However, railway network resilience shares a similar concept and methodology to network vulnerability analysis. Compared to vulnerability analysis, resilience analysis mainly provides a much broader sociotechnical framework to cope with infrastructure threats and disruptions, including preparedness, response, recovery, and adaptation stages [35]. Meanwhile, Hollnagel [36] emphasizes four cornerstones of resilience, including knowing what to do, what to look for, what to expect, and what has happened. Vulnerability analysis deals primarily with knowing what to expect, which is an important prerequisite for adequate proactive actions. The framework indicates the role of vulnerability studies in contributing to the overall goal of strengthening the resilience of a transport system.

Compared to road systems, the vulnerability and resilience of a railway system share a similar relationship. It is known that the majority of railway system analyses are situated on the vulnerability analysis, focusing on methodology developments by measuring consequences on network performance as shown by Stage 1 of the network performance curve in Fig. 2.1 [33]. Current vulnerability measures are mostly rooted in network topology and graph theory from the supply side, which neglects impacts on rail passengers from the demand side. However, as indicated in Fig. 2.1, network performance curves under incidents change not only in Stage 1 but also in Stages 2 and 3; network vulnerability analysis addresses only one part of the performance curves and thus has rare implications for the other two stages that are important for the recovery of a rail network. Moreover, network resilience analysis with a topological approach could hardly describe the recovery capability and rapidity of the network, that is, Stage 3, especially the accumulation and dispersion of delayed passengers. As indicated by Stage 2 of the curves in Fig. 2.1, which most resilience literature assumes as a horizontal line, network vulnerability analyses did not capture this stage. Once incidents occur, the network performance would still change with the time duration at Stage 2 instead of being unchanged. This unchanged or overlooked assumption would underestimate the consequences of incidents and overestimate the network performance.

In short, a railway system could be more vulnerable and less resilient under disruptions because of its low network redundancy but large daily passenger flow,

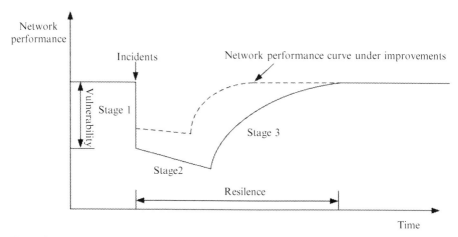

Fig. 2.1 Rail network performance curve under incidents.
Reproduced with permission based on: Q.C. Lu, Modeling network resilience of rail transit under operational incidents, Transp. Res. A Policy Pract. 117(11) (2018) 227–237.

especially in populated countries. The importance of a robust and reliable railway system from economic and passengers traveling perspectives has led to considerable research to understand the mechanisms and interrelationships of the system under disruptions.

2.2 Methodologies for railway vulnerability and resilience

2.2.1 Accessibility-based rail system vulnerability methodology

A large body of research work has contributed to the development and improvement of accessibility for various purposes [37,38]. Accessibility is also proven to be an important measure for transportation network vulnerability analysis [7,11,39], which was mostly developed for applications on road networks. The accessibility of public transport network has recently attracted much attention and has become an important field of research [40]. Therefore, a proposed accessibility-based rail system vulnerability method is being developed for failures of stations, links, and/or lines, while including land use characteristics around stations. Normally, stations are more exposed and vulnerable to disruptions, which have more complex passenger activities and land use features, as well as a higher probability of becoming terrorists' targets rather than links. The mathematical construct of the proposed methodology starts with a station-based accessibility measure as follows:

- *Station-based rail network accessibility*

Under emergent railway system disruptions, some passengers may reroute to unaffected lines to reach their destinations, whereas cannot find alternative routes except for other public transport services such as ground bus and taxi within walkable distance. The mobility to choose alternative routes/modes and the accessibility to arrive at destinations are both reduced for rail transit passengers whose routes are affected. Thus, one proposed method measured the accessibility of these two situations separating the affected and unaffected stations.

Let N be the set of all stations in a rail network. $N_W \subseteq N$ is the subset of working stations after disruption, and $N_D = N - N_W$ denotes the subset of station(s) disrupted. The location-based accessibility of station i could be formulated as:

$$A_i = \alpha SI_i^0 \left(\frac{\sum_{j \in N_W} AR_{ij} FR_{ij}}{\sum_{j \in N_W} AR_{ij}^0 FR_{ij}^0} \right)^{-1} + (1-\alpha) SI_i^0 \left(\sum_{j \in N_D} BR_j \left(\frac{AR_{ij} FR_{ij}}{AR_{ij}^0 FR_{ij}^0} \right) \right)^{-1} \quad (i \neq j, i \in N_w) \quad (2.1)$$

where A_i denotes the accessibility of rail station i, SI_i^0 represents station importance of rail station i among all the stations, AR_{ij}^0 is the passenger volume from rail station j to station i before network disruption, AR_{ij} denotes the passenger volume from rail station j to station i after network disruption, FR_{ij}^0 shows the travel cost of rail transit from station j to station i before network disruption, FR_{ij} represents the travel cost of rail transit from station j to station i after network disruption, FB_{ij} is the travel cost of a ground bus from rail station j to station i after network disruption, BR_j denotes the bus capacity ratio in terms of total passenger volume within walkable distance of rail station j, α is the balance factor of j between working stations and disrupted stations, and $\alpha=1$, if $j \in N_W$; $\alpha=0$, if $j \in N_D$.

The station importance is described as the total passenger flow using a station before network disruption; the more passengers who use this station, the more important the station becomes. Station importance here is defined as passengers arriving and departing a station and is constructed as:

$$I_i^0 = DE_i^0 + AR_i^0 \quad (i \in N) \tag{2.2}$$

$$SI_i^0 = \frac{I_i^0}{\sum_{i \in N} I_i^0} \quad (i \in N) \tag{2.3}$$

where DE_i^0 is the total departure passenger volume at rail station i before network disruption representing the generation of station i, AR_i^0 denotes the total arrival passenger volume at rail station i before network disruption showing the attractiveness of station i, and I_i^0 represents the total passenger volume of station i.

The bus capacity ratio is defined as the ratio of number of bus stops within walkable distance of station j divided by the total passenger volume of this station to the maximum bus stops each passenger has within the study area, which is calculated as Eq. (2.4).

$$BR_j = \frac{BS_j/I_j^0}{\text{Max}\left(BS_j/I_j^0\right)} \quad (j \in N) \tag{2.4}$$

where BS_j is the total bus stations within walkable distance of rail station j. This measures the relative capacity of buses to serve the rail transit passengers under station disruptions. This ratio measures the capacity of bus services accommodating delayed rail passengers.

As a result, the overall accessibility of working rail stations A_W could be calculated as:

$$A_w = \sum_{i \in N_W}$$

$$A_i = \sum_{i \in N_W} \alpha SI_i^0 \left(\frac{\sum_{j \in N_W} AR_{ij} FR_{ij}}{\sum_{j \in N_W} AR_{ij}^0 FR_{ij}^0}\right)^{-1}$$

$$+ (1-\alpha) SI_i^0 \left(\sum_{j \in N_D} BR_j \left(\frac{AR_{ij} FR_{ij}}{AR_{ij}^0 FR_{ij}^0}\right)^{-1}\right) \quad (i \neq j) \tag{2.5}$$

where A_W is the overall accessibility of working stations.

For station $i \in N_D$, passengers could not reroute to unaffected rail transit lines but instead have to take other transit modes such as buses to reach station i. The accessibility of station i is then proposed as:

$$A_i = SI_i^0 BR_i \left(\frac{\sum_{j \in N} AR_{ij} FR_{ij}}{\sum_{j \in N} AR_{ij}^0 FR_{ij}^0} \right)^{-1} \quad (i \neq j, i \in N_D) \tag{2.6}$$

The accessibility of closed stations is equal to the accessibility summation of all the disrupted stations, which is

$$A_D = \sum_{i \in ND} A_i = \sum_{i \in ND} SI_i^0 BR_i \left(\frac{\sum_{j \in N} AR_{ij} FR_{ij}}{\sum_{j \in N} AR_{ij}^0 FR_{ij}^0} \right)^{-1} \quad (i \neq j) \tag{2.7}$$

where A_D is the total accessibility of disrupted stations.

- *Rail network accessibility under station(s) failure*

Based on the interpretation of the previously discussed station-based accessibility, the network accessibility consists of the accessibility of both working stations and disrupted stations. Thus, if rail station(s) d is disrupted, the rail network accessibility could be formulated as:

$$A^d = A_w^d + A_D^d \tag{2.8}$$

where A^d is the rail network accessibility when station(s) d is closed, A_w^d denotes the accessibility of all the working stations when station(s) d is closed, A_D^d describes the accessibility of all the disrupted stations when station(s) d is closed, and d represents the station or set of stations failed and is nonnegative integer.

Eq. (2.8) calculates the rail transit network accessibility under station failure while including the availability and capacity of ground bus transportation for closed stations. This calculation allows no station failure scenario, that is, $d=0$ as well as single and multiple station disruptions. Partial station failure could also be included in this methodology when separating arrival and departure passenger volumes of closed and working lines at a station. Another advantage of this method is that, if $d=N$, most current vulnerability measures would have network performance values of zero since all the stations are disrupted, but in reality this is not the case and passengers could still travel using other transit modes, which could be addressed with the proposed accessibility measure.

- *Rail network accessibility under link(s) failure*

In the case of disruptions on a link of the rail network, if stations at the two ends of the link are also included, the network accessibility under link failure is constructed as:

$$A^L = A^d, \quad DC_d = \text{Max}(DC_m, DC_n) \quad (m, n \in N_L) \tag{2.9}$$

where A^L denotes the rail network accessibility when link L failed, DC_m represents the degree centrality of disrupted station m, which is calculated as the number of direct connections of m, with other stations, m and n are disrupted stations of link l, and N_L represents the station set of link l. If only link disruption is considered, the network accessibility under link failure is equal to the accessibility of station failure, which only belongs to the line containing this link. The network accessibility under link failure could be described as:

$$A^L = A^d(d \in N_R \text{ and } d \notin N_{U-R}, LR) \tag{2.10}$$

where N_R is the set of all stations of line R, N_L represents the set of stations on link L, and U denotes all the lines of a rail network.

- *Rail network accessibility under line(s) failure*

Rail line is a constitution of stations and links whose failure could be described by simultaneous failures of the stations and links of a line. Based on network accessibility under link failure, rail network accessibility under line failure could be measured as:

$$A^R = \sum A^d \tag{2.11}$$

where A^R denotes the rail network accessibility when line(s) R failed, $d \in N_R$ if stations on line R are disrupted, and $d \in N_R$ and $d \notin N_{U-R}$ if stations on line R failed, but transfer stations are still working for other lines.

- *Rail network accessibility including land use characteristics*

The proposed accessibility index is aimed to measure station accessibility under incidents, and the importance of each rail station and the independency of land use on rail network are both accounted for the improvement of the accessibility index. It is defined as the potential opportunities for interaction among stations in which the opportunities are weighted by the land use characteristics around stations, the number of people arriving or departing at stations, the number of people traveling between two stations, and a ratio of impedance function before and after incidents. This could be constructed as:

$$RA_i = C_i A_i \tag{2.12}$$

where RA_i represents the accessibility of rail transit station i under incidents; C_i denotes the independency of surrounding land use on rail transit within walkable distance of station i, and is only applied when there are incidents on the network; and A_i is a location-based accessibility index of rail transit station i.

The independency of land use on rail transit C_i could then be formulated as:

$$C_i = L_i \left(n_i^c w_i^c + n_i^s w_i^s \right) \tag{2.13}$$

where L_i describes the degree of mixed land use within walkable distance of rail transit station i; n_i^c and n_i^s show the alternative weight for car and ground bus around station i, respectively; w_i^c is the availability of car alternative on the land use around station i;

and w_i^s represents the availability of ground bus alternative within walkable distance of station i. Thus, the change of the dependency and share of rail transit before and after rail incidents could be captured by C_i.

The degree of mixed land use index L_i around station i is calculated as:

$$L_i = 1 - \left\{ \frac{\left|\frac{r_i}{L} - \frac{1}{4}\right| + \left|\frac{m_i}{L} - \frac{1}{4}\right| + \left|\frac{c_i}{L} - \frac{1}{4}\right| + \left|\frac{o_i}{L} - \frac{1}{4}\right|}{3/2} \right\} \quad (2.14)$$

where r_i shows the residential land area around station i, m_i denotes the manufacturing land area around station i, c_i is the commercial and office land area around station i, and o_i is the area of other land use types around station i.

The 3/2 in the denominator of Eq. (2.14) is used for normalizing the land use mix degree L_i from 0 to 1. As a result, generally speaking, when different land use types are more optimally mixed, the mixed degree of land use L_i is closer to 1, and land use would have weaker dependency on rail. It is almost impossible to have only one type of land use around a station since there may always be multiple land use types such as roads and green space, hence, the value of the degree of mixed land use could be calculated within (0,1].

$$w_i^c = \frac{N_i^c}{N_{\max}^c} \quad (2.15)$$

$$w_i^s = \frac{N_i^s}{N_{\max}^s} \quad (2.16)$$

The calculations of two travel alternative modes around a station are shown in Eqs. (2.15), (2.16). The car alternative availability is measured as the car ownership ratio of the people living or working on land within walkable distance of station i. Specifically, w_i^c is calculated as the car alternative availability of station i divided by the maximum value of car availability in the study area. In this way, the car alternative availability is normalized to [0, 1]. The ground bus alternative availability within walkable distance of station i, w_i^s, is measured with the number of bus stops within walkable distance of a station and is proposed similarly as w_i^c.

Therefore, the mixed land use index and alternative availability indices are all positively correlated with the independency degree index C_i. When there are incidents, the higher the value of C_i, the greater accessibility will be, and the maximum value of this independency is 1.

- *Accessibility-based rail system vulnerability*

As for rail transit accessibility, A_i is the transitional accessibility index and is constructed as:

$$A_i = w_i^A \sum_{j=1}^{n-1} w_j^G \left(\frac{f(t_{ij})}{f^0(t_{ij})} \right)^{-1} \quad (j \neq i) \quad (2.17)$$

where w_i^A is the importance of station i, and is calculated as the total number of people departing and arriving at this station, denoted by N_i^{da}, divided by the total number of people departing and arriving at all the stations of a study area, denoted by N_i^{da}; w_j^G is the weight of trip generation of station j ($j \neq i$) traveling to station i, and is calculated as the ratio of the number of people departing from station j to station i, denoted by N_{ji}, and the total number of people traveling to station i denoted by N_i; $f^0(t_{ij})$ is the travel cost between stations i and j without station failure; $f(t_{ij})$ is the travel cost between stations i and j after station failure; and n represents the number of rail stations in the study area.

$$w_i^A = \frac{N_i^{da}}{N^{da}} \tag{2.18}$$

$$w_j^G = \frac{N_{ji}}{N_i} \tag{2.19}$$

The main advantages of the prior proposed station-based accessibility index could be explained in that it considers the impacts of surrounding land use on a railway system and quantifies the land use independency on rail network accessibility.

$$DD^d = \sum_{i=1}^{n} A_i^0 - \sum_{i=1}^{n} RA_i^d \tag{2.20}$$

where DD^d denotes the degree of railway system degradation if station(s) d fails, A_i^0 is the accessibility of rail station i without rail network degradation, and RA_i^d represents the accessibility of station i if rail network is degraded under failure of station(s) d.

Specifically, the accessibility changes for individual station under failure(s) could also be calculated to identify stations mostly affected. The accessibility change rate for each station under station(s) d failure is calculated as reduction ratio of the proposed accessibility (RAR):

$$RAR_i^d = \frac{A_i^0 - RA_i^d}{A_i^0} \tag{2.21}$$

where RAR_i^d is the accessibility reduction ratio of rail station i under the failure of station(s) d.

As a result, the railway system vulnerability could be measured as:

$$V_d = prob_d \times DD^d \tag{2.22}$$

where V_d denotes the vulnerability of railway system under incident(s) d and $prob_d$ is the probability of rail network incident d.

2.2.2 Rail system resilience modeling

Based on functions and structure features of the rail system, an importance-impedance-based rail network resilience method integrating topological and passenger flow characteristics was modeled. Particularly, the approach looks into the changing resilience of a rail network with the time duration of incidents, addressing the effects on delayed travel demand. Based on failures of stations, this model explicitly accounts for the effects of accumulatively affected passengers on network performance, quantifying the resilience of the rail system with different incidents' duration time.

- *OD-based importance-impedance network degradation measures under incidents*

The impacts of station incidents on the network would be different depending on affected passengers and increased travel time. To measure these impacts, an importance-impedance rail network performance model based on origin and destination (OD) travels was proposed.

Let N be the set of stations in a rail network. $N_W \subseteq N$ is the subset of working stations after incidents, and $N_C = N - N_W$ denotes the subset of station(s) closed and disconnected due to incidents, as some stations on rail network legs would be separated from the main network, although they are not affected directly. N_{OD} is the node set on the shortest path between an OD pair including origin node, destination node, and passing through nodes, and $N_{OD} \in N$. The importance of a rail system without any incident could be measured as:

$$I_M = \sum_{OD}(I_O B_O + I_D B_D) \quad N_{OD} \in N \tag{2.23}$$

where I_M is the importance of rail system M represented by the summary of importance of all OD pairs; I_O denotes the node importance of origin station and represented by the origin passenger volume (persons per hour), which is similarly proposed by Jenelius et al. [4]; I_D shows the node importance of destination station and is represented by the destination passenger volume (persons per hour); B_O represents the betweenness centrality of origin station; and B_D is the betweenness centrality of destination station. The impedance of an urban rail network is:

$$D_M = \sum_{OD} T_{OD} I_{OD} \quad N_{OD} \in N \tag{2.24}$$

where D_M is the impedance of rail system M being equal to the summary of impedance between all OD pairs, T_{OD} is the travel time between OD stations without network disruption (minutes), and I_{OD} denotes the importance of OD travel without network disruption (persons per hour), equal to the total passenger volume on the OD travel. The importance-impedance-based network performance P_M could then be equated as:

$$P_M = I_M / D_M \tag{2.25}$$

A higher value of P_M indicates better performance of rail network M, and a lower value shows poorer rail network performance. This performance index is typically used for before and after analyses comparing the P_M values before and after incidents.

The previous metric includes structural and functional measurements of a rail network. The structural measurement calculates the importance of OD travel consisting of topological and passenger volume importance of the OD stations. The functional measurement is the integration of travel time and total passengers between OD stations.

$$B_n = \sum_{n \neq O \neq D} \frac{\sigma_{OD(n)}}{\sigma_{OD}} \quad OD \in N, n \in N \tag{2.26}$$

where B_n is the betweenness centrality of node n, σ_{OD} denotes the total number of shortest paths from station O to station D, and $\sigma_{OD(n)}$ is the number of shortest paths between O and D stations passing through station n.

Travel time of the functional measurement, consisting of waiting time, in-vehicle travel time, and transfer time measured as minutes, is formulated as follows:

$$T_{OD} = T_W + T_{IV} + T_{TR} \tag{2.27}$$

where T_W is the waiting time for rail transit, T_{IV} denotes the in-vehicle travel time, and T_{TR} is the transfer time at transfer stations.

Performance of a rail system after incident(s) is then formulated as:

$$P_M^f = I_M^f / D_M^f = \frac{\sum_{OD(NOD \in NW)} I_O B_O^f + I_D B_D^f}{\sum_{OD(\forall NOD \in NW, \exists NOD \in NC)} T_{OD}^f I_{OD}} \tag{2.28}$$

where P_M^f is the importance-impedance-based performance of rail system M after incident(s) f, I_M^f is the importance of rail system M after incident(s) f, D_M^f shows the impedance of rail system M after incident(s) f, B_O^f denotes the betweenness centrality of origin station after incident(s) f, B_D^f is the betweenness centrality of destination station after incident(s) f, and T_{OD}^f is the travel time between OD stations after incident(s) f.

As shown in Eq. (2.29), travel time could fall into three categories after incidents, which are (I) the same as that before incidents; (II) changed because of rerouting; or (III) changed because of waiting for recovery. Waiting time under incidents is further divided into waiting time after arriving at the station before recovery and dispersion time until getting on the train after recovery.

$$T_{OD}^f = \begin{cases} T_{OD}, & \text{if OD travel is not affected}, N_{OD} \in N_W \\ T'_{OD}, & \text{if OD travel is rerouted}, N'_{OD} \in N_W \\ T_f - t + \left\lceil \frac{V + A \cdot t}{C} \right\rceil T_{FR} + T_{IV} + T_{TR}, t \in [0, T_f], & \text{if OD travel is delayed}, N_{OD} \in N_C \end{cases} \tag{2.29}$$

where $N_{OD}{'}$ indicates the updated working OD node set after incident(s), $T_{OD}{'}$ ($OD \in N_{OD}{'}$) is the updated travel time as a result of rerouting after incident(s) (minutes), T_f denotes the duration time of incident(s) (minutes), t ($t \in [0, T_f]$) represents the arrival time of passengers at station(s) during the incident(s) beginning with 0 if passengers just arrive when the incident(s) occurs, V is the number of passengers waiting in the station when incident(s) happens, A shows the average passenger arrival rate at stations (persons per minute), C is the capacity of each train (number of passengers), and T_{FR} denotes the time interval between trains (minutes), that is, train headway after recovery.

$T_f - t$ calculates the waiting time of passengers who arrive at time t before recovery. $\left\lceil \frac{V+A \cdot t}{C} \right\rceil$ returns the ceiling value of $\frac{V+A \cdot t}{C}$ and calculates the number of trains needed to transport $V + A \cdot t$ passengers, and thus $\left\lceil \frac{V+A \cdot t}{C} \right\rceil \cdot T_{FR}$ denotes the time needed to get on board after recovery for passengers who arrive at the station at time t. As a result, the total waiting time for passengers to get on the train after recovery could be formulated as $T_f - t + \left\lceil \frac{V+A \cdot t}{C} \right\rceil T_{FR}$.

$$E_M^f = \frac{P_M - P_M^f}{P_M} \tag{2.30}$$

where E_M^f denotes the degree of network degradation of rail network M under operational incidents f.

- *Rail system resilience measurement*

Based on Eqs. (2.25), (2.28), the numerator of Eq. (2.28) would decrease as some stations would be closed on the degraded rail network, and the denominator of the equation would increase with the increase of travel time. Both of these would result in a lower network performance under incidents than that without network degradation. As a result, the resilience of a rail network under incidents is defined as the speed of rail network recovery from the worst network performance under incidents to its original state. It could be measured as:

$$R_M^f = \frac{E_M^f}{\max\left\{T_f + \left\lceil \frac{V+A \cdot t}{C} \right\rceil T_{FR}\right\}} = \frac{P_M - P_M^f}{\max\left\{T_f + \left\lceil \frac{V+A \cdot t}{C} \right\rceil T_{FR}\right\} P_M} \tag{2.31}$$

where R_M^f calculates the degree of rail network resilience M under incidents f.

In the case of multiple simultaneous station incidents, Eq. (2.31) includes a maximum station recovery time, that is, $\max\left\{T_f + \left\lceil \frac{V+A \cdot t}{C} \right\rceil T_{FR}\right\}$ representing recovery time of the network. Station(s) whose failure would have the most impact on the resilience of a rail network could also be identified based on the changes of network resilience values.

The previous rail system resilience model could be not only used for assessing single and multiple station failures but also applicable to rail segment failures under operational incidents. Under rail link failures, the structural part of network resilience could be calculated based on the network without the link(s); the functional part could be updated with the changed travel time between OD pairs if the link(s) failed.

2.3 Railway vulnerability and resilience practices in Chinese cities

2.3.1 Shanghai Metro Network vulnerability and resilience practice

Shanghai Metro operates 14 metro lines with a total length of 617 km ranking the first all over the world in 2016. There is 1 circular line and 13 radial lines constituting the rail transit network of Shanghai. The average daily passenger flow of the Shanghai Metro system is more than 8 million and reaches over 10 million on peak days, thus the rail transit of Shanghai is regarded as one of the busiest metro systems in the world. There are 10 more lines under construction with a further length of 216 km, and the total length of metro lines in Shanghai has exceed 800 km with 500 stations by the end of 2020. The data used for analysis include the Shanghai Metro Network with 281 stations and 51 transfer stations, and peak hour OD data from Shanghai Public Transportation Card for the year of 2015 provided by Shanghai Metro. The number of passengers arriving and departing at each station was extracted from the OD data.

It is expected that the Shanghai Metro system will have even more passengers in the future with the continuous metro construction and increasing metro travel demand. With such a large ridership, the metro system of Shanghai is susceptible to disruptive events such as train breakdown, signal control system breakdown, communication or electrical power system failure, and so on. From January 2016 to May 2017, a total of 167 operational incidents were reported for Shanghai Metro, and the types and proportions of the incidents are shown in Fig. 2.2A. The top three incidents with the most occurrence frequency are train breakdown, signal control failure, and doors breakdown. The highest operation risk is train breakdown accounting for 50% in major operational incidents. Based on statistics of the above 167 incidents, the average duration time and corresponding standard deviations of the incidents are different varying from minutes to more than 1 h as indicated in Fig. 2.2B. Although the probability of electrical system failures is only 0.05%, its average working time loss for the rail system is close to 100 min, which implies that the potential risk of electrical system failure has significant effects on the rail system. Impacts of these incidents on passengers would be delayed travel, unsatisfied travel demand, and even accidents resulting from crowded people, all of which would lead to losses of travel time and business and work opportunities. As the financial center of China, Shanghai would bear many economic losses from frequent delays on its big rail transit network with such large passenger flow.

- *Shanghai Metro Network vulnerability analysis*

To investigate the impacts on the Shanghai Metro Network under station(s) or link(s) failures, 20 incident scenarios including 15 station and 5 link failures were studied.

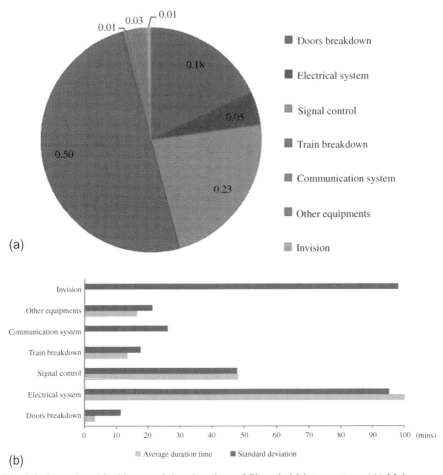

Fig. 2.2 Operational incidents and duration time of Shanghai Metro system: (A) Major operational incidents and occurrence frequency of Shanghai Metro system; (B) Average duration time and corresponding standard deviations of major incidents.
Reproduced with permission based on: Q.C. Lu, Modeling network resilience of rail transit under operational incidents, Transp. Res. A Policy Pract. 117(11) (2018) 227–237.

Daily passenger volume of each station during morning peak hours (7:00–9:30 AM) is also illustrated in Fig. 2.3. The 20 scenarios are proposed based on the stations or links and characterized either by large passenger flow (A), multiline transfer (B), intersected with transport hub such as airport and railway station (C), areas with many big events (D) or suburban stations with relative high passenger flow (E), and urban poor areas (F) as shown in Table 2.1. Further scenarios of simultaneous failures of more than one station and/or link are also evaluated. The probability of each scenario was supposed to be 0.25, 0.50, and 0.75, respectively, and then the metro network vulnerability is computed in the table.

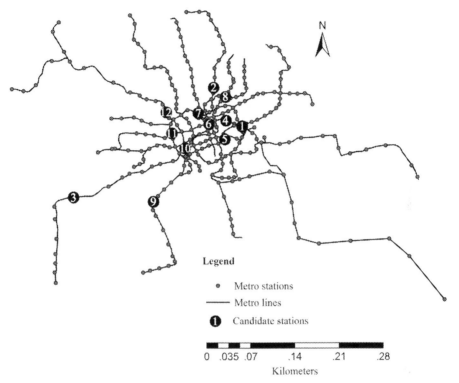

Fig. 2.3 Shanghai Metro Network and candidate stations for analysis.

As indicated in Fig. 2.4, stations or links whose failure would result in high network accessibility reduction are mainly distributed inside or close to the inner circle road. However, failures of several stations in the suburban areas would also result in relatively high network-wide accessibility reductions, which should be paid more attention; once failed, fewer metro network redundancy and alternative modes are available in these areas. Transfer metro stations are usually reported to be more important stations both in urban and suburban areas determined by their topological characteristics of connecting abundant transfer lines.

To investigate the impacts on a rail system under different infrastructure failures, 20 incident scenarios including 15 stations and 5 links failures are suggested. As shown in Table 2.1, under the proposed 20 single, 6 multiple incident scenarios, as well as 2 section incident scenarios, the overall network accessibility was calculated. Under single station or link failure, it could be found that the overall network accessibility reduction ratios under stations No. 15 and No. 17 failures are extremely small since the travel flow of these two stations are the least compared to all the other selected stations. It is also indicated that less passenger flow of a station would contribute to less station importance and thus fewer impacts on network performance once the station is disrupted. Except for passenger volume, the network accessibility could also be affected by land use characteristics around stations. Therefore, from the

Table 2.1 Vulnerability of Shanghai Metro Network under different scenarios.

Station ID	Station(s) or link(s)	Classification	RAR (%)	Importance rank	Metro network vulnerability (10^{-3}) Failure probability 0.25	0.50	0.75
Single station failure scenarios							
1	Xujiahui	A & B	85.3593	10	213.3983	426.7965	640.1948
3	Shanghai Railway Station	A & C	85.7028	6	214.2570	428.5140	642.7710
4	Lujiazui	A	85.2348	11	213.0870	426.1740	639.2610
5	Xinzhuang	A	85.0722	13	212.6805	425.3610	638.0415
6	Century Avenue	B	85.5257	8	213.8143	427.6285	641.4428
8	Pudong Airport	C & E	84.2345	24	210.5863	421.1725	631.7588
9	Shanghai South Railway Station	C	85.7409	5	214.3523	428.7045	643.0568
10	West Shanghai Railway Station	C	84.9598	15	212.3995	424.7990	637.1985
11	Longyang Road	B & E	85.3744	9	213.4360	426.8720	640.3080
12	Oriental Sports Center	B & F	85.1669	12	212.9173	425.8345	638.7518
14	Jufeng Road	E	84.5719	20	211.4298	422.8595	634.2893
15	Songjiang South Railway Station	C & F	84.2086	26	210.5215	421.0430	631.5645
17	Shuyuan	F	84.2083	27	210.5208	421.0415	631.5623
18	China Art Museum	D	84.2841	23	210.7103	421.4205	632.1308
20	Jinshajiang Road	B & E	84.7945	16	211.9863	423.9725	635.9588
Single link failure scenarios							
2–24	People's Square —Huangpi Road (S)	A & B	84.5678	21	211.4195	422.8390	634.2585
7–21	Hongqiao Railway Station —Hongqiao Airport Terminal 2	C & E	84.2023	28	210.5058	421.0115	631.5173
19–23	Shanghai Indoor Stadium —Yishan Road	B	84.2215	25	210.5538	421.1075	631.6613
13–25	Siping Road—Tongji University	E	84.7311	17	211.8278	423.6555	635.4833
22–16	Jiading Xincheng—Shanghai Circuit	F	84.2908	22	210.7270	421.4540	632.1810

Table 2.1 Continued

Station ID	Station(s) or link(s)	Classification	RAR (%)	Importance rank	Metro network vulnerability (10^{-3}) Failure probability		
					0.25	0.50	0.75
Multiple failure scenarios
—	Xujiahui & People's Square	—	86.7385	2	216.8463	433.6925	650.5388
—	Jiading Xincheng & Dongchuan Road	—	84.6751	18	211.6878	423.3755	635.0633
—	Chifeng Road & Nanjing Road (E)	—	85.8835	4	214.7088	429.4175	644.1263
—	Xujiahui & Nanjing Road (W)	—	86.2587	3	215.6488	431.2935	646.9403
—	Shanghai University & Zhouhai Road & Lingzhao Xincun	—	85.0614	14	212.6535	425.3070	637.9605
—	Xujiahui & Century Park—Longyang Road	—	85.6910	7	214.2275	428.4550	642.6825
Section failure scenarios
—	Jinshajiang Road & Longde Road & Changshou Road & Zhenping Road & Caoyang Road	—	86.8761	1	217.1902	434.3805	651.5707
—	West Gaoke Road & South Yanggao Road & Jinxiu Road & Dongming Road & Linyi Xincun & Shanghai Children's Medical Center	—	84.6530	19	211.6325	423.2650	634.8975

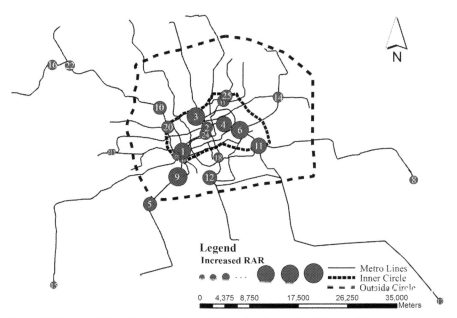

Fig. 2.4 Shanghai Metro Network accessibility reduction under different disruptions.

perspective of land use independency, failure of station No. 10 will have greater impacts on rail transit network than that of station No. 20, while the impacts on the rail transit network performance will be less when compared to station No.9. Although located in the suburban area with fewer passengers, the link between stations No. 22 and No. 16 has relatively serious impacts on the network accessibility when disrupted. In short, the larger passenger volume a station has and the heavier dependency of nearby land use, greater impacts on rail transit network performance could be observed.

In addition, this practice also addressed the breakdown of multiple elements on the Shanghai Metro Network, including multiple station failures, multiple link failures, combinations of station and link failures, and section incident scenarios. Results are shown in Table 2.1. It is obvious that multiple incidents generally have greater impacts on the whole network than single incidents, and the impacts of section incidents can be either greater or smaller than other types of incidents, thus disasters at a large scale don't always mean greater impacts on the rail transit network compared to small scale incidents such as one or several stations and links out of operation. As a result, the impacts on a rail transit network would depend on the importance of the stations or links affected instead of the scale of impacts. The impacts of multiple element failures, including multiple incidents and section incidents, depend not only on which and how many stations and links are disrupted but also the passenger flow and surrounding land use characteristics of those closed stations and links.

From the prior analysis, it can be found that if incidents occur inside and close to the inner circle road of central business areas, the whole network will generally show relative high vulnerability. The disruptions of suburban stations may not pose high impacts on the network vulnerability. However, some suburban stations may affect the vulnerability of the metro network than downtown stations, which may have many more alternatives of rail transit routes and other travel modes.

- *Shanghai Metro Network resilience analysis*

The proposed methodology was used for the evaluation of the resilience of the Shanghai Metro Network under operational incidents. The incidents were assumed to occur on 12 stations over the network with different duration times representing various types of incidents. The 12 stations were selected for network resilience analysis either due to their important roles on the topology of the Shanghai Metro Network or because of their heavy passenger flows. Locations of these selected stations could be identified in Fig. 2.2. Based on statistics of the operational incidents, the duration time of the analyses were assumed to be 5, 10, 20, and 40 min for each of the 12 stations, corresponding to occurrence of small incidents, door breakdowns, train and communication system breakdowns, and control system failures, respectively.

As shown in Fig. 2.5, the longer the incidents continue, the lower the network resilience becomes, and the rail network receives low resilience values under incidents on nontransfer stations but demonstrates high capability of recovery under transfer station disruptions. The network is the most resilient when Station 7 is closed followed by Station 10. The network also shows high resilience values if transfer stations such as Stations 1 and 6 failed. This might be because transfer stations have more redundant metro lines and could rapidly accommodate impacts of the incidents; however, people have to travel with one metro line at nontransfer stations, which have limited capacity to absorb the degradation impacts. The network resilience degrees change differently among the proposed incidents. These changes in network resilience under different station disruptions should call for attention of rail managers and decision makers. For example, if operational incidents occur on Station 7, it is better to fix the

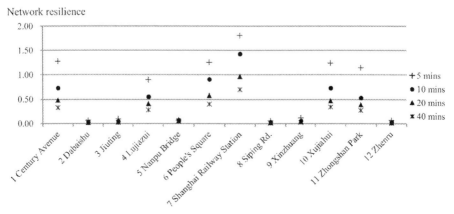

Fig. 2.5 Degree of network resilience under different operational incidents.

disruptions within 10 min as the resilience values would be more reduced if the incidents continue for 20 min or more.

Among all the incidents, the network keeps the lowest resilience when Station 8 is closed. Station 8 is located at the transfer station of two network legs, and incidents on this station would interdict travels of stations on both legs resulting in lower resilience than those of nontransfer stations and transfer stations with one network leg. It is worth noting that with the increase of incidents' duration time, network resilience is reduced to the same value under failures of Stations 8 and 12. Thus, if operational incidents occur on stations like Station 12, it should be fixed as soon as possible because if the incidents last for a longer time, it would be even harder for the network to recover by itself. Low network resilience values could also be found at failures of Station 5, but the resilience values remain similar among 10-, 20-, and 40-min incidents on this station.

Fig. 2.6 shows the degrees of network degradation and resilience under incidents of 40 min on the proposed stations. Based on Fig. 2.6A, stations that cause high network degradation could be either transfer stations or nontransfer stations on legs. However, as demonstrated in Fig. 2.6B, the network shows higher resilience under incidents on stations in the central area of the city. The network becomes much less resilient when stations on network legs are closed under incidents. These might be because the rail network in the city center is more redundant, and people have more alternatives to choose from when stations failed. Stations in the suburban area or legs of the network are less redundant and show low capability of recovery after incidents. Another interpretation would be that there are many subcity centers in Chinese big cities such as Shanghai, and people would live and work near metro stations in these subcenters because of the convenience of either land use diversity or commuting to the city center. As a result, travel demand at these stations would be even higher than that of many stations in the city center. Both the high travel demand and low rail redundancy of suburban stations would result in high network degradation and low network resilience under incidents.

2.3.2 Shenzhen rail network vulnerability within multimodal transit network

With the third-largest public transport network in China, Shenzhen has a transit ridership of over 10.5 million per day. Shenzhen bus transit operates 919 bus routes with a daily ridership of 5.9 million. Consisting of 5 lines and 118 stations including 13 transfer stations, Shenzhen urban rail transit network (SURTN) has a length of 178.0 km and ranks sixth in the country. However, it has the fourth rail transit ridership in China with nearly 3.0 million passengers per day after Beijing, Shanghai, and Guangzhou. The peak day rail ridership in Shenzhen would reach 3.5 million. With another 6 lines under construction, SURTN will be extended to 11 lines with a total length of 434.9 km in 2020. As a job-immigrant city in China, Shenzhen's rail transit is expected to play a more and more important role in people's daily travel. Therefore, the network and passenger flow data of SURTN were used. The urban rail transit network of Shenzhen with stations and links is shown in Fig. 2.7 for the year of 2013. The

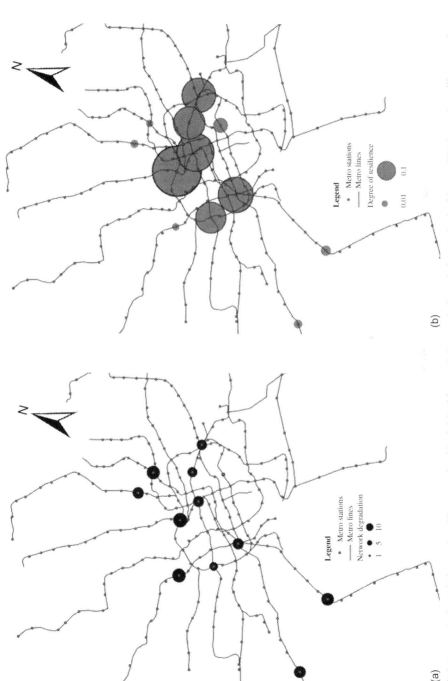

Fig. 2.6 Network degradation and resilience under 40min' incident: (A) Network degradation under incidents; (B) Network resilience under incidents.

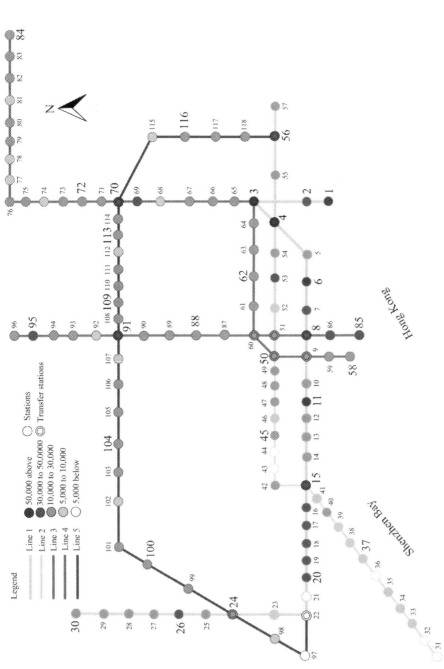

Fig. 2.7 Shenzhen urban rail transit network and passenger volume distribution. Reproduced with permission based on: Q.C. Lu, S. Lin, Vulnerability analysis of urban rail transit network within multi-modal public transport networks, Sustain. For. 11(7) (2019) 2109-1-14.

network for analysis also contains the Shenzhen bus transit network that year. As shown in Fig. 2.7, passenger flow on SURTN is mainly distributed on the southeastern part of the network. The station data include average daily OD trip matrix between rail stations in 2013 and the number of bus stops within 600 m of rail stations [41]. The candidate rail stations for analysis are classified based on the station OD and the station location on the network, that is, normal stations (NS), stations with large passenger volume (LPS), transfer stations (TS), and stations far away from city center but having a relative high ridership (FCPS) are chosen as candidate stations for analysis. The travel time of rail transit includes in-vehicle travel time, transfer time, and vehicle time headway. Ground buses share the same transit network with rail transit, and passengers of closed stations would choose buses based on the shortest-path principle on the transit network.

Due to different types of rail system infrastructure, this practice captures different types of stations of the Shenzhen metro network within the multimodal public transport network. Different types of stations were assumed to be disrupted individually, and the station-based network accessibility was calculated under each disruption. The consequence of a station failure was measured as network-wide accessibility reduction compared with its original value of 1. Multiple station failures were then analyzed for the network. The results are shown as follows:

- *Measuring network accessibility under individual station failures*

Network accessibility of SURTN was calculated for each of the 30 station failure scenarios, and results are shown in Fig. 2.8. Among all the candidate stations, the failure of Station 4 would result in the most network accessibility reduction. Located in the central business district, Station 4 has the most passenger volume among all the rail stations in Shenzhen but does not have the highest bus availability. Following Station 4, Stations 3, 11, 26, 6, and 1 also cause high accessibility reduction if disrupted. All of these stations are on Line 1, which was built the earliest and goes through the most developed area of Shenzhen. Station 62 is proved to be the least important station whose failure causes the least network accessibility reduction. Substantial accessibility reductions could be observed for most of the transfer stations as most of them have relative high ridership. Station 30 has the most importance increase under failure, compared with the rank based on passenger volume. One reason would be the pretty low bus availability around the station, while another could be explained by its widely distributed OD flows, which could be seriously affected for long-distance bus alternative travels if Station 30 is disrupted. It is worth noting that all the disrupted stations are still accessible but have substantial reductions (more than 80%) in their accessibility.

The top 15 stations whose failures result in the most accessibility reductions are shown in Table 2.2. The network accessibility values do not vary too much under these station failures; however, the station importance rank is different from the rank of original passenger volume. Stations 4 and 3 are shown to be the most important stations, and no differences are reported under both ranks. The top 5 most affected working stations under each station failure are also identified in Table 2.2. The five most affected working stations mainly belong to the same rail line as the disrupted stations, which demonstrates that most indirect impacts of a station failure are imposed on

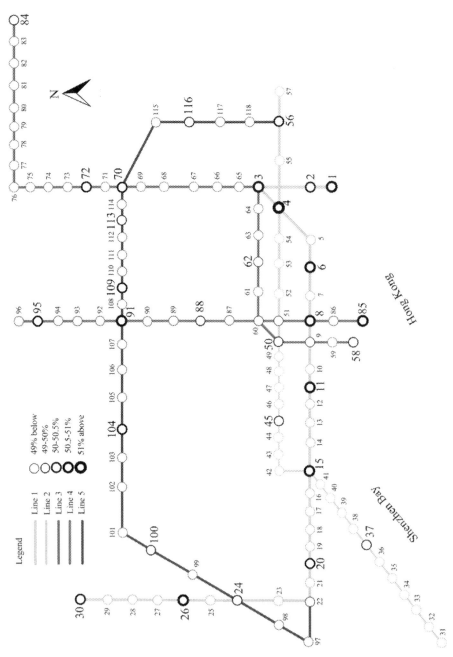

Fig. 2.8 Network-wide accessibility reduction under failures of 30 stations.

Table 2.2 Top 15 stations with most impacts on network accessibility.

Station ID	Station types	Passenger volume rank	Network accessibility after station disruption (%)	Accessibility reduction rank	Most affected working stations
4	TS	1	88.623	1	5, 18, 28, 94, 116
3	TS	2	89.058	2	18, 71, 94, 90, 28
6	LPS	3	89.204	5	8, 91, 7, 30, 18
1	LPS	4	89.337	6	28, 18, 94, 5, 3
11	LPS	5	89.297	3	18, 91, 12, 116, 28
8	TS	6	89.381	7	86, 6, 18, 90, 91
15	TS	7	89.789	10	86, 30, 18, 90, 5
26	FCS	8	89.307	4	18, 28, 91, 30, 116
85	FCS	9	89.285	8	91, 18, 90, 86, 94
70	TS	11	89.681	15	91, 118, 116, 117, 3
91	TS	12	89.348	9	116, 86, 90, 94, 71
109	NS	13	89.664	12	91, 118, 117, 18, 28
20	LPS	14	89.380	11	18, 28, 30, 17, 3
72	NS	17	89.726	14	91, 18, 66, 117, 28
30	FCS	18	89.633	13	28, 18, 116, 91, 24

stations sharing the same line. Moreover, the most affected working stations include not only surrounding stations of the disrupted station but also far away stations with large passenger flow. Station 86 is shown to be the most affected station under failures of Stations 8 and 15, as the OD trips of Station 86 to and from Stations 8 and 15 are the most among other stations. Besides, Stations 28 and 91 are also shown to be in the top 5 most affected stations under many station failures.

- *Network accessibility under multiple station failures*

The network accessibility under simultaneous multiple station failures is calculated for each scenario in Table 2.3. The scenarios with large passenger volume do not necessarily cause high network accessibility reduction. The top 5 most affected working stations under each scenario are included in the results. Similar to individual station failures, the most affected working stations are mainly those on the same line with disrupted stations. This could be explained by the high OD trip distributions of Station 18 on stations of Line 1 and those in the southeast of the network; when stations in these areas failed, the accessibility of Station 18 would be importantly affected.

Table 2.3 Network accessibility under multiple station failures.

Station ID	Station types	Passenger volume rank	Network accessibility after disruptions (%)	Accessibility reduction rank	Most affected working stations
3, 4	TS, LPS	1	71.766	3	18, 94, 5, 24, 90
11, 45, 70	TS, LPS	2	64.802	1	19, 40, 91, 116, 118
8, 30, 50	TS, FCS	3	64.822	2	19, 86, 91, 90, 92
1, 20	FCS, LPS	4	72.626	4	18, 28, 25, 24, 3
56, 91	TS, LPS	5	72.880	5	116, 117, 118, 54, 114
24, 26	TS	6	73.043	6	91, 30, 28, 18, 113

Together with results in Table 2.2, it could be reached that Stations 116, 117, and 118 are vulnerable to station failures on the same line especially to disruptions of transfer station since they show high accessibility reductions under these failures.

2.4 Conclusions

In this chapter, the vulnerability and resilience of a railway system is discussed comprehensively from conceptual, methodological, and practical perspectives. According to the reviews and analyses of the current literature, the relationship of railway system vulnerability and resilience is presented. This chapter then elaborates an accessibility-based vulnerability method to explicitly account for passenger flow features and land use characteristics around stations under conditions of station(s), links(s), and line(s) failures. Meanwhile, the methodology measuring the resilience of a railway system under different operational incidents is also captured, which identifies critical stations according to network resilience dynamics. The main contributions of the vulnerability and resilience methods fall into two aspects. On one hand, it interprets the importance of stations while considering alternative public transport systems and various land use characteristics around stations, and thus helps to analyze the effects of passenger flow distribution characteristics and land use dependencies on the vulnerability of the railway system. On the other hand, the proposed methodology addresses not only the topological impacts but also the effects on people's travel including rerouted and delayed passengers changing with time, and the system resilience changes with the increase of incidents duration time resulting in criticality changes of the stations.

The methodologies for the vulnerability and resilience of railway system are not only presented and discussed but also applied to railway systems of two cities in China. There are important implications for the management and decision making of railway systems under disruptions. First, based on the analyses of railway system vulnerability and resilience, practical measures could be implemented by rail planners and operators to design and manage resilient railway systems. The vulnerability of the network could be reduced with network plan and design measures during the railway network expansion, especially for critical stations. The analysis of railway system vulnerability based on land use characteristics could provide a scientific basis at the planning stage for the coordination between land use development and rail network planning to avoid potential risks. Second, rail managers should pay more attention to the relationship between rail systems and other transport systems under different disruptive events. Thus, bus and other transit systems should be planned in coordination with railway systems to enhance the resilience of public transportation system as a whole. For example, when incidents occur on an urban rail transit network, bridging bus design and operation decisions could be made based on the identified impacts of disrupted stations on the system. Additional measures, such as passenger evacuation plans, could also be adopted for the most affected stations. Finally, rail planners should manage stations based on the resilience changes of the network and prepare countermeasures according to different duration time of incidents. For some stations, rail managers should ensure incidents will be fixed within a short time but not necessarily put too many resources on other stations if multiple stations simultaneously fail. Besides, failure of a circular line would have the worst impact on a rail system, but a station closure on the circular line might not cause the worst impact. Thus, rail planners should manage stations on circular lines independently to avoid line failure. As for important transfer stations, more redundant rail networks should be planned and designed around these stations, and rail planners might also increase train frequency after recovery to ensure a resilient railway system.

Based on the these practices and conclusions, the analyses of railway system vulnerability and resilience are useful to effectively understand the mechanism of rail network performance varying under different disruptions, thus providing insights for the management and decision making of railway systems. However, limitations still exist. For example, people's travel behavior under different rail incidents should be surveyed and analyzed with discrete choice models so as to give more insights into the mode share of different travel alternatives within a multimodal public transport network. All of these limitations should be addressed with more efforts in the future.

Acknowledgments

These research works are funded by the National Natural Science Foundation of China (71971029) and Natural Science Basic Research Program of Shaanxi (No. 2021JC-28). The support of Huo Yingdong Education Foundation (No. 171069) is acknowledged.

References

[1] A. Cox, F. Prager, A. Rose, Transportation security and the role of resilience: a foundation for operational metrics, Transp. Policy 18 (2) (2011) 307–317.
[2] B.P.Y. Loo, K.Y.K. Leung, Transport resilience: the occupy central movement in Hong Kong from another perspective, Transp. Res. A Policy Pract. 106 (2017) 100–115.

[3] K. Berdica, An introduction to road vulnerability: what has been done, is done and should be done, Transp. Policy 9 (2002) 117–127.
[4] E. Jenelius, T. Petersen, L.G. Mattsson, Importance and exposure in road network vulnerability analysis, Transp. Res. A Policy Pract. 40 (7) (2006) 537–560.
[5] M.A.P. Taylor, G.M.D. D'Este, Critical Infrastructure, Springer, Berlin, 2007.
[6] X.Z. Chen, Q.C. Lu, Z.R. Peng, J.E. Ash, Analysis of transportation network vulnerability under flooding disasters, Transp. Res. Rec. 2532 (2015) 37–44.
[7] J. Sohn, Evaluating the significance of highway network links under the flood damage: an accessibility approach, Transp. Res. A Policy Pract. 40 (6) (2006) 491–506.
[8] M.G.H. Bell, U. Kanturska, J.D. Schmöcker, A. Fonzone, Attacker-defender models and road network vulnerability, Phil. Trans. R. Soc. A 366 (1872) (2008) 1893–1906.
[9] A. Chen, C. Yang, S. Kongsomsaksakul, M. Lee, Network-based accessibility measures for vulnerability analysis of degradable transportation networks, Netw. Spat. Econ. 7 (3) (2007) 241–256.
[10] Q.C. Lu, Z.R. Peng, J. Zhang, Identification and prioritization of critical transportation infrastructure: a case study of coastal flooding, J. Transp. Eng. 141 (3) (2015). 04014082-1-8.
[11] M.A.P. Taylor, G.M. D'Este, S.V.C. Sekhar, Application of accessibility based methods for vulnerability analysis of strategic road networks, Netw. Spat. Econ. 3 (6) (2006) 267–291.
[12] O. Cats, M. Yap, N. Van Oort, Exposing the role of exposure: public transport network risk analysis, Transp. Res. A Policy Pract. 88 (6) (2016) 1–14.
[13] V.L. Knoop, M. Snelder, H.J. Van Zuylen, S.P. Hoogendoorn, Link-level vulnerability indicators for real-world networks, Transp. Res. A Policy Pract. 46 (5) (2012) 843–854.
[14] B.Y. Chen, W.H.K. Lam, A. Sumalee, Q. Li, Z.C. Li, Vulnerability analysis for large-scale and congested road networks with demand uncertainty, Transp. Res. A Policy Pract. 46 (3) (2012) 501–516.
[15] L.-G. Mattsson, E. Jenelius, Vulnerability and resilience of transport systems—a discussion of recent research, Transp. Res. A Policy Pract. 81 (11) (2015) 16–34.
[16] Z. Wang, A.P. Chan, J. Yuan, B. Xia, M. Skitmore, Q. Li, Recent advances in modeling the vulnerability of transportation networks, J. Infrastruct. Syst. 21 (2) (2014). 06014002-1-9.
[17] S. Derrible, C. Kennedy, Network analysis of world subway systems using updated graph theory, Transp. Res. Rec. 2112 (2009) 17–25.
[18] Y. Yang, Y. Liu, M. Zhou, F. Li, C. Sun, Robustness assessment of urban rail transit based on complex network theory: a case study of the Beijing Subway, Saf. Sci. 79 (2015) 149–162.
[19] O. Cats, E. Jenelius, Vulnerability analysis of public transport networks: a dynamic approach and case study for Stockholm, in: The 5th International Symposium on Transportation Network Reliability (INSTR2012), Hong Kong, China, 2012.
[20] R. Jiang, Q.C. Lu, Z.R. Peng, A station-based rail transit network vulnerability measure considering land use dependency, J. Transp. Geogr. 66 (2018) 10–18.
[21] Q.C. Lu, S. Lin, Vulnerability analysis of urban rail transit network within multi-modal public transport networks, Sustain. For. 11 (7) (2019). 2109-1-14.
[22] E. Rodríguez-Núñez, J.C. García-Palomares, Measuring the vulnerability of public transport networks, J. Transp. Geogr. 35 (2014) 50–63.
[23] X. Li, Y. Liu, Z. Gao, D. Liu, Linkage between passenger demand and surrounding land-use patterns at urban rail transit stations: a canonical correlation analysis method and case study in Chongqing, Int. J. Transp. Sci. Technol. 5 (1) (2016) 10–16.
[24] C. Bhat, J.Y. Guo, A comprehensive analysis of built environment characteristics on household residential choice and auto ownership levels, Transp. Res. B Methodol. 41 (5) (2007) 506–526.

[25] R. Faturechi, E. Miller-Hooks, Measuring the performance of transportation infrastructure systems in disaster: a comprehensive review, J. Infrastruct. Syst. 21 (1) (2015). 04014025-1-15.
[26] A. Reggiani, Network resilience for transport security: some methodological considerations, Transp. Policy 28 (2013) 63–68.
[27] S. Derrible, C. Kennedy, The complexity and robustness of metro networks, Phys. A: Stat. Mech. Appl. 389 (17) (2010) 3678–3691.
[28] A. De-Los-Santos, G. Laporte, J.A. Mesa, F. Perea, Evaluating passenger robustness in a rail transit network, Transp. Res. C 20 (1) (2012) 34–46.
[29] M. D'Lima, F. Medda, A new measure of resilience: an application to the London underground, Transp. Res. A Policy Pract. 81 (2015) 35–46.
[30] E. Miller-Hooks, X. Zhang, R. Faturechi, Measuring and maximizing resilience of freight transportation networks, Comput. Oper. Res. 39 (2012) 1633–1643.
[31] X. Zhang, E. Miller-Hooks, K. Denny, Assessing the role of network topology in transportation network resilience, J. Transp. Geogr. 46 (2015) 35–45.
[32] O. Cats, G.-J. Koppenol, M. Warnier, Robustness assessment of link capacity reduction for complex netwokrs: application for publich transport systems, Reliab. Eng. Syst. Saf. 167 (2017) 544–553.
[33] Q.C. Lu, Modeling network resilience of rail transit under operational incidents, Transp. Res. A Policy Pract. 117 (11) (2018) 227–237.
[34] A. Reggiani, P. Nijkamp, D. Lanzi, Transport resilience and vulnerability: the role of connectivity, Transp. Res. A Policy Pract. 81 (2015) 4–15.
[35] K.E. Worton, Using socio-technical resilience frameworks to anticipate threat, in: 2012 Workshop on Socio-Technical Aspects on Security and Thrust (STAST), Cambridge, Massachusetts, USA, 2012.
[36] E. Hollnagel, Prologue: the scope of resilience engineering, in: E. Hollnagel, J.P. Dédale, D. Woods, J. Wreathall (Eds.), Resilience Engineering in Practice: A Guidebook, Ashgate, UK, 2011.
[37] C. Bhat, S. Handy, K. Kockelman, H. Mahmassani, A. Gopal, I. Srour, L. Weston, Development of an Urban Accessibility Index: Formulations, Aggregation, and Application, Center for Transportation Research, The University of Texas at Austin, Austin, USA, 2002.
[38] T. Litman, Evaluating Accessibility for Transportation Planning, Victoria Transport Policy Institute, Victoria, British Columbia, 2016.
[39] Q.C. Lu, Z.R. Peng, Vulnerability analysis of transportation network under scenarios of sea level rise, Transp. Res. Rec. 2263 (2011) 174–181.
[40] N. Nassir, M. Hickman, A. Malekzadeh, E. Irannezhad, A utility-based travel impedance measure for public transit network accessibility, Transp. Res. A Policy Pract. 88 (2016) 26–39.
[41] M.J. Jun, K. Choi, J.E. Jeong, K.H. Kwon, H.J. Kim, Land use characteristics of subway catchment areas and their influence on subway ridership in Seoul, J. Transp. Geogr. 48 (2015) 30–40.

Rail resilience to climate change: Embedding climate adaptation within railway operations

Emma J.S. Ferranti[a,b], Andrew D. Quinn[a,b], and David J. Jaroszweski[b,c]
[a]School of Engineering, University of Birmingham, Birmingham, United Kingdom, [b]Birmingham Centre for Railway Research and Education (BCRRE), University of Birmingham, Birmingham, United Kingdom, [c]School of Geography, Earth and Environmental Science, University of Birmingham, Birmingham, United Kingdom

3.1 Introduction

3.1.1 Introduction to climate change

Climate change is one of the greatest challenges of our time. Anthropogenic emissions of greenhouse gases have changed the radiation balance within the earth's atmosphere causing the average global temperature to increase and changing our climate. In 2019, climate change was brought to fore by the global climate strikes initiated by Greta Thunberg, action by environmental movements such as Extinction Rebellion, and campaigning by other international influencers such as Sir David Attenborough. This has increased global awareness of the issue, bringing it to the attention of governments, organizations, and publics across the world. For example, Climate Action is one of the United Nations' 2015 Sustainable Development Goals, the Organization for Economic Cooperation and Development (OECD) undertakes a portfolio of work to support low-emission climate-resilient pathways [1], and from 2021 the World Bank will double the funding available for countries to take ambitious climate action equating to $200 billion over 5 years [2]. These initiatives build on earlier work led by the United Nations Framework Convention on Climate Change (UNFCCC) including the Kyoto Protocol, which set the first global emissions targets in 1997, the Cancun Adaptation Framework, which recognized that adaptation should be given the same priority as mitigation in 2010, and the 2015 Paris Agreement, that produced the most recent global emissions targets.

Under the 2015 Paris Climate Agreement, 186 states and the European Union committed to reduce their emissions to limit global temperature rise to 1.5°C above industrial levels. Limiting global average temperature increase to 1.5°C will reduce the risks to human and natural systems [3]. That said, 2019 was already 0.95°C warmer than preindustrial levels, and although the level of preindustrial warming varies annually (e.g., 2018 was 0.83°C warmer), experts believe that current policies and practices will not deliver on the Paris Agreement and that a global warming of 2.5–2.9°C by 2100 is more realistic [4,5] (Fig. 3.1). The financial implications of climate change

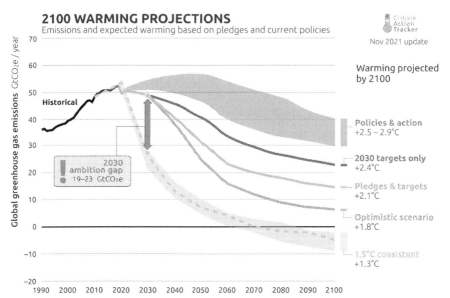

Fig. 3.1 The amount of global warming projected for different emissions pathways linked to current policies, pledges, and targets.
From Climate Action Tracker, 2100 Warming projections: emissions and expected warming based on pledges and current policies, November 2021. Available at: https://climateactiontracker.org/global/temperatures/. Copyright © 2021 by Climate Analytics and NewClimate Institute. All rights reserved.

are stark; without action, the cost of climate impacts is expected to rise to between 1% and 3.3% of global GDP (Gross Domestic Product) by 2060, with a central projection of 2% [6]. The OECD notes that changes in crop yield and labor productivity have the greatest negative impacts, followed by sea-level rise from the 2015s onwards. Negative impacts are particularly large in Africa and Asia.

The global scientific evidence base for climate change is collated by the Intergovernmental Panel on Climate Change (IPCC) via its series of regular Assessment Reports. The most recent assessment, the Fifth Assessment (AR5) was completed in 2014, and the reports describe the physical science basis for climate change [7], the impacts, adaptation, and vulnerability (global [8]; regional [9]) and aspects of mitigation [10]. The next assessment reports are due in 2022. The IPCC also produces other specialist reports such as the special report on 1.5°C warming, or climate change and the land [11]. Other reputed sources of scientific information on climate change include the United State National Oceanic and Atmospheric Administration (NOAA) produce an annual State of the Climate report (e.g., Blunden and Arndt [12]), the World Meteorological Organization (e.g., WMO [13]), Climate Impact Lab [14], and numerous national weather and climate services, globally, such as the UK Met Office, who provide global and regional (Europe and United Kingdom) climate change projections via the UK Climate Impacts Program [15].

3.1.2 Climate change and the railway sector and key definitions

Although climate change varies regionally, there are several generalizable climate impacts of relevance to the railway industry. These include an increase in the frequency of extreme weather such as droughts, heavy rainfall, or heatwaves, an increased frequency of wildfires, and rising sea levels [3]. Railway organizations are accustomed to dealing with poor weather conditions as part of daily operations and typically have established procedures to reduce delays and disruption to customers, and the broader socio-economic consequences such as loss of productivity (e.g., MOWE-IT [16]). Operational resilience forms a sound basis for climate resilience, not least as future climate impacts may be an increase in the frequency or intensity of existing weather events. However, other climate impacts are more gradual, and therefore, must be considered in long-term strategic planning to ensure that new and existing infrastructure is suitable for future climate. This is particularly relevant when considering long-life infrastructure that may be operating in a different climate to that in which it was designed or constructed, or in a region that may experience sea-level rise, or an increase in the frequency or intensity of wildfires or extreme weather events. Failure to consider climate change can lead to increased costs or delays in the future; Dobney et al. [17] estimate that without action, the cost of buckling due to high temperatures would double on the UK railway network by 2080s under a high emissions scenario. In a study on the London Underground, Greenham et al. [18] describe how without action, the frequency and length of heat-related days will increase by 29% and 30%, respectively, under a high emissions scenario. See Dawson et al. [19] for other climate impacts for the UK railway sector. In the United States, delays caused by increased temperatures could be $45 to $60 billion more by 2100 under a high emissions scenario [20].

Climate Adaptation describes the ongoing process by which organizations make changes to enable climate resilience, both now and in the future. This may include physical changes to infrastructure, such as increasing the stress-free temperature of continuous welded rail or retrofitting overhead lines to include auto-tensioning to reduce line sag. It may also include organizational changes in the way services are provided, or changes in approach to asset management, such as developing adaptive management approaches (Box 3.1). Climate change adaptation should not be a special project undertaken by a separate group but should be integrated with standard operational and planning procedures as part of business as usual [28]. The railway sector has an established track record of dealing with risk; climate change is another risk, that brings short-term risks (e.g., increase in intensity and frequency of extreme weather events that we already experienced) and longer-term risks (e.g., sea-level rise) to railway infrastructure and operations.

By adapting to the challenges of climate change, railway infrastructure and operations can become climate-resilient. There are multiple definitions and applications of resilience, across the fields of academia. These include definitions stemming from engineering resilience, i.e., the ability of a system to be consistent, efficient, and predictable, and ecological resilience, i.e., amount of disturbance a system can take before it changes to a new steady state (after Holling [29]). Wang [30] builds on these

Box 3.1 Adaptive management

Adaptive management describes a stepped process to adaptation, whereby each phase of adaption reduces risk to an acceptable level as the environment changes, and adaptation measures and options are regularly reviewed and assessed as part of an iterative process. The progression to a new step of adaptation is not determined by fixed times frames, as in traditional project management, but, condition monitoring, maintenance needs and inspection, and updated climate information, brought together through frequent risk reassessment. Moreover, all options for adaption to a range of potential future climates are considered at the outset, to ensure that early adaptation does not prevent future adaptation measures, and contributes to the long-term plan of adaptation. This approach delivers the best cost-benefits over the lifespan of the infrastructure asset or system, not just in the short-term.

This approach to climate adaptation produces a range of flexible "adaptive pathways" that permit dynamic planning over time that can response to the uncertainties of climate change, such as the uncertainty of the global emissions pathway, or the uncertainty associated with climate change projections. When visualized, these pathways look like a metro map, showing different routes to reach the same destination [21,22]. In practice, adaptation pathways and the process of adaptive management have been adopted in New York, following the impact of Hurricane Sandy [23], for the Thames Barrier, London, United Kingdom [24], and for river and coastal management in the Netherlands [25], Australia [26] and New Zealand [27], and more.

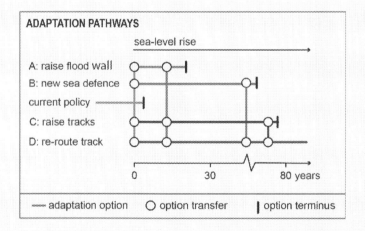

> The figure shows a series of potential options to adapt to sea-level rise and coastal flooding. In this scenario, the current policy is ineffective, with regular inundation of the tracks representing an unacceptable level of risk. Option A will address the problem in the short-term, but options B and C offer longer-term protection at greater expense, but cannot be undertaken together. Only option D offers long-term adaptation under high emissions scenarios, but Option D can be selected after options A to C are no longer viable. The railway organization should decide which pathway to follow, and use environmental indicators (e.g., height of storm surge, frequency of flooding) to decide when to switch pathway. Diagram adapted for rail sector from Ref. [22].

concepts and defines "comprehensive transportation in resilience" as a system that has the qualities of "recovery, reliability and sustainability." Other studies (e.g., Reggiani et al. [31] and Mattsson and Jenelius [32]) have examined the relationship between resilience and vulnerability across a spectrum of transport research. In this best practice handbook, we use a practitioner-focused definition of resilience: "the ability of the transport network to withstand the impacts of extreme weather, to operate in the face of such weather and to recover promptly from its effects" [33].

3.1.3 Rail adapt: Best practice from the International Union of Railway

Acknowledging the need for climate change adaptation within the rail sector, the International Union of Railways (UIC) commissioned the Rail Adapt Project in 2016/17. The project built on previous projects such MOWE-IT (Management of Weather Events in the Transport System; [16]); WEATHER2 (Weather Extremes: Impacts on Transport Systems and Hazards for European Regions); EWENT (Extreme Weather impacts on European Networks of Transport); and UK-based Tomorrow's Railway and Climate Change Adaptation [34]. The Rail Adapt Project had a dual purpose; firstly, to inform and update UIC members about climate change matters, and secondly, to work with members and other stakeholders to develop a framework for climate adaptation for use by the railway sector. The project ran workshops in Vienna (October 2016), London (April 2017), China (June 2017), and Morocco (September 2017), that were attended by over 50 organizations from 20 countries, mainly by delegates from the rail sector, but also from other transport organizations, academia, and consultancy [28]. These workshops facilitated knowledge exchange between UIC members and the other stakeholders of initiatives, experiences, and best practices in the field of climate resilience. This wealth of tacit knowledge was combined with a review of international best practices to develop a two-sided framework for climate adaptation. The Rail Adapt Framework is suitable for any transport organization, or division of an organization, to increase resilience to a changing climate, and its associated hazards such as extreme weather and sea-level rise. The Rail Adapt Framework is presented in Section 3.3, and the detailed Rail Adapt report [35] is available to download for free from the UIC website. Quinn et al. [28] provide greater detail on the co-creation process of the framework for adaptation, including stakeholder engagement and feedback.

3.1.4 Chapter structure

This introduction has provided an overview of the global context of climate change. The current trajectory of global emissions indicates that regional climates will continue to change, global sea levels will rise, and extreme events such as droughts, heavy rainfall, high temperatures, and wildfires will continue to increase in their frequency of occurrence [3]. As such, it is imperative that the railway sector, whose infrastructure and strategic planning can span decades into the future, considers climate change adaptation as part of business as usual in order to be climate-resilient. Section 3.2 reviews best practice in climate adaptation from across the infrastructure sector, including multinational, national, regional, and local examples. Many of the examples are from the United Kingdom, United States, and Europe, where climate change adaptation is currently practiced more widely. Section 3.3 describes the Rail Adapt Framework, which can be used by any rail organization, regardless of their current level of climate change awareness and preparedness, in order to increase their climate resilience. Section 3.4 concludes this study and provides future directions of research and further sources of information. Two Boxes contain information that cross-cuts the different sections of this report on Adaptive Management (Box 3.1), and Interdependencies (Box 3.2).

Box 3.2 Interdependencies

When one infrastructure system is dependent on another in order to operate, they are considered to be interdependent. Indeed, most infrastructure sectors rely on power to operate some, if not all of their assets [19]. The railway sector needs power to run the trains, and to operate signage, and ICT is essential for effective communication. Therefore identifying which other infrastructure sectors that assets depend on is essential to understanding asset vulnerability and is an important part of the assessment process (Section 3.3.2.2). Failure to consider interdependencies can give a false impression of total system resilience [36].

The TRaCCA Project examined the potential impacts of interdependencies on the railway system [34]. TRaCCA identified several elements that could be interdependent including; people, materials and/or buildings, machinery, fuel and power, and communications and data, and considered how a failure in an external system (i.e., the supply of power), for example due to flooding, could impact railway operations.

External system	Impact of flooding	Potential impact on railway
Electrical power	Power station shut down by flooding	Power shortages for traction, signage, communications, depots, stations
Water systems, e.g., drainage, sewerage	Flooding overloads drainage. Disruption to clean water supply	Lack of clean water/sanitation problems at stations
Flood defenses	Flooding of houses Bridge damage	Staff shortages due to flooded homes Flooded stations /rail underpasses Bridge damage requires inspection
Other transport systems	Flooding closes roads	Staff/passengers cannot reach or leave railway stations Machinery for rail works cannot be moved
Fuel supply	Flooded roads prevents fuel supply	No fuel for railway operations

After Rail Safety and Standards Board (RSSB), Tomorrow's Railway and Climate Change Adaptation: Work Package 1 Summary Report. The Arup TRaCCA WP1 Consortium (Arup, CIRIA, JBA Consulting, the Met Office and the University of Birmingham) in Collaboration with the RSSB Project Team, RSSB, London, UK, 2016 and A.D. Quinn, A. Jack, S. Hodgkinson, E. J.S. Ferranti, J. Beckford, J. Dora, Rail Adapt: Adapting the Railway for the Future: A Report for the International Union of Railways (UIC), UIC, Paris, France, 2017.

In order to address the complexity of interacting and cross-cutting risks, research in support of the UK Climate Change Risk Assessment (CCRA), developed a series of systems maps that depict inter and cross sectoral interactions for different climate drivers and sectors [37]. Modeling based on these systems maps showed that different climatic effects can compound each other, increasing the total overall risk. Other approaches to studying interdependencies include Pescaroli & Alexander [38], who developed a framework that classifies interdependencies into four types: compound, interconnected, interacting, and cascading. Murdock et al. [39] outline a method for quantifying disruption caused by different failures that could be replicated by practitioners, and visualize the information using interdependency circle diagrams. For a review see Val et al. [40].

3.2 Best practice in climate adaptation and resilience

3.2.1 Multinational examples

There are several approaches to climate adaptation and resilience that span countries and continents. From 2013 to 2019, the 100 Resilient Cities program led by The Rockefeller Foundation provided financial and technical support to increase urban resilience for a global network of member cities thereby providing a valuable resource of global best practice, including several tools and publications [41]. Although the initiative did not focus solely on climate change adaptation, climate change, coastal flooding, and transportation are featured resilience challenges. The European Union has a Climate Change Adaptation Strategy, which was adopted in 2013. This has three main priorities; (1) to encourage action by member states; (2) to support better-informed decision making; and (3) to promote adaptation in key EU vulnerable sectors, of which infrastructure is considered a sector [42]. As part of mainstreaming climate change adaptation across Europe, the strategy requires that infrastructure providers undertake resilience assessments before obtaining funding from European Regional Development Fund and Cohesion Fund. For example, in Slovakia, climate change risks and vulnerabilities, specifically floods and future extreme weather, have been included in the modernization of the key passenger and freight corridor connecting Bratislava, Žilina, and Košice which is under development with finance for the EU [43]. With the exception of Croatia, Bulgaria, and Latvia, all member states have a National Adaptation Plan, many of which include adaptation measures for infrastructure. For a review of the quality of national adaptation plans, see Woodruff and Regan [44] and references therein. The EU also supports the Climate-Adapt portal [43], which collates information on climate change impacts, vulnerability, and adaptation to support decision making. This includes a database of case studies with 286 sources of data for transport adaptation including reports, case studies, tools, and links to ongoing research. The Organization for Economic Cooperation and Development [45] also provides financial and policy support for climate-resilient infrastructure (e.g., Vallejo and Mullan [46]).

Within the infrastructure sector, World Road Association (PIARC) and the World Association for Waterborne Transport Infrastructure (PIANC) both developed materials for climate adaptation for their members. PIARC has developed a framework for climate change adaptation that supports the transition to climate-resilient infrastructure via four steps; (i) identifying scope, variables, risks, and data; (ii) assessing and prioritizing risk; (iii) developing and selecting adaptation options; and (iv) integrating outcomes of (i)-(iii) into decision making [47]. PIANC also advocate a four-stage methodology to increase resilience to climate change, namely; (i) understanding the impact of climate change on assets, operations systems and identifying who needs to be involved; (ii) identifying the climate information required; (iii) assessing the risk to waterborne transport assets; and (iv) presenting a portfolio of potential measures (structural, operational, institutional) to be considered when developing an adaptation pathway [48].

3.2.2 National examples

In the United Kingdom, infrastructure adaptation to climate change is legislated under the 2008 Climate Change Act via the Adaptation Reporting Power (ARP) process, which includes Climate Change Risk Assessments (CCRAs), which feed into the National Adaptation Program (NAP). Under the ARP, organizations of a public nature, such as major infrastructure owners and operators, or other bodies such as the Environment Agency should report how they are addressing the impact of climate change. The CCRA is a governmental-level risk assessment of the impacts of climate change, with a chapter specifically on infrastructure [19,49]. The CCRA process determines the urgency with which adaptation actions should be prioritized for a number of key risks areas, for instance, risks to transport infrastructure from extreme heat. This prioritization is based on an assessment of the magnitude of the current and projected future climate impacts in that area (with a consideration on how the system or sector may change in the future), and the extent to which current and future planned adaptation will address this risk. As with most national assessments, this is primarily achieved through a review of published evidence, with supplementary quantitative work carried out where necessary (e.g., national flood risk modeling to support sectoral assessments). Both the ARP and CCRA feed into the NAP, and the process is repeated every 5 years. The NAP lays out high-level adaptation objectives and actions for government, businesses, and society. Studies of the two previous rounds of the ARP (2010–12 and 2013–16) are positive, with 78% of reporting organizations reporting that the process led to organizational change in climate change adaptation [50], with benefits including corporate-level awareness and action of climate change risk and adaptation, and integration of climate change adaptation as business as usual [51]. Progress toward meeting the actions laid out in the NAP is similarly reviewed every 5 years by the Commission on Climate Change (CCC), with any identified adaptation shortfall feeding back into the CCRA in a cyclical process (e.g., CCC [52]). As an example, the most recent review indicated good progress in adaptation planning in the rail sector, while identifying a need for improved recording of indicators to give a better sense of how vulnerability to climate risk is changing.

The most recent United States National Climate Assessment [53] presents a review of published evidence on climate risk in 15 sectoral chapters (including transport). The assessment is more narrative and descriptive in nature compared with the methodological urgency scoring framework presented in the CCRA. The assessment centers on awareness-raising through "key messages," as well as a focus on sectoral engagement through workshops and a greater involvement of practitioners in the chapter team. In parallel with the sector-based chapters, a further 10 chapters explore regional climate risk on a holistic/whole-systems basis, developing structures based on their particular circumstances. For instance, the North East [54] identified five key focus areas including "maintaining urban areas and communities and their interconnectedness." This regional basis allowed engagement with a diverse range of decision makers and a forum to discuss interdependencies between sectoral risks, such as failure in energy supply cascading through to critical systems supporting transport operations, such as subway signals.

Other studies examine national adaptation plans. For example, Bauer and Steurer [55] contrast national adaptation in Germany and the Netherlands within the context of long-term water management. These countries have contrasting approaches to adaptation. Germany (like the United Kingdom) has a multisectoral approach to adaptation with national measures that are followed at provincial level and by other organizations. The Netherlands uses a sectoral approach, namely the Dutch Delta Program [56]. In Finland, The Climate Policy Program for the Ministry of Transport and Communications' administrative sector 2009–20 requires a series of adaptation measures, with annual progress reporting. The Finnish Transport Agency is implementing rail-specific adaptation measures such as strengthening and protecting structures, developing weather warning and monitoring systems with weather service providers, and improving rescue services and safety information (from Ref. [35]). In another rail case study, Lingren et al. [57] interviewed key personnel within the Swedish Rail Administration to understand their approaches to, and understanding of, climate change adaptation. Their conclusions, which align with the Rail Adapt Project [35] are relevant to all rail organizations, namely; (i) that systematic mapping of climate change vulnerabilities and consequences is essential; (ii) climate change needs to be considered early within the planning processes and included in risk and vulnerability assessments, which may need modification to incorporate future-orientated goals; and (iii) when designing adaptation goals, other requirements should be considered in order to prevent counterproductive measures, and also to look for mutual benefits, e.g., aligning climate and environmental goals [57].

3.2.3 Regional and local examples

Within the United Kingdom, the ARP process has mandated infrastructure organizations such as Network Rail and Highways England to develop climate change adaptation plans. Network Rail has developed Weather Resilience and Climate Change Adaptation plans (2014–19) for each of their eight routes, that currently experience different weather-related challenges, and will experience regionally different future climates [58]. For example, the southeast of the United Kingdom will be warmer and drier, particularly in summer, posing problems such as signaling equipment overheating, and also for embankments, which are predominantly composed of moisture-sensitive soil types [59]. The northwest of the United Kingdom will experience more frequent heavy rainfall events, especially in winter, leading to impacts such as landslips and flooding. These WRCCA plans are currently being updated for the 2019–24 planning period. Network Rail also contributed to the Tomorrow's Railway and Climate Change Adaptation Program that developed a set of decision-making tools and information for railway resilience [34,60]. Highways England has developed a framework for climate adaptation, and via the ARP process has identified the potential climate hazards, vulnerabilities, options for adaptation, and an adaptation timeline, which is currently under review as the part of the next round of ARP [61].

At local and regional levels, several projects have applied adaptive management as part of their climate adaptation process (see Box 3.1). This approach was pioneered by the Thames Estuary 2100 Project, which developed a 100-year plan to manage flood risk in and around London, United Kingdom. In order to address the uncertainty of future climate projections, the project came up with a series of future possible climates, including a low-probability high-end climate change scenario, and then developed a series of adaptation options for the different climate impacts associated with the different scenarios. These were combined into adaptation pathways, and environmental indicators (in this case, rate of sea-level rise, peak flow heights, peak surge heights) will determine which pathway is followed and therefore which adaptation measures are required (see Ranger et al. [24]). Building on this work, New York City began developing adaptation pathways to increase resilience to coastal flooding, using environmental indicators, and regular 3-yearly reviews as part of its climate change adaption strategy [62]. These adaptation pathways initially focused on avoiding disruption to current systems, but following the damage caused by Hurricane Sandy in 2012 (> US$19 billion), which took the level of risk to an unacceptable level, these evolved into pathways that delivered transformative change on a regional level [23]. San Francisco has also utilized adaptation pathways to manage the impact of future sea-level rise. The San Francisco Bay area has critical road and rail networks located in a region prone to flooding. As part of the Adapting to Rising Tides project [63], Dynamic Adaptive Plans were developed to protect transport infrastructure, including the Capital Corridor intercity rail route from different levels of climate change and associated sea-level rise [64].

3.3 Rail Adapt Framework for climate change adaptation

3.3.1 Introduction

The Rail Adapt Framework (Fig. 3.2) was co-created with railway stakeholders [28], specifically for the rail sector, although the flexible design makes it suitable for other businesses and organizations in the transport sector, and more broadly. Through an iterative circular process, it enables organizations or divisions of organizations to increase their preparedness for climate change, by adapting their infrastructure and operations and thereby increasing their climate resilience. This circular, iterative process is particularly important to address the inherent uncertainty of climate change [65], such as the uncertainty of the climate impacts (e.g., the change to heatwaves or droughts), and the uncertainty of the socio-economic factors that determine the emissions pathway, and the response to a changing climate (see Section 3.1.1).

The Rail Adapt Framework also encourages links with stakeholders, in different parts of the same organizations, and in other sectors. Railway infrastructure rarely operates in isolation and has interdependencies (Box 3.2) with the energy sector, who provide power for railway operations, and the ICT sector (Information and Communications Technology) enables communications. Infrastructure is only as strong as its weakest link, and failure at a critical node, sometimes referred to as a "single point of failure" (e.g., Dft [33]) can lead to cascade failures across multiple infrastructure sectors. For example, the flooding of an electricity supply substation in Lancaster in December 2015 left the city without power for over 30h and affected all critical

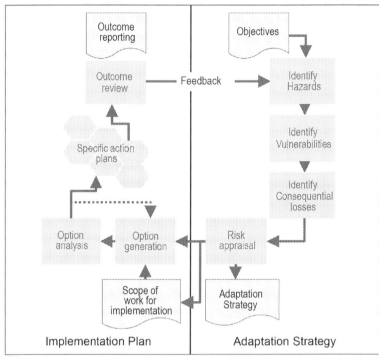

Fig. 3.2 The two-sided Rail Adapt Framework.
Modified from A. Quinn, E. Ferranti, S. Hodgkinson, A. Jack, J. Beckford, J. Dora, Adaptation becoming business as usual: a framework for climate-change-ready transport infrastructure, Infrastructures 3(2) (2018) 10.

infrastructure including road and rail transport, ICT, water supply, and emergency services [66]. Moreover, the impact of a hazard such as flooding or landslip can have a potentially far greater impact where infrastructure is co-located, such as electricity cables built into bridges (e.g., Booth et al. [67]). Accordingly, working with other stakeholders and across infrastructure sectors to identify interdependencies and those weakest links with the potential to cause disproportionate impact across multiple infrastructure sectors is crucial.

3.3.2 Developing an adaptation strategy

3.3.2.1 Defining objectives

Setting the objectives is the first step toward developing an adaptation strategy. The strategy may be needed to comply with external requirements, such as the UK's Adaptation Reporting Power (ARP), or to draw down finance from an international fund such as the European Union, or the World Bank, for national/regional transport planning, or part of internal strategic planning processes, or all, or more. Understanding the different contexts for adaptation, and setting the objectives accordingly, is crucial to prevent conflicts between the requirements for different stakeholders, and prevent duplication of effort. The objectives should also link to and support the existing practices for risk management within an organization. In many respects, climate change is just another risk, and should therefore be considered within existing risk assessment and management procedures, as part of business as usual, rather than a separate project. Climate change adaptation may be considered alongside existing International Organization for Standardization (ISO) standards for asset management (ISO55000), risk management for safety and financial planning (ISO31000), organizational resilience (ISO22316), quality assurance (ISO14000/9000), or social responsibility (ISO26000). The ISO is currently developing standards specifically for climate change adaptation. The first, Adaptation to climate change—Principles, requirements, and guidelines (ISO14090) were published in June 2019.

3.3.2.2 Hazards, vulnerabilities, consequences

Fig. 3.3 expands the Rail Adapt Framework to permit consideration of the hazards, vulnerabilities, and losses, which forms the next steps toward developing an adaptation strategy. Climate change is associated with a range of hazards, from changes in

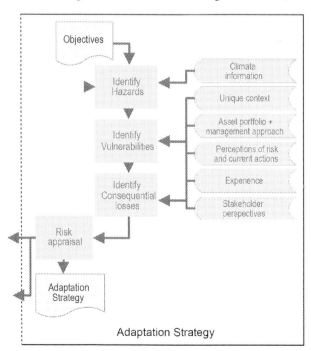

Fig. 3.3 Expanded view of the adaptation strategy from the Rail Adapt Framework.
Modified from A. Quinn, E. Ferranti, S. Hodgkinson, A. Jack, J. Beckford, J. Dora, Adaptation becoming business as usual: a framework for climate-change-ready transport infrastructure, Infrastructures 3(2) (2018) 10.

the frequency or intensity of extreme weather events such as heavy rainfall, droughts, or heatwaves, and nonmeteorological hazards such as wildfires or sea-level rise. These vary regionally and will vary depending upon the future emissions pathway (Section 3.1.1), with the greatest impacts associated with the higher emissions pathways, i.e., a greater degree of global warming compared to preindustrial levels [3]. It is also important to consider sequences of meteorological events, such as a period of drought followed by heavy rainfall. For example, the power loss in Lancaster in 2015 was caused by a combination of 8 weeks of rainfall, an incoming tide, and an unusually heavy 2-day rainfall event [66], and rail delays in July 2015 were a combination of high temperatures, heavy rainfall and lightning strikes [68]. These are called multihazards, cascade hazards, or compound events (e.g., AghaKouchak et al. [69]), and there is limited understanding of how the frequency of such multihazard meteorological events will change in the future.

Identifying the future climate change hazards for a regionally or nationally operating rail or railway organization is complex, and will likely require information and possibly external support from weather and climate service providers [35]. Information is available from Copernicus Climate Change Service [70], which provides data from European Union's Earth Observation Unit and from the IPCC Assessment Reports that describe the potential impact of climate change on human and natural systems at a regional scale [9]. The UK Met Office also provides global climate change projections, including 12 km resolution projections for Europe, as part of the UK Climate Projections (UKCP) program, for a range of different emissions scenarios [71]. There are multiple UKCP18 products available for the United Kingdom, including high-resolution climate projections (2.2 km) and observational datasets (e.g., Fig. 3.4). It is important to link the timescales of the climate information to infrastructure lifespans. For example, a regional increase in temperature is likely to be gradual, and therefore will have limited impact on short-life assets such as telecommunications, but may be relevant for long-life infrastructure such as underground or tunnel infrastructure. An adaptive management approach is particularly important for long-life infrastructure (Box 3.1). As railway stakeholders may not have a background in climate science, they may require external guidance on which projections or emissions scenarios to use, or support with different data formats, perhaps from a specialized agency or consultancy [72].

Understanding the vulnerability or susceptibility of railway infrastructure and operations to future changes is also complex. Known weather impacts may occur more frequently (if extreme events occur more frequently), or with greater intensity, and such changes in frequency or intensity can be adapted accordingly (e.g., Dobney et al. [17]). In some rail organizations, the vulnerability of infrastructure and operations to climate change hazards may be unknown. In this step of the framework, combining observational datasets with railway datasets on performance or faults (e.g., Ferranti et al. [68,74]) can quantify known links between weather and performance. However, this relationship may not be straightforward as infrastructure failure can be a combination of multiple causes such as extreme weather and poor maintenance, or multiple combinations of extreme weather, for example, drought conditions followed by heavy rainfall. Another approach is to use analogs to understand how infrastructure

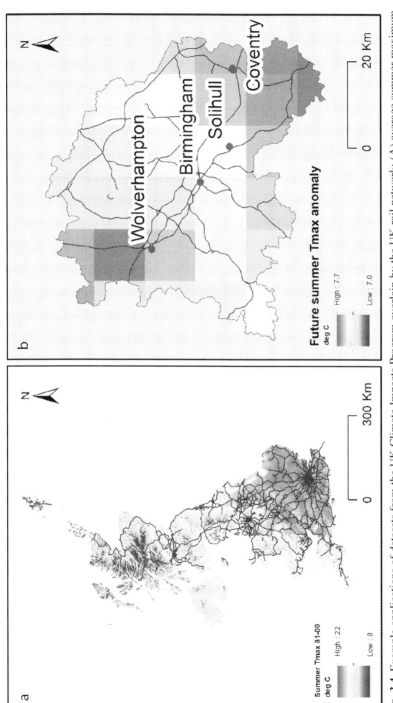

Fig. 3.4 Example applications of datasets from the UK Climate Impacts Program overlain by the UK rail network; (A) average summer maximum temperature in °C between 1981 and 2000 [73], and (B) for the West Midlands, probabilistic projections of the future maximum temperature anomaly for summer 2080–99, as compared to 1981–2000, as produced by the 13th ensemble member for RCP8.5, equivalent to global average temperature rise of 4.3°C by 2100. Produced under an open government license. http://www.nationalarchives.gov.uk/doc/open-government-license/version/3/.

may be vulnerable to future climates. Sanderson et al. [75] developed climate and railway analogs for the United Kingdom, i.e., those countries that already have a climate similar to that projected for the United Kingdom in the future, and have railway infrastructure comparable to that used within the United Kingdom. To support decision making with these uncertainties, the TRaCCA Project [34] developed a series of diagrams that summarized the links between extreme weather and specific problems, and whether action could be taken, or whether this was a knowledge gap, in order to prioritize next actions. Where data is limited, a traffic-light approach can be used to define whether vulnerability is high, medium, or low, for the different assets, processes, or geographical information, to enable the vulnerability assessment to take place. The iterative nature of the Rail Adapt Framework ensures that as knowledge increases, this can be incorporated into the adaptation strategy.

Once the hazards and vulnerabilities have been identified, the consequential losses can be estimated. These will include the direct repair costs of failure and compensation to affected passenger or freight users. For example, Network Rail (United Kingdom) estimate that weather management costs between £100–£200 million per year, although this is likely an underestimate as attributing equipment failure to extreme weather is not always possible, and the estimate does not include the cost of managing a weather event [76]. The broader socio-economic costs and interdependent stakeholders should also be considered, particularly if the infrastructure can be considered a "single point of failure" [33], such as a railway line serving a remote community. After coastal flooding destroyed the railway track at Dawlish, a coastal town in southwest England located along a line that provided important freight and passenger connections along a peninsula in the United Kingdom, the track repair cost an estimated £35 million [77]. However, the broader socio-economic impact of the line closure on tourism, fishing, was considered to be between £60 million and £1.2 billion, with some businesses reporting losses of £100–1000 each day the line was closed [78].

3.3.2.3 Risk appraisal

The risk appraisal prioritizes the next steps for climate adaptation. The information collated on hazards, vulnerabilities, and consequential losses to service provides the evidence base for a risk appraisal. This identifies the short-term risks that the organization needs to adapt to in the near future, such as extreme weather, and the longer-term risks, such as rising sea-level or gradual temperature increase that will be relevant when designing new infrastructure with a long lifespan. It is wise to consider short and long-term risks together to avoid maladaptation or retrofitting, i.e., where action taken in the short-term creates a problem that needs to be solved in the future. There are multiple approaches to dealing with risk, and organizations often have an existing methodology as part of routine risk management and assessment. The UK Treasury provides general advice on how to appraise and evaluate policies, projects, and programs [79] and the CCRA process has a bespoke system for assessing and scoring risk [19,80].

Part of the risk assessment is a financial appraisal that takes into consideration the beneficial costs of adaptation action, as compared to the costs of inaction, and the available resources within the organization. This should include the costs to the railway organization, and broader socio-economic costs, as discussed in Section 3.3.2.2. Stakeholder engagement via the Rail Adapt Project raised several important points:

- Adaptation costs can be mitigated via existing procedures or strategies for asset management, thereby integrating it as part of business as unusual. This prevents adaptation from being considered an optional extra, or an extra cost, within a separate activity stream, and thereby reduces the potential for siloisation or duplication of effort.
- Although technically possible from an engineering perspective, the cost of delivering a seasonally agnostic railway (i.e., one which is operational in all weather conditions) may exceed the benefits to passengers or freight users and may create unacceptable risks to railway employees who work during extreme weather.
- It is important to acknowledge and act when there becomes a financial crossover where the cost of additional adaptation outweighs the benefits derived. For example, adaptation to prevent the repeated flooding of a coastal section of railway in an area of rising sea-level will 1 day become financially unsustainable.
- By considering the whole life value of an asset, and the range of operating conditions it may experience as the climate changes, will deliver the best value solution over time (see Box 3.1 on Adaptive management).

3.3.3 Implementing adaptation plans

3.3.3.1 Option generation and analysis

The right-hand side of the Rail Adapt Framework describes how to develop an adaptation strategy, including setting objectives, identifying hazards, vulnerabilities and consequences, and prioritizing areas for adaptation, via the risk appraisal (Section 3.3.2). The left-hand side of the framework (Fig. 3.5), the implementation plan, draws on this work to generate adaptation options for each identified risk, that take into account technical, social, environmental, and financial constraints. These

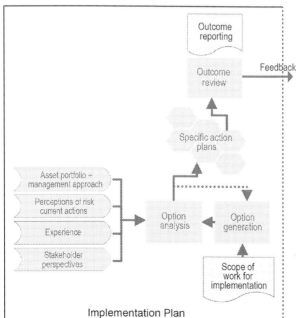

Fig. 3.5 Expanded view of the implementation plan from the Rail Adapt Framework. From A. Quinn, E. Ferranti, S. Hodgkinson, A. Jack, J. Beckford, J. Dora, Adaptation becoming business as usual: a framework for climate-change-ready transport infrastructure, Infrastructures 3(2) (2018) 10.

feed into an action plan, the outcomes of which are reviewed in due course, feeding forwards into the next iteration of the adaptation strategy.

Developing a diverse range of options to enable a genuine options analysis is crucial. Financial constraints can be a concern for railway organizations, but this should not inhibit the desired outcome. Costs should also be considered of asset lifespan, particularly for longer life assets that may experience greater and more frequent climate impacts in the future. There may be multiple approaches to increasing resilience to the identified risks, including engineering solutions, operational planning, and organizational enhancements. Moreover, approaches should consider short and long-term actions within the context of the changing climate and the infrastructure lifespan. For example, replacing an asset that has failed due to extreme high temperatures with one of identical specifications will quickly restore operations. However, if high temperatures are occurring or projected to occur more frequently, then replacing the asset with a more heat-resistant alternative may be more appropriate. Raising the height of assets can reduce coastal flood risk in the short-term, but depending on the rate of sea-level rise, the long-term relocation of assets may be more appropriate. Considering the long-term options prevents maladaptation whereby short-term solutions prevent long-term requirements that can lead to costly retrofitting. The MOWE-IT project [16] provides a range of options for reducing the impact of weather on transport operations. Similarly, the TRaCCA Project produced practitioner-focused fact sheets on managing winter conditions, flooding, and heat [34]. Option generations should also consider external stakeholders; managing flood risk may be mutually beneficial for other infrastructure managers with co-located assets that are similarly vulnerable. Lastly, the other multifunctional benefits of the options should be also considered such as the opportunity to reduce carbon emissions or increase capacity.

Adopting an adaptive management approach should be central to option generation and analysis. By developing adaptation pathways (e.g., Box 3.2), the relative merits and lifespans of the different adaptation options can be collated and compared to support decision making (e.g., Ranger et al. [24]). The iterative and circular nature of the Rail Adapt Framework supports adaptive management by triggering a review of the hazards, vulnerabilities, and consequences at regular interviews, and thereby consideration of whether moving to a new phase of adaptation is appropriate.

Railway organizations will have existing processes for options appraisal and the development of action plans as part of their asset management, and these should be applied, where possible, to the different options for climate change adaptation in order to integrate this into business as usual. However, it is important to note that the potential disruption of potential weather events that occur at an unknown frequency at some point in the future can be troublesome to incorporate into existing tools such as Cost Benefit Analysis (CBA), which tend to have a short-term focus. Other methods, such as Multi-Criteria Analysis (MCA) that supports decision making where factors other than costs are important may be appropriate for adaptation and option assessment. Dodgson et al. [81] provide a manual of approaches for MCA; the World Road Association (PIARC) compares CBA and MCA within the context of infrastructure investment within their framework for adaptation [47]. Developing adaptation pathways with costs calculated over the lifespan of the infrastructure asset can also assist economic appraisal (e.g., Haasnoot et al. [21]).

3.3.3.2 Implementation of specific action plans

The implementation of the action plans created by the options analysis will vary dependent on the organization, and the action plan itself. Some actions, such as upgrading overhead lines to reduce heat-related line sag may start immediately, but take place over several months and years, dependent on the size of the railway network. Other actions may be completed within a shorter time frame. Regardless of the time frame or type of action plan, there are four underpinning principles to contemplate:

- Infrastructure or design decisions should reflect the climate of the future, as appropriate for the lifespan of the system.
- The future socio-economic environment, and how the usage of the system may change, must be considered (e.g., Jaroszweski et al. [82]).
- Implementation should utilize the best information available from both internal and external sources.
- Implementation should integrate where appropriate with existing procedures of assessment management and investment, i.e., become part of business as usual.

3.3.3.3 Outcome review

Reviewing the outcomes of the implementation actions, and feeding this forward into the next iteration of Adaptation planning is the final step of the implementation plan. The approaches used to evaluate the implementation actions and improvements to climate resilience can be quantitative or qualitative and could include metrics such as performance to schedule or lost customer minutes, or passenger surveys on system changes. The review process should also recognize changes in organizational capacity, for example, increased knowledge of climate change, or new organizational skills such as climatological analysis of data to create the evidence base for risk appraisal. The review outcomes should also be shared with other relevant stakeholders, such as those who helped shape the initial objectives, those who will be involved in future iterations of the adaptation framework, or other organizations with interdependent infrastructure impacted by the climate adaptation strategy. Broad dissemination of the challenges and achievements of climate adaptation is crucial for developing tacit knowledge within the organization, thereby embedding it in business as usual for the future.

3.3.4 Managing and supporting adaptation

For successful climate adaptation and short and long-term climate resilience, an organization must have the capacity to develop the climate adaptation strategy, and implement the necessary actions, and iteratively review and evaluate the adaptation process through time, albeit with the support of external experts (e.g., climate service providers) where appropriate. Fig. 3.6 simplifies a typical organizational hierarchy and climate adaptation awareness, knowledge, and experience are required by individuals and divisions at all these levels. Those at government or regulatory level must use their climate awareness to set objectives and rules governing long-term investments that link to international and national drivers for climate change adaptation such as the UN Sustainable Development Goals, the Paris Climate Agreement, National

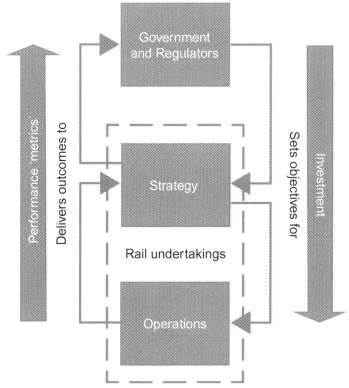

Fig. 3.6 Typical organization hierarchy.
From A.D. Quinn, A. Jack, S. Hodgkinson, E.J.S. Ferranti, J. Beckford, J. Dora, Rail Adapt: Adapting the Railway for the Future: A Report for the International Union of Railways (UIC), UIC, Paris, France, 2017.

Adaptation Plans, and more (see Section 3.2). Those responsible for strategic planning within railway organizations need to understand these international drivers, to ensure that adaptation within their organization aligns with and meets the external requirements such as the UK Adaptation Reporting Power, and also to access sources of funding for climate adaptation, such as the Adaptation Fund administered by the World Bank [83]. They can also implement adaptation in strategic planning and asset management processes. At local or operational level, the individuals and teams must be climate change aware to recognize how service failures or disruption caused by extreme weather or other hazards such as wildfires links to current or future climate hazards. The tacit knowledge derived from resolving daily operational issues within a region is invaluable to identify hazards, vulnerabilities, and losses, to support the risk appraisal, and also to generate options for adaptation.

Capability (or capacity) assessments can be undertaken to evaluate an organization's level capacity to adapt to climate change. This could be achieved by undertaking "maturity assessments" as part of ISO55000, for which the UIC [84] has prepared practical guidance for railway organizations. The CaDD (Capacity Diagnosis and) Development software [85] is designed specifically for organizations to measure and improve their performance to adapt to climate impacts and undertake carbon managers. This tool was pioneered by infrastructure stakeholders who took part in the Rail Adapt Project, including HS2 and Network Rail [28]. The CaDD analyses provided baseline capacity assessments alongside a targeted action plan to deliver capacity targets.

3.4 Conclusions

This chapter has explored a framework within which decision makers in railway undertakings can design effective climate adaptation actions appropriate to their organizations. Climate change is unequivocal, and despite actions through civil society, government, and intergovernmental actors to reduce emissions, it is acknowledged that continued anthropogenic climate change is unavoidable, with adaptation to this change increasingly given equal importance alongside mitigation. Although national adaptation programs create a cyclical process where adaptation can be assessed and steered at a national level, including the steps of adaptation reporting (such as the UK's ARP), climate risk assessments (such as the UK's CCRA and the USA's NCA) and resultant national adaptation plans (e.g., the UK's NAP), organization-level adaptation requires its own decision-making process.

This chapter focused on the UIC-sponsored Rail Adapt, which builds on previous international transport-related adaptation projects to both understand the risks that climate change poses for railways, and to work with UIC members and other stakeholders to create a framework within which effective adaptation actions can be determined. Utilizing an iterative and circular process, the framework allows stakeholders to define their objectives (often partially determined by government or standards-mandated requirements), understand the expected hazards, vulnerabilities, and consequences associated with climate change, appraise the risk these poses, generate and analyze options to counter these risks, implement action plans based on these options and finally to review these actions, feeding the outcomes back into the adaptation process on that organization (and those organizations that share interdependencies). The framework operates at multiple timeframes, allowing stakeholders to identify short-term risks that need to be adapted to immediately, and longer-term risks such as sea-level rise, which can be designed into the specification of infrastructure renewal. As the process takes the organization as its unit of study, eliciting organizational (tacit) knowledge in the improvement of resilience, and measuring and improving organizational capacity is of high importance.

Finally, as adaptation takes on increased importance and decisions are made at increasingly long timeframes, it is essential that frameworks such as that described in this chapter take into consideration the rapid changes that are also taking place in the sector which may change the sector's exposure and vulnerability to climate hazards. It has generally been observed that concurrent changes in how a sector or system develops alongside climatic change are often lacking when exploring climate impacts and adaptation [86]. It is already acknowledged that the railways are experiencing a rapid period of change, encompassing developments that will affect both the vulnerability of infrastructure and operations to climate hazards (the move to moving block signaling and automation, increasing use of remote condition monitoring, improved downscaled weather forecasting, electrification and the introduction of alternative energy systems such as hydrogen) as well as its exposure (for instance the European Commission's 2011 White Paper targets for a modal shift from road to rail of 50% by 2050). The speed and extent of these changes are uncertain and will be joined by additional developments not currently anticipated, all driven by the underlying socio-economic development of the country or region in question (itself influenced by wider global trends).

Jaroszweski et al. [82,87] explored the integration of concurrent changes into climate impact assessment in a transport context. Scenario development and the exploration of qualitative storylines and quantitative indicators of what the future rail system may look like is a useful way of testing the robustness of adaptation decisions. By developing sector-specific scenarios from existing generic scenarios such as the Shared Socio-economic Pathways [88] decision makers can explore future visions of the rail sector paired to specific climate emission scenarios (Representative Concentration Pathways (RCPs). The use of consistent overlying scenarios also allows integration with similar impact and adaptation studies developed for linked sectors such as energy and the built environment. Further work is needed to explore how these scenarios should be utilized in frameworks such as Rail Adapt. Ideally, this should take whole systems and multisectoral approach to ensure adaptation is planned in an efficient way and that maladaptation is avoided.

References

[1] OECD, OECD Work in Support of Climate Action, OECD Environment Directorate, Paris, France, 2019. December http://www.oecd.org/env/cc/OECD-work-in-support-of-climate-action.pdf.

[2] World Bank, 2019 Press Release No: 2019/CCG/112, 2019. https://www.worldbank.org/en/news/press-release/2019/01/15/world-bank-group-announces-50-billion-over-five-years-for-climate-adaptation-and-resilience.

[3] IPCC, Summary for policymakers, in: V. Masson-Delmotte, P. Zhai, H.-O. Pörtner, D. Roberts, J. Skea, P.R. Shukla, T. Waterfield (Eds.), Global Warming of 1.5°C. An IPCC Special Report on the Impacts of Global Warming of 1.5°C Above Pre-industrial Levels and Related Global Greenhouse Gas Emission Pathways, in the Context of Strengthening the Global Response to the Threat of Climate Change, Sustainable Development, and Efforts to Eradicate Poverty, World Meteorological Organization, Geneva, Switzerland, 2018. 32 pp.

[4] NOAA National Centers for Environmental Information, State of the Climate: Global Climate Report for Annual 2019, 2020, Retrieved on 16 January 2020 from https://www.ncdc.noaa.gov/sotc/global/201913.

[5] CAT Climate Action Tracker, 2021. https://climateactiontracker.org/global/temperatures/. (Accessed 29 April 2021).

[6] OECD, The Economic Consequences of Climate Change, OECD Publishing, Paris, 2015, https://doi.org/10.1787/9789264235410-en.

[7] IPCC, Climate change 2013: the physical science basis, in: T.F. Stocker, D. Qin, G.-K. Plattner, M. Tignor, S.K. Allen, J. Boschung, P.M. Midgley (Eds.), Contribution of Working Group I to the Fifth Assessment Report of the Intergovernmental Panel on Climate

Change, Cambridge University Press, Cambridge, United Kingdom and New York, NY, USA, 2013, p. 1535, https://doi.org/10.1017/CBO9781107415324.
[8] IPCC, Climate change 2014: impacts, adaptation, and vulnerability. Part A: global and sectoral aspects, in: C.B. Field, V.R. Barros, D.J. Dokken, K.J. Mach, M.D. Mastrandrea, T.E. Bilir, L.L. White (Eds.), Contribution of Working Group II to the Fifth Assessment Report of the Intergovernmental Panel on Climate Change, Cambridge University Press, Cambridge, United Kingdom/New York, NY, USA, 2014. 1132 pp.
[9] IPCC, Climate change 2014: impacts, adaptation, and vulnerability. Part B: regional aspects, in: V.R. Barros, C.B. Field, D.J. Dokken, M.D. Mastrandrea, K.J. Mach, T.E. Bilir, L.L. White (Eds.), Contribution of Working Group II to the Fifth Assessment Report of the Intergovernmental Panel on Climate Change, Cambridge University Press, Cambridge, United Kingdom/New York, NY, USA, 2014, p. 688.
[10] IPCC, Climate change 2014: mitigation of climate change, in: O. Edenhofer, R. Pichs-Madruga, Y. Sokona, E. Farahani, S. Kadner, K. Seyboth, J.C. Minx (Eds.), Contribution of Working Group III to the Fifth Assessment Report of the Intergovernmental Panel on Climate Change, Cambridge University Press, Cambridge, United Kingdom/New York, NY, USA, 2014.
[11] H.O.P. Mbow, A. Reisinger, J. Canadell, P. O'Brien, Special Report on Climate Change, Desertification, Land Degradation, Sustainable Land Management, Food Security, and Greenhouse Gas Fluxes in Terrestrial Ecosystems (SR2), IPCC, Ginevra 650 (2017).
[12] J. Blunden, D.S. Arndt, State of the climate in 2018, Bull. Am. Meteorol. Soc. 100 (9) (2019) Si–S305, https://doi.org/10.1175/2019BAMSStateoftheClimate.1.
[13] WMO, WMO Provisional Statement on the State of the Global Climate in 2019, World Meteorological Organization (WMO), Geneva, Switzerland, 2019. https://library.wmo.int/doc_num.php?explnum_id=10108.
[14] Climate Impacts Lab, 2019. http://www.impactlab.org/. (Accessed 6 January 2020).
[15] J.A. Lowe, D. Bernie, P. Bett, L. Bricheno, S. Brown, D. Calvert, R. Clark, K. Eagle, T. Edwards, G. Fosser, F. Fung, UKCP18 Science Overview Report, Met Office Hadley Centre, Exeter, UK, 2018.
[16] MOWE-IT, Guidelines and Recommendations for Reducing the Impact of Weather on Rail Operations, 2014, Available from: http://mowe-it.eu.
[17] K. Dobney, C. Baker, L. Chapman, A. Quinn, The future cost to the United Kingdom's railway network of heat-related delays and buckles caused by the predicted increase in high summer temperatures owing to climate change, Proc. Inst. Mech. Eng. Part F J. Rail Rapid Transit. 224 (2010) 25–34, https://doi.org/10.1243/09544097JRRT292.
[18] S.V. Greenham, E.J.S. Ferranti, A.D. Quinn, K. Drayson, The impact of high temperatures and extreme heat to delays on the London, Meteorol. Appl. 27 (3) (2020) e1910.
[19] D.J. Jaroszweski, R. Wood, L. Chapman, A. Bell, S. Bell, J. Berman, G. Darch, E.J.S. Ferranti, S. Gosling, I. Haigh, P. Hughes, D. Lombardi, J. McCullough, A. Netherwood, E. Palin, K. Paulson, C. Payne, M. Pregnolato, G. Watson, D. White, UK Climate Risk Independent Assessment Technical Report: Chapter 4, Infrastructure, Report prepared for the Adaptation Sub-Committee of the Committee on Climate Change, London, 2022. https://www.ukclimaterisk.org/wp-content/uploads/2021/06/CCRA3-Chapter-4-FINAL.pdf.
[20] P. Chinowsky, J. Helman, S. Gulati, J. Neumann, J. Martinich, Impacts of climate change on operation of the US rail network, Transp. Policy 75 (2019) 183–191.
[21] M. Haasnoot, M. van Aalst, J. Rozenberg, K. Dominique, J. Matthews, L.M. Bouwer, J. Kind, N.L. Poff, Investments under non-stationarity: economic evaluation of adaptation pathways, Clim. Chang. 161 (2019) 1–13.

[22] M. Haasnoot, J.H. Kwakkel, W.E. Walker, J. ter Maat, Dynamic adaptive policy pathways: a method for crafting robust decisions for a deeply uncertain world, Glob. Environ. Chang. 23 (2) (2013) 485–498.
[23] C. Rosenzweig, W. Solecki, Hurricane Sandy and adaptation pathways in New York: lessons from a first-responder city, Glob. Environ. Chang. 28 (2014) 395–408.
[24] N. Ranger, T. Reeder, J. Lowe, Addressing 'deep' uncertainty over long-term climate in major infrastructure projects: four innovations of the Thames Estuary 2100 Project, EURO J. Decis. Process. 1 (3–4) (2013) 233–262.
[25] M. Haasnoot, J. Schellekens, J.J. Beersma, H. Middelkoop, J.C.J. Kwadijk, Transient scenarios for robust climate change adaptation illustrated for water management in the Netherlands, Environ. Res. Lett. 10 (10) (2015) 105008.
[26] J. Barnett, S. Graham, C. Mortreux, R. Fincher, E. Waters, A. Hurlimann, A local coastal adaptation pathway, Nat. Clim. Chang. 4 (12) (2014) 1103–1108.
[27] J. Lawrence, R. Bell, P. Blackett, S. Stephens, S. Allan, National guidance for adapting to coastal hazards and sea-level rise: anticipating change, when and how to change pathway, Environ. Sci. Pol. 82 (2018) 100–107.
[28] A. Quinn, E. Ferranti, S. Hodgkinson, A. Jack, J. Beckford, J. Dora, Adaptation becoming business as usual: a framework for climate-change-ready transport infrastructure, Infrastructures 3 (2) (2018) 10.
[29] C.S. Holling, Engineering resilience versus ecological resilience, in: National Academy of Engineering (Ed.), Engineering Within Ecological Constraints, vol. 31, The National Academies Press, Washington, DC, 1996, p. 32. https://doi.org/10.17226/4919.
[30] J.Y. Wang, 'Resilience thinking' in transport planning, Civ. Eng. Environ. Syst. 32 (1–2) (2015) 180–191.
[31] A. Reggiani, P. Nijkamp, D. Lanzi, Transport resilience and vulnerability: the role of connectivity, Transp. Res. A Policy Pract. 81 (2015) 4–15.
[32] L.G. Mattsson, E. Jenelius, Vulnerability and resilience of transport systems—a discussion of recent research, Transp. Res. A Policy Pract. 81 (2015) 16–34.
[33] DfT, Transport Resilience Review: A Review of the Resilience of the Transport Network to Extreme Weather Events, HMSO, London, 2014, p. 166.
[34] Rail Safety and Standards Board (RSSB), Tomorrow's Railway and Climate Change Adaptation: Work Package 1 Summary Report. The Arup TRaCCA WP1 Consortium (Arup, CIRIA, JBA Consulting, the Met Office and the University of Birmingham) in Collaboration with the RSSB Project Team, RSSB, London, UK, 2016.
[35] A.D. Quinn, A. Jack, S. Hodgkinson, E.J.S. Ferranti, J. Beckford, J. Dora, Rail Adapt: Adapting the Railway for the Future: A Report for the International Union of Railways (UIC), UIC, Paris, France, 2017.
[36] Q. Mao, N. Li, Assessment of the impact of interdependencies on the resilience of networked critical infrastructure systems, Nat. Hazards 93 (1) (2018) 315–337.
[37] WSP, Interacting risks in infrastructure and the built and natural environments, Research in Support of the UK's Third Climate Change Risk Assessment Evidence Report, WSP, UK, 2020. April 2020 https://www.ukclimaterisk.org/wp-content/uploads/2020/07/Interacting-Risks_WSP.pdf. (Accessed 30 April 2021).
[38] G. Pescaroli, D. Alexander, Understanding compound, interconnected, interacting, and cascading risks: a holistic framework, Risk Anal. 38 (11) (2018) 2245–2257.
[39] H. Murdock, K. de Bruijn, B. Gersonius, Assessment of critical infrastructure resilience to flooding using a response curve approach, Sustainability 10 (10) (2018) 3470.
[40] D.V. Val, D. Yurchenko, M. Nogal, A. O'Connor, Climate change-related risks and adaptation of interdependent infrastructure systems, in: Climate Adaptation Engineering, Butterworth-Heinemann, 2019, pp. 207–242.

[41] 100 Cities, 2019. http://www.100resilientcities.org/. (Accessed 27 January 2020).
[42] European Commission, The EU Strategy on Adaptation to Climate Change, 2013. https://ec.europa.eu/clima/sites/clima/files/docs/eu_strategy_en.pdf. (Accessed 27 January 2020).
[43] Climate Adapt, Incorporating climate change risks in planning the modernization of the railway corridor in Slovakia, in: Case Study in the EU Climate-Adapt Portal, 2020. https://climate-adapt.eea.europa.eu/metadata/case-studies/incorporating-climate-change-risks-in-planning-the-modernization-of-the-railway-corridor-in-slovakia. (Accessed 15 February 2021).
[44] S.C. Woodruff, P. Regan, Quality of national adaptation plans and opportunities for improvement, Mitig. Adapt. Strateg. Glob. Chang. 24 (1) (2019) 53–71.
[45] OECD, Adaptation to Climate Change, 2020. http://www.oecd.org/env/cc/adaptation.htm. (Accessed 29 January 2020).
[46] L. Vallejo, M. Mullan, Climate-resilient infrastructure: getting the policies right, OECD Environment Working Papers, No. 121, OECD Publishing, Paris, 2017, https://doi.org/10.1787/02f74d61-en.
[47] C. Toplis, M. Kidnie, A. Marchese, C. Maruntu, H. Murray, R. Sebille, S. Thomson, International Climate Change Adaptation Framework for Road Infrastructure (No. 2015R03EN), PIARC World Road Association, Paris, France, 2015.
[48] PIANC, Climate Change Adaptation Planning for Ports and Inland Waterways. EnviCom WG Report N° 178 - 2020. The World Association for Waterborne Transport Infrastructure, PIANC Secrétariat Général, Belgium, 2020. https://www.pianc.org/shop/download/12611.
[49] R.F. Warren, R.L. Wilby, K. Brown, P. Watkiss, R.A. Betts, J.M. Murphy, J.A. Lowe, Advancing national climate change risk assessment to deliver national adaptation plans, Philos. Trans. R. Soc. A Math. Phys. Eng. Sci. 376 (2121) (2018) 20170295.
[50] S.R. Jude, G.H. Drew, S.J. Pollard, S.A. Rocks, K. Jenkinson, R. Lamb, Delivering organisational adaptation through legislative mechanisms: evidence from the adaptation reporting power (Climate Change Act 2008), Sci. Total Environ. 574 (2017) 858–871.
[51] R.B. Street, S. Jude, Enhancing the value of adaptation reporting as a driver for action: lessons from the UK, Clim. Pol. 19 (10) (2019) 1340–1350.
[52] CCC, Progress in Preparing for Climate Change—2019 Progress Report to Parliament, 2019. https://www.theccc.org.uk/publication/progress-in-preparing-for-climate-change-2019-progress-report-to-parliament/.
[53] J.M. Jacobs, M. Culp, L. Cattaneo, P. Chinowsky, A. Choate, S. DesRoches, S. Douglass, R. Miller, Transportation, in: D.R. Reidmiller, C.W. Avery, D.R. Easterling, K.E. Kunkel, K.L.M. Lewis, T.K. Maycock, B.C. Stewart (Eds.), Impacts, Risks, and Adaptation in the United States: Fourth National Climate Assessment, vol. II, U.S. Global Change Research Program, Washington, DC, USA, 2018, pp. 479–511, https://doi.org/10.7930/NCA4.2018.CH12.
[54] L.A. Dupigny-Giroux, E.L. Mecray, M.D. Lemcke-Stampone, G.A. Hodgkins, E.E. Lentz, K.E. Mills, E.D. Lane, R. Miller, D.Y. Hollinger, W.D. Solecki, G.A. Wellenius, P.E. Sheffield, A.B. MacDonald, C. Caldwell, Northeast, in: D.R. Reidmiller, C.W. Avery, D.R. Easterling, K.E. Kunkel, K.L.M. Lewis, T.K. Maycock, B.C. Stewart (Eds.), Impacts, Risks, and Adaptation in the United States: Fourth National Climate Assessment, vol. II, U.S. Global Change Research Program, Washington, DC, USA, 2018, pp. 669–742, https://doi.org/10.7930/NCA4.2018.CH18.
[55] A. Bauer, R. Steurer, National Adaptation Strategies, what else? Comparing adaptation mainstreaming in German and Dutch water management, Reg. Environ. Chang. 15 (2) (2015) 341–352.

[56] S.H. Verduijn, S.V. Meijerink, P. Leroy, How the Second Delta Committee set the agenda for climate adaptation policy: a Dutch case study on framing strategies for policy change, Water Altern. 5 (2) (2012) 469.

[57] J. Lindgren, D.K. Jonsson, A. Carlsson-Kanyama, Climate adaptation of railways: lessons from Sweden, Eur. J. Transp. Infrastruct. Res. 9 (2) (2009), https://doi.org/10.18757/ejtir.2009.9.2.3295. https://journals.open.tudelft.nl/ejtir/article/view/3295. (Accessed 30 January 2020).

[58] Network Rail, 2020. https://www.networkrail.co.uk/communities/environment/climate-change-and-weather-resilience/climate-change-adaptation. (Accessed 29 January 2020).

[59] Network Rail, SOUTH EAST 2019–2024 Route CP6 Weather Resilience and Climate Change Adaptation Plan, Network Rail, Milton Keynes, UK, 2019. https://cdn.networkrail.co.uk/wp-content/uploads/2019/10/South-East-CP6-WRCCA-Plan.pdf.

[60] E.J. Palin, et al., Future projections of temperature-related climate change impacts on the railway network of Great Britain, Clim. Chang. 120 (1–2) (2013) 71–93.

[61] Highways England, Highways England Climate Adaptation Risk Assessment Progress Update—2016, Highways England, London, UK, 2016. https://assets.publishing.service.gov.uk/government/uploads/system/uploads/attachment_data/file/596812/climate-adrep-highways-england.pdf. (Accessed 30 January 2020).

[62] C. Rosenzweig, W. Solecki, Introduction to climate change adaptation in New York City: building a risk management response, Ann. N. Y. Acad. Sci. 1196 (1) (2010) 13–18.

[63] A. Nguyen, B. Dix, W. Goodfried, J. LaClair, L. Lowe, S. Yokoi, R. Fahey, Adapting to rising tides—transportation vulnerability and risk assessment pilot project, 2011. http://www.adaptingtorisingtides.org/wp-content/uploads/2015/04/RisingTides_Technical Report_sm.pdf. Technical Report.

[64] T.A. Wall, W.E. Walker, V.A. Marchau, L. Bertolini, Dynamic adaptive approach to transportation-infrastructure planning for climate change: San-Francisco-Bay-Area case study, J. Infrastruct. Syst. 21 (4) (2015) 05015004.

[65] OECD, Adapting to the Impacts of Climate Change—Policy Perspectives, 2015, Available from: https://www.oecd.org/environment/cc/Adapting-to-the-impacts-ofclimate-change-2015-policy-Perspectives-27.10.15%20WEB.Pdf.

[66] E. Ferranti, L. Chapman, D. Whyatt, A perfect storm? The collapse of Lancaster's critical infrastructure networks following intense rainfall on 4/5 December 2015, Weather 72 (1) (2017) 3–7.

[67] J. Booth, M. Drye, D. Whensley, P. McFarlane, S. McDonald, Future of flood resilience for electricity distribution infrastructure in Great Britain, CIRED Open Access Proc. J. 2017 (1) (2017) 1158–1161.

[68] E. Ferranti, L. Chapman, S. Lee, D. Jaroszweski, C. Lowe, S. McCulloch, A. Quinn, The hottest July day on the railway network: insights and thoughts for the future, Meteorol. Appl. 25 (2) (2018) 195–208.

[69] A. AghaKouchak, L.S. Huning, F. Chiang, M. Sadegh, F. Vahedifard, O. Mazdiyasni, H. Moftakhari, I. Mallakpour, How do natural hazards cascade to cause disasters? Nature 561 (2018) 458–460, https://doi.org/10.1038/d41586-018-06783-6.

[70] Copernicus Climate Change, 2020. https://climate-adapt.eea.europa.eu/knowledge/adaptation-information/climate-services. (Accessed 2020).

[71] J.M. Murphy, G.R. Harris, D.M.H. Sexton, E.J. Kendon, P.E. Bett, R.T. Clark, K.E. Eagle, G. Fosser, F. Fung, J.A. Lowe, R.E. McDonald, R.N. McInnes, C.F. McSweeney, J.F.B. Mitchell, J.W. Rostron, H.E. Thornton, S. Tucker, K. Yamazaki, UKCP18 Land

Projections: Science Report. Met Office Technology Report, Met Office, Bracknell, 2018, p. 191 pp. https://www.metoffice.gov.uk/pub/data/weather/uk/ukcp18/science-reports/UKCP18-Land-report.pdf.
[72] De Buck, Climate Data Needed to Address Resilience to Climate Change in Standards for Infrastructures, 2018. https://www.nen.nl/web/file?uuid=58972028-a29e-46fe-80af-ef01cfc48f8c&owner=441f3990-cd12-49c7-8d35-a371ae1ace23.
[73] Met Office, D. Hollis, M. McCarthy, M. Kendon, T. Legg, I. Simpson, HadUK-Grid Gridded and Regional Average Climate Observations for the UK, Centre for Environmental Data Analysis, 2018. http://catalogue.ceda.ac.uk/uuid/4dc8450d889a491ebb20e724debe2dfb. (Accessed 29 January 2020).
[74] E. Ferranti, L. Chapman, C. Lowe, S. McCulloch, D. Jaroszweski, A. Quinn, Heat-related failures on Southeast England's railway network: insights and implications for heat risk management, Weather Clim. Soc. 8 (2) (2016) 177–191.
[75] M.G. Sanderson, H.M. Hanlon, E.J. Palin, A.D. Quinn, R.T. Clark, Analogues for the railway network of Great Britain, Meteorol. Appl. 23 (4) (2016) 731–741.
[76] Network Rail, NR Weather Analysis Report Issue 1.0 March 2015, Network Rail, London, 2015. 55 pp.
[77] HoCTC, House of Commons Transport Committee investing in the railway, in: Seventh Report of Session 2014–15, House of Commons London, 2015.
[78] DMF, Holding the line? Reviewing the impacts, responses and resilience of people and places in Devon to the winter storms of 2013/2014, 2015. A Summary Report from the Devon Maritime Forum.
[79] HM Treasury, The Green Book: Appraisal and Evaluation in Central Government, 2016, Available from: https://www.gov.uk/government/publications/the-green-book-appraisaland-evaluation-in-central-governent.
[80] Defra, Method for undertaking the CCRA part II—detailed method for stage 3: assess risk, 2010. Project deliverable number D.2.1.1 September 2010. Contractors: HR Wallingford Ltd, The Met Office, Alexander Ballard Ltd, Collingwood Environmental Planning, Entec Ltd UK, Paul Watkiss Associates, Metroeconomica http://randd.defra.gov.uk/Document.aspx?Document=GA0204_9587_TRP.pdf.
[81] J.S. Dodgson, M. Spackman, A. Pearman, L.D. Phillips, Multi-Criteria Analysis: A Manual, Department for Communities and Local Government, London, UK, 2009. January.
[82] D. Jaroszweski, L. Chapman, J. Petts, Assessing the potential impact of climate change on transportation: the need for an interdisciplinary approach, J. Transp. Geogr. 18 (2) (2010) 331–335.
[83] World Bank Adaptation Fund, 2020. https://www.adaptation-fund.org/apply-funding/. (Accessed 29 January 2020).
[84] UIC, UIC Railway Application Guide-Practical Implementation of Asset Management through ISO 55001, Internatioanl Union of Railways, Paris, France, 2016. November. Available from: https://uic.org/IMG/pdf/iso_55000_implementation_guidelines_on_rail ways_infrastructure_organisations.pdf.
[85] CADD, The CaDD (Capacity Diagnosis and) Development Software, 2020. https://www.cadd.global/. (Accessed 27 January 2020).
[86] K.W. Steininger, B. Bednar-Friedl, H. Formayer, M. König, Consistent economic cross-sectoral climate change impact scenario analysis: method and application to Austria, Clim. Serv. 1 (2016) 39–52, https://doi.org/10.1016/j.cliser.2016.02.003.
[87] D. Jaroszweski, E. Hooper, L. Chapman, The impact of climate change on urban transport resilience in a changing world, Prog. Phys. Geogr. 38 (4) (2014) 448–463.

[88] K. Riahi, D.P. van Vuuren, E. Kriegler, J. Edmonds, B.C. O'Neill, S. Fujimori, N. Bauer, K. Calvin, R. Dellink, O. Fricko, W. Lutz, A. Popp, J.C. Cuaresma, K.C. Samir, M. Leimbach, L. Jiang, T. Kram, S. Rao, J. Emmerling, K. Ebi, T. Hasegawa, P. Havlik, F. Humpenöder, L.A. Da Silva, S. Smith, E. Stehfest, V. Bosetti, J. Eom, D. Gernaat, T. Masui, J. Rogelj, J. Strefler, L. Drouet, V. Krey, G. Luderer, M. Harmsen, K. Takahashi, L. Baumstark, J.C. Doelman, M. Kainuma, Z. Klimont, G. Marangoni, H. Lotze-Campen, M. Obersteiner, A. Tabeau, M. Tavoni, The shared socioeconomic pathways and their energy, land use, and greenhouse gas emissions implications: an overview, Glob. Environ. Chang. 42 (2017) 153–168.

Rail transport resilience to demand shocks and COVID-19

Erik Jenelius
Department of Civil and Infrastructural Engineering, KTH Royal Institute of Technology, Stockholm, Sweden

4.1 Introduction

Transport system resilience can be defined as the ability to prepare for and to withstand, absorb and adapt to shocks, and to recover from the consequences in a timely and efficient manner [1]. Compared to, e.g., road transport, rail transport is characterized by relatively high levels of regulation and control. While these factors allow the system to run efficiently under normal conditions, they also contribute to make the system inflexible and sensitive to disruptions. Responses such as timetable replanning and train rerouting are often difficult to achieve in a short time.

The rail transport system can conceptually be divided into three layers (Fig. 4.1): First, there is the technical and physical infrastructure including the rails, signals, bridges, vehicles, etc. Second, there are the services built on top of the infrastructure in terms of train routes, departures, etc. Third, there are the users, i.e., travelers and transporters, who use the system for trips and shipments.

In economic terms, the first (infrastructure) and second (services) layers can be regarded as the supply side of the system, while the third (users) represents the demand side. Fourth, there are the actors responsible for planning, operating, and maintaining the infrastructure and services and regulating the interactions between supply and demand. From the management's perspective, strategic goals are typically focused on the demand layer, such as reducing travel times and increasing punctuality, while the available actions are mainly concentrated on the supply layers, including new investments, maintenance, traffic control, etc.

The infrastructure layer provides a basic level of capacity in terms of how many trains the system can facilitate, their speed, etc. Changes in the infrastructure are typically long-term strategic decisions. Given the constraints set by the infrastructure, the service layer determines the practical capacity for different origin-destination combinations in terms of number of departures, available space, etc. Adjustments can be accommodated in the short to medium time frame through operational and tactical planning decisions. Ideally, infrastructure and service capacity is designed to meet the requirements from the demand side. Even if this is the case under typical conditions, disruptions in any of the system layers can lead to severe degradation of the system function.

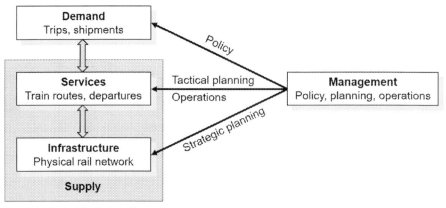

Fig. 4.1 A model of the rail transport system.

Infrastructure disruptions are caused by, e.g., technical failures (bridge collapses, power outages, vehicle malfunctions, etc.), extreme operating conditions (floods, snowstorms, etc.), or deliberate attacks. Disruptions in the service layer can occur from events such as human errors or crew shortages due to sickness or labor conflicts. Disruptions in the management may arise from, for example, policy shifts or budget cuts. Demand shocks, finally, can occur due to special events or various forms of societal upheaval and crises. Most attention in practice and in research has been given to supply cuts such as rail segment closures or service delays (see Bešinović [2] for a recent literature review). In contrast, this chapter focuses on disruptions in the demand layer with particular attention to the COVID-19 pandemic.

4.2 Rail transport demand shocks

Like most transport systems, rail passenger and freight demand are subject to systematic and random variations over time. Systematic variations include long-term market trends and seasonal, within-week, and within-day cycles, which can be forecasted and incorporated into planning and operations. Moderate random fluctuations can be accommodated through spare capacity in the system. Sometimes, however, larger unexpected events lead to demand shocks that cannot be fully absorbed through the standard planning processes but call for resilience planning. Table 4.1 classifies demand shocks along two dimensions: whether demand increases or decreases and the time frame of the event.

Some demand shocks both develop and dissipate within a relatively short time period. Examples include special events such as concerts, games, and conventions. Such events can cause unusually high levels of travel demand that exceed the nominal service supply. When demand exceeds supply a range of negative consequences can occur, including crowding, sold out tickets and denied boarding which in turn often lead to longer travel times. These impacts can cascade through the network as passengers and transporters seek alternative routes. The increased pressure on the supply may spill over into train delays, which can further exacerbate the situation [3].

Table 4.1 Classification of rail transport demand shocks.

	Short-term	Long-term
Demand increase	Special event (e.g., concert, sports event, convention)	Societal trend (e.g., sustainable mobility)
Demand decrease	Threat (e.g., terrorism, weather alert)	Crisis (e.g., pandemic, economy, conflict)

Other events can cause a sudden temporary drop in demand. Tragically, almost all modes of transport, including the rail system, have been subject to deadly terrorist attacks. While the purpose of such attacks may be to disrupt people's lives, evidence from the 2005 bombings of the London Underground suggests that travel demand returned relatively quickly to nominal levels [4]. If the focus is restricted to the rail system rather than society in general, the loss of function from such events is thus limited as long as the infrastructure and service layers are intact.

Unanticipated demand changes, both positive and negative, can be persistent over longer periods of time. Demand surges may arise from societal movements such as the increased focus on climate change and sustainable mobility. As an example, Sweden has seen a recent dramatic increase in demand for rail travel to the European continent. Due to numerous technical and regulatory restrictions, however, supply has been slow to adapt, which has led to fully booked departures long in advance and unsatisfied demand. Recent political initiatives aim to have procured night train services in place by August 2022 [5].

Long-lasting drops in travel and transport demand can be initiated by various societal processes and crises, such as economic downturns and political conflicts. The introduction of a competing transport mode can also cause a dramatic reduction in the market share for the rail system in a particular segment. Reductions in demand generally have limited impacts on system function in the short term; in fact, the level of service for remaining customers may increase as a result of faster response times and less crowding. Over time, however, the loss of customers is likely to result in a loss of revenues for the service providers, system operators, and managers. This will in turn force such actors to increase prices, reduce or, in the worst cases, completely shut down traffic. If demand eventually recovers the process of reinstating the transport services may lag significantly, leading to a severe loss of system function. The next section focuses on the mobility restrictions in response to the COVID-19 pandemic, which provides a prime example of a long-term negative demand shock.

4.3 Impacts of COVID-19 on rail demand and supply

The COVID-19 pandemic has had unprecedented cascading effects on the economy and the transport system. The rapid global spread is partly due to the high level of mobility offered by the transport system [6,7]. The rail transport system contributed to spread of the virus [8,9], but also assisted in the distribution of essential health care supplies [10]. Most countries closed borders and restricted mobility as well as locked down activities and businesses with the aim of suppressing the spread. These measures led to a drastic reduction in both local and international travel and transport demand. The effects were most pronounced for the aviation industry, where a corresponding reduction in supply soon followed. Public transport, including rail, was also greatly affected, with ridership losses ranging from 30%–35% in Hong Kong and South Korea to 70%–80% in the United States and United Kingdom, and complete shutdown in India [11]. The transport infrastructure, meanwhile, was left intact.

Data from Sweden are used to illustrate the impacts on the rail transport system. The Swedish Public Health Agency increased the risk level of COVID-19 spreading in Sweden from "low" to "moderate" on 2 March 2020 and to "very high" on 10 March. From this time, Swedish citizens were advised to stay at home if feeling sick in any way and work from home if possible. Meetings involving more than 50 people were banned, and high schools, colleges, and universities were closed for students. Since early April 2020, people were advised to travel by public transport only if necessary. These recommendations were still in place as of July 2020 [12].

Fig. 4.2 shows the weekly supply of passenger train services in Sweden during the first half of 2019 and 2020, respectively, separated into short-distance (commuter and airport trains), medium-distance (regional trains), and long-distance (interregional, night, and high-speed trains). From week 3 to week 12 the supply was similar between years. During weeks 13 and 14 the traffic decreased rapidly to levels far below those from 2019. Between weeks 13 and 26, the traffic was 10%–30% lower in 2020 than in 2019. Train kilometers for long-distance trains have decreased the most, 43% in week 26. The decrease for short-distance trains and medium-distance trains is 12% and 15%, respectively. The drop in supply also occurred the fastest for the long-distance traffic.

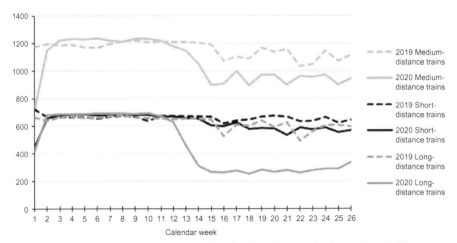

Fig. 4.2 Weekly passenger train supply in Sweden by train type, in thousand train kilometers. Data from Swedish Transport Administration, diagram produced by Transport Analysis, available at: https://www.trafa.se/en/rail-traffic/rail-traffic-9442/ (Accessed 8 July 2020).

Fig. 4.3 shows the daily demand for commuter train trips in Stockholm from February to May 2019 and 2020, respectively. A dramatic drop in ridership occurred in mid-March 2020 when threat levels were raised and stay-at-home recommendations were implemented. From around 10 April, when the demand loss was around 60%, until at least 31 May ridership has recovered slowly but consistently. Combined with Fig. 4.2 it can be concluded that the sharp drop in short-distance train demand was not preceded or coupled with drop in supply of the same magnitude.

Fig. 4.4 shows the weekly freight rail traffic in Sweden for the same weeks in 2019 and 2020. Unlike the passenger train traffic, there is no clear difference between the 2 years. Thus, at least in this time frame, the impacts of COVID-19 are mainly found on the passenger transport side.

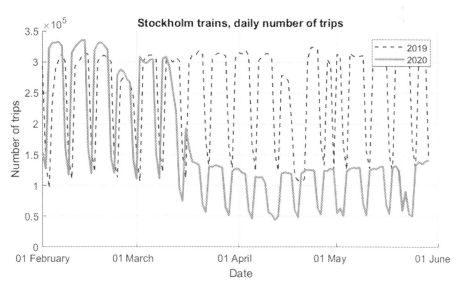

Fig. 4.3 Daily commuter train ridership in Stockholm, in number of trips. Ticket validation data provided by Region Stockholm Transport Administration.

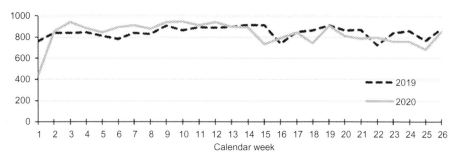

Fig. 4.4 Weekly freight rail traffic in Sweden, in million gross-gross tonne kilometers hauled. Data from Swedish Transport Administration, diagram produced by Transport Analysis, available at: https://www.trafa.se/en/rail-traffic/rail-traffic-9442/ (Accessed 8 July 2020).

4.4 Rail system resilience

Resilience is commonly illustrated as in Fig. 4.5, where the function of the system is monitored over time. A shock will initiate a degradation of system function, possibly followed by a period of consolidation, and an eventual recovery of the nominal function. Higher resilience can be achieved by reducing the magnitude of function degradation, or reducing the duration of the consolidation and recovery periods.

Quantitative resilience analysis requires some measure of system function. Reggiani et al. [13] emphasize the close relation between resilience and network connectivity. Simple measures of network connectivity include the relative size of the largest connected component of the network, i.e., the largest number of nodes that can be reached from each other through at least one network route, and the average distance between all node pairs. However, such indicators are mainly developed for the analysis of infrastructure disruptions such as closures of rail segments.

This chapter formulates a framework for transport system resilience analysis that incorporates both supply and demand shocks. In this framework, system function loss corresponds to a shortage of supply in relation to the demand. This can occur due to either a reduction in supply or an increase in demand. The lack of supply can manifest itself in different ways, such as reduced number of departures, higher prices, longer travel times, and increased crowding.

4.5 Supply losses

Fig. 4.6 illustrates the most commonly considered case, where a disruption causes a sudden supply cut while the demand remains relatively unchanged. The lack of resilience is represented by the total loss of function until supply is restored to the baseline level.

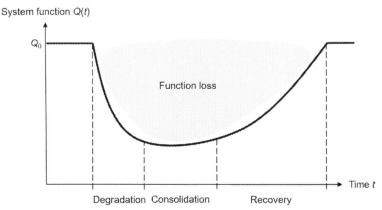

Fig. 4.5 The system resilience curve highlighting the stages following a disruption: function degradation, consolidation, and recovery.

The literature on rail network resilience to supply disruptions is extensive; only a few examples can be presented here. Most studies have been of a network topological nature, which requires only limited data about the infrastructure and transport system. Recent applications include Bhatia et al. [14] who study the Indian rail network and Zhang et al. [15] who compare the high-speed rail networks of United States, China, and Japan. In the area of urban rail transit, Yang et al. [16], Chopra et al. [17], and Sadaat et al. [18] focus on the Beijing, London, and Washington DC metro networks, respectively. Xing et al. [19] extend the purely topological analysis by using passenger flows as link weights in a representation of the Shanghai rail transit system. Sun et al. [20] use a similar weighted network as basis for a dynamic model of cascading failures due to flow overloads following an initial disruption of a station. The model is applied to the Beijing metro network, where the results indicate that the larger the initial perturbation, the earlier all stations will fail.

Ouyang et al. [21] assess the vulnerability of the Chinese Railway System to single and multiple random node failures, with network function measured as the average number of stations reachable from each start station and the share of trains that can complete their trips. Three different routing models are compared: first, a simple connectivity-based model in which a train trip can be completed as long as all served stations are connected with at least one path; second, a shortest path model in which a trip can only be completed if the shortest path connecting all served stations is intact; and third, a more realistic model in which the trains are routed according to their real-world time tables. The authors find that the shortest path model produces similar resilience results to the timetable path model, while the purely connectivity-based model deviates significantly. This highlights some of the limitations of the general complex network theory when applied to specific cases.

On the freight transport side, Peterson and Church [22] develop a network vulnerability modeling framework. The authors apply a slight generalization of the classical shortest path algorithm to find the best way of routing a shipment in the original network and rerouting it after a certain link has been disrupted. Using this as the basis for the estimation of the detour costs of a disrupted link, they analyze the impacts for rail

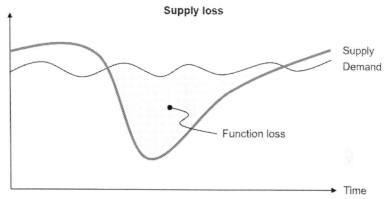

Fig. 4.6 System function loss in case of a supply loss.

freight to and from the State of Washington in the United States due to the disruption of an important bridge. Bababeik et al. [23] extend the model to include the optimal rescheduling of trains during the disruption. Gedik et al. [24] formulate an interdiction problem where an antagonist seeks to maximize the sum of train transport and delay costs by disrupting a limited number of nodes in the network given that an operator reroutes the trains on the remaining network to minimize these costs. The authors apply the model to a real coal transport network in the United States and find that increasing the number of disruptions escalates the delay cost dramatically, whereas the transport cost remains more or less the same. Khaled et al. [25] propose a more elaborate mixed integer-programming model for forming trains out of railcars and routing the trains through the network. Train forming occurs at yards, a subset of nodes with limited capacities of trains and train cars that can be handled per day. The model is applied to a US rail network, where the importance of nodes and links are evaluated as the difference in total travel time before and after the element is removed.

4.6 Demand spikes

Fig. 4.7 illustrates the case where a spike in transport or travel demand brings it above the supply capacity. The excess demand can lead to functional loss in terms of increased travel times, crowding, denied boarding, and unsatisfied demand.

The literature on transport system resilience against demand shocks is limited; on the other hand, the topic is closely related to decision-making and planning under high uncertainty, which has a long history of research [26]. Some attention has been given to transport planning for large events such as Olympic Games, which attract high volumes of people during a limited time. While the events are known and planned for long in advance, uncertainties regarding passenger volumes on specific lines, etc. are high due to the lack of experience from similar circumstances. Hensher and Brewer [27] evaluate the transport planning for the 2000 Summer Olympics in Sydney, Australia. Overall, the authors find that the city handled the surge in travel

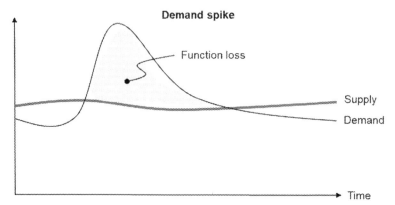

Fig. 4.7 System function loss in case of a demand spike.

demand well through new and reinforced train and bus lines. With a few exceptions, passenger loads on the new trains were lower than forecasted whereas the regular train services were largely unaffected. Furthermore, volunteers on the stations helped increase the passenger throughput substantially. Kassens-Noor [26] describes the transport planning process behind Boston's bid for the 2024 Olympic Games, characterizing it as planning for "grand opportunities." In other words, the aim was to not only solve the short-term transport needs during the games but to take the chance of making large long-term transformations of the transport system.

Noursalehi et al. [28] focus on adaptive control strategies for smaller events such as local football games. They propose a real-time short-term prediction methodology for passenger arrivals at public transport stations. Ensemble predictions are used to combine the outputs from multiple models based on their respective accuracy. The method is applied to the Central Line of the London Underground, with special attention to modeling the effect of planned events. Similarly, Chen et al. [29] use time series models to predict passenger flows during special events, focusing on two stations near the Olympic Sports Center, Nanjing, China.

Compared to supply cuts, demand shocks are not as intuitive to represent in the framework of network topology. Many existing resilience and vulnerability studies consider demand at the network route level as a function of supply, but not the other way around. Capturing the impacts of a short-term demand spike on system function requires a more sophisticated model of the interactions between the rail infrastructure, vehicles, travelers and/or goods, and decision makers. A number of candidate models have been introduced in recent years, although they have been applied primarily to assess the impacts of supply disruptions.

Cats and Jenelius [3] introduce a dynamic, stochastic, and multimodal notion of public transport network vulnerability, accounting for interactions between supply and demand and the accumulated effect of disruption on system performance. The authors also study the mitigating impact of real-time information provision and find that it may have a significant positive influence, although counter-examples also exist due to cascading effects. Cats and Jenelius [30] study the possibility of reducing vulnerability by increasing the capacity on lines that can serve as alternatives when critical links are disrupted, and propose a methodology for identifying the lines where capacity increases are the most effective.

Shen et al. [31] consider cascading failures in metro networks and propose a flow redistribution model during station failures. The authors apply data from Nanjing, China to the model in order to analyze the cascading failure process and optimize the robustness of the system. Hong et al. [32] propose a spatiotemporal framework for analyzing the vulnerability of a railway system with heterogeneous train flows. The rail system is modeled as a physical layer and a service layer, where the former incorporates damaged components, occurrence times, and durations, and the latter captures time-related attributes of trains and passengers. The framework is applied to assess the passenger delays in the Chinese Railway System under various disruption scenarios. Szymula and Bešinović [33] introduce an optimization model for assessing the vulnerability of rail passenger networks and adjusting train routes and timetables in response to network disruptions and adaptive passenger flows. The framework is applied to a part of the Dutch railway network, where the results indicate that link criticality is a function of passenger demand rather than a static network topological property.

4.7 Demand losses

Substantial losses of transport demand which are not caused by transport supply disruptions have not received much attention in the resilience literature. In part, the reason may be that such events do not have immediate negative impacts on system function in terms of travel times, accessibility, etc. Furthermore, drops in transport demand may be manifestations of events for which the most severe impacts occur externally to the transport system. Within the transport system, the direct consequences are mainly concentrated on the service suppliers who lose revenues from ticket sales and shipment contracts. Thus, resilience to demand disruptions has been studied primarily within the field of supply chain risk management (e.g., Ali et al. [34]).

The COVID-19 pandemic has highlighted the need to understand and manage the long-term transport system effects of demand drops. As Fig. 4.8 illustrates, an extended decrease in demand may force suppliers to reduce traffic as a response to revenue losses. When demand eventually recovers, the market may have changed so dramatically that it cannot quickly recover to meet the demand. Thus, the main losses of system function may occur as late as several years after the first demand drop depending on how long mobility restrictions are maintained.

Ivanov [35] characterizes epidemic outbreaks such as COVID-19 as a distinct type of supply chain disruption risk, involving long durations, cascading effects, and high uncertainty. A simulation-based methodology is used to predict short-term and long-term impacts of COVID-19 on supply chain performance. The study finds that the timing of the closing and opening of facilities, lead time, speed of epidemic propagation, and disruption durations are important factors that determine the impacts. Wang et al. [36] use an agent-based simulation model to study the transport system effects of

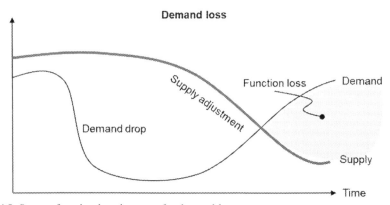

Fig. 4.8 System function loss in case of a demand loss.

different public transport reopening strategies for New York City. The study shows that even a reopening to full public transport capacity would reach only 73% of pre-COVID ridership due to behavioral inertia. At the same time, car trips would increase up to 142% of pre-COVID levels.

Forecasting the long-term effects of a demand disruption on transport supply is in general a difficult problem, considering the heterogeneous range of transport providers with varying roles and incentives. Typically, transport system models treat supply as exogenous to facilitate appraisals of the benefits of investments and policy changes. Economic models of trade and transport may provide insights at an aggregate level (e.g., Brancaccio et al. [37]).

4.8 Increasing demand loss resilience

Our notion of transport resilience suggests that resilience to demand losses can be enhanced in multiple ways. Table 4.2 lists three basic strategies and examples of actions that have been taken in various places around the world to reduce system function loss during COVID-19.

The first approach, illustrated in Fig. 4.9, is to limit the reduction in demand through various adaptations of the transport system. By mitigating the demand loss, the need to cut down supply is reduced, which increases the capability of handling a subsequent recovery. During COVID-19, many public transport and rail providers have sought to maintain business by addressing health risk concerns, including improving cleaning, ventilation, and crowding management in vehicles and in stations.

The second approach, illustrated in Fig. 4.10, is to increase the responsiveness of supply, i.e., to adjust supply to meet demand faster in all phases of a shock. This approach requires that infrastructure, resources, and services are agile, which is a significant challenge for planning and operations. If successfully implemented, however, the approach can increase system resilience without substantial costs. Possible actions include reducing service frequencies and operating trains with fewer cars. On the other hand, a pandemic such as COVID-19 has conflicting health risks associated with reducing capacity due to increased crowding. In Stockholm, the public transport

Table 4.2 Demand loss resilience enhancement strategies and examples implemented during COVID-19.

Resilience enhancement strategy	Example actions during COVID-19
Demand loss mitigation	Increased cleaning of vehicles and stations, enforced social distancing, crowd management, fleet allocation
Supply responsiveness	Adjusted service frequencies and vehicle capacities, fleet allocation
Supply tenacity	Governmental support

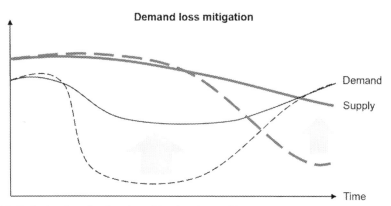

Fig. 4.9 Reducing function loss by mitigating demand loss.

administration and operators agreed in late March 2020 to reduce service frequencies but returned to nearly nominal service levels in early April following media reports of high crowding levels.

Fig. 4.11 illustrates the third approach, which is to increase the tenacity of supply, i.e., to prevent supply from dropping in response to the demand cut. This means that supply will be able to accommodate the demand once it returns to its original levels. Furthermore, the system can guarantee that critical societal functions (e.g., access to emergency healthcare and supplies) will be maintained at all times and that health risks do not substantially deteriorate. This approach will generally require some form of governmental support or insurance in order to finance the operations during the period of revenue loss.

During COVID-19, countries have supported rail traffic in different ways. In an international survey, WSP [11] note that most studied countries support urban rail transit. However, while aviation receives direct support in all studied countries, the same kind of support for long-distance rail transport is uncommon. In the United Kingdom, funding equivalent to 37 million USD were allocated to support trams

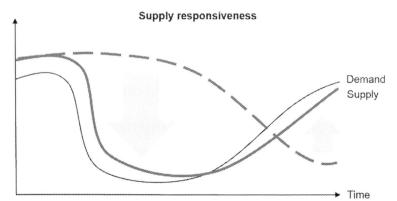

Fig. 4.10 Reducing function loss in case of a demand cut by increasing supply responsiveness.

Rail transport resilience to demand shocks and COVID-19

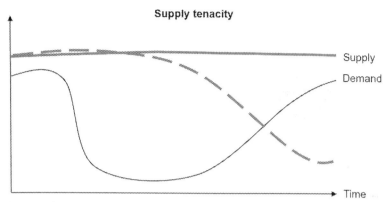

Fig. 4.11 Reducing function loss in case of a demand cut through supply tenacity.

and metros in certain areas. Further, revenue and cost risks were transferred from operators to the government for 6 months, rail tickets bought before the end of March were refunded, and the continued operation of certain regionally important rail lines was ensured. In the United States, the national rail sector received support and 25 billion USD was allocated to public transportation. In Sweden, the equivalent of 300 million USD were allocated to support public transportation. Private rail operators receive no specific support but could apply for nonsector specific governmental support [11].

4.9 Concluding remarks

This chapter has introduced a framework for understanding rail transport resilience that incorporates supply shocks as well as positive and negative demand shocks. While most attention tends to fall on supply disruptions (e.g., infrastructure failures), the rail system must also cope with short-term and long-term demand surges (e.g., due to special events and societal trends) and losses (e.g., due to threats and crises). The mobility restrictions motivated by the COVID-19 pandemic are a stark illustration of the need to consider the rail system's ability to withstand, adapt to and recover from long-term negative demand shocks. From the perspective of transport system resilience, a significant challenge of long-term demand losses lies in ensuring that supplied infrastructure and services are sufficient once demand recuperates. The chapter identified demand loss mitigation, supply responsiveness, and supply tenacity as three strategies for increasing resilience. In practice, a combination of multiple actions is likely to be the most effective.

As of 2022, the full long-term effects of COVID-19 on the rail transport system are yet unknown. How to continue the development toward a sustainable and resilient transport system under the new circumstances will be a main focus for practitioners and researchers in the following years.

References

[1] E. Jenelius, L.-G. Mattsson, Resilience of transport systems, in: R. Vickerman (Ed.), International Encyclopedia of Transportation, 1st, 7, Elsevier, 2021, pp. 258–267.

[2] N. Bešinović, Resilience in railway transport systems: a literature review and research agenda, Transp. Rev. 40 (4) (2020) 457–478.

[3] O. Cats, E. Jenelius, Dynamic vulnerability analysis of public transport networks: mitigation effects of real-time information, Netw. Spat. Econ. 14 (3–4) (2014) 435–463.

[4] F. Prager, G.R. Beeler Asay, B. Lee, D. von Winterfeldt, Exploring reductions in London underground passenger journeys following the July 2005 bombings, Risk Anal. 31 (5) (2011) 773–786.

[5] Sweden Ministry of Infrastructure, Night trains to Europe will now be procured, 2020. Press release 24 July 2020, Available from: *https://www.government.se/press-releases/2020/07/night-trains-to-europe-will-now-be-procured/*. (Accessed 28 July 2020).

[6] C. Musselwhite, E. Avineri, Y. Susilo, Editorial JTH 16—the coronavirus disease COVID-19 and implications for transport and health, J. Transp. Health 16 (2020) 100853.

[7] N.C. Peeri, N. Shrestha, S. Rahman, R. Zaki, Z. Tan, S. Bibi, M. Baghbanzadeh, N. Aghamohammadi, W. Zhang, U. Haque, The SARS, MERS and novel coronavirus (COVID-19) epidemics, the newest and biggest global health threats: what lessons have we learned? Int. J. Epidemiol. 49 (3) (2020) 717–726.

[8] M. Hu, H. Lin, J. Wang, C. Xu, A.J. Tatem, B. Meng, X. Zhang, Y. Liu, P. Wang, G. Wu, H. Xie, S. Lai, Risk of Coronavirus disease 2019 transmission in train passengers: an epidemiological and modeling study, Clin. Infect. Dis. 72 (4) (2021), ciaa1057.

[9] J. Zhen, C. Chan, A. Schoonees, E. Apatu, L. Thabane, T. Young, Transmission of respiratory viruses when using public ground transport: a rapid review to inform public health recommendations during the COVID-19 pandemic, S. Afr. Med. J. 110 (6) (2020) 478–483.

[10] C. Hendrickson, L.R. Rilett, The COVID-19 pandemic and transportation engineering, J. Transp. Eng. 146 (7) (2020) 01820001.

[11] WSP, Rail and the effects of the COVID-19 pandemic, 2020. White Paper. Available from: *https://www.wsp.com/en-SE/insights/rail-and-the-effects-of-the-covid-19-pandemic*. (Accessed 7 July 2020).

[12] E. Jenelius, M. Cebecauer, Impacts of COVID-19 on public transport ridership in Sweden: Analysis of ticket validations, sales and passenger counts, Transp. Res. Interdiscip. Perspect. 8 (2021) 100242.

[13] A. Reggiani, P. Nijkamp, D. Lanzi, Transport resilience and vulnerability: the role of connectivity, Transp. Res. A 81 (2015) 4–15.

[14] U. Bhatia, D. Kumar, E. Kodra, A.R. Ganguly, Network science based quantification of resilience demonstrated on the Indian railways network, PLoS One 10 (11) (2015) e0141890.

[15] J. Zhang, F. Hu, S. Wang, Y. Dai, Y. Wang, Structural vulnerability and intervention of high speed railway networks, Physica A 462 (2016) 743–751.

[16] Y. Yang, Y. Lui, M. Zhou, F. Li, C. Sun, Robustness assessment of urban rail transit based on complex network theory: a case study of the Beijing Subway, Saf. Sci. 79 (2015) 149–162.

[17] S.S. Chopra, T. Dillon, M.M. Bilec, V. Khanna, Network-based framework for assessing infrastructure resilience: a case study of the London metro system, J. R. Soc. Interface 13 (2016) 20160113.

[18] Y. Sadaat, B.M. Ayyub, Y. Zhang, D. Zhang, H. Huang, Resilience of metrorail networks: quantification with Washington, DC as a case study, ASCE-ASME J. Risk Uncertainty Eng. Syst., Part B: Mech. Eng. 5 (2019) 041011.

[19] Y. Xing, J. Lu, S. Chen, S. Dissanayake, Vulnerability analysis of urban rail transit based on complex network theory: a case study of Shanghai Metro, Public Transp. 9 (2017) 501–525.

[20] L. Sun, Y. Huang, Y. Chen, L. Yao, Vulnerability assessment of urban rail transit based on multi-static weighted method in Beijing, China, Transp. Res. A 108 (2018) 12–24.

[21] M. Ouyang, L. Zhao, L. Hong, Z. Pan, Comparisons of complex network based models and real train flow model to analyze Chinese railway vulnerability, Reliab. Eng. Syst. Saf. 123 (2014) 38–46.

[22] A.K. Peterson, R.L. Church, A framework for modeling rail transport vulnerability, Growth Chang. 39 (4) (2008) 617–641.

[23] M. Bababeik, N. Khademi, A. Chen, Increasing the resilience level of a vulnerable rail network: the strategy of location and allocation of emergency relief trains, Transp. Res. E 119 (2018) 110–128.

[24] R. Gedik, H. Medal, C. Rainwater, E.A. Pohl, S.J. Mason, Vulnerability assessment and re-routing of freight trains under disruptions: a coal supply chain network application, Transp. Res. E 71 (2014) 45–57.

[25] A. Khaled, M. Jin, D.B. Clarke, M.A. Hoque, Train design and routing optimization for evaluating criticality of freight railroad infrastructures, Transp. Res. B 71 (2015) 71–84.

[26] E. Kassens-Noor, Transportation planning and policy in the pursuit of mega-events: Boston's 2024 Olympic bid, Transp. Policy 74 (2019) 239–245.

[27] D. Hensher, A.M. Brewer, Going for gold at the Sydney Olympics: how did transport perform? Transp. Rev. 22 (4) (2002) 381–399.

[28] P. Noursalehi, H.N. Koutsopoulos, J. Zhao, Real time transit demand prediction capturing station interactions and impact of special events, Transp. Res. C 97 (2018) 277–300.

[29] E. Chen, Z. Ye, C. Wang, M. Xu, Subway passenger flow prediction for special events using smart card data, IEEE Trans. Intell. Transp. Syst. 21 (3) (2020) 1109–1120.

[30] O. Cats, E. Jenelius, Planning for the unexpected: the value of reserve capacity for public transport network robustness, Transp. Res. A 81 (2015) 47–61.

[31] Y. Shen, G. Ren, B. Ran, Cascading failure analysis and robustness optimization of metro networks based on coupled map lattices: a case study of Nanjing, China, Transportation 48 (2021) 537–553.

[32] L. Hong, B. Ye, H. Yan, H. Zhang, M. Oyang, X. He, Spatiotemporal vulnerability analysis of railway systems with heterogeneous train flows, Transp. Res. A 130 (2019) 725–744.

[33] C. Szymula, N. Bešinović, Passenger-centered vulnerability assessment of railway networks, Transp. Res. B 136 (2020) 30–61.

[34] S.M. Ali, M.H. Rahman, T.J. Tumpa, A.A. Moghul Rifat, S.K. Paul, Examining price and service competition among retailers in a supply chain under potential demand disruption, J. Retail. Consum. Serv. 40 (2018) 40–47.

[35] D. Ivanov, Predicting the impacts of epidemic outbreaks on global supply chains: a simulation-based analysis on the coronavirus outbreak (COVID-19/SARS-CoV-2) case, Transp. Res. E 136 (2020) 101922.

[36] D. Wang, B.Y. He, J. Gao, J.Y.J. Chow, K. Ozbay, S. Iyer, Impact of COVID-19 behavioral inertia on reopening strategies for New York City transit, Int. J. Transp. Sci. Tech. 10 (2) (2021) 197–211.

[37] G. Brancaccio, M. Kalouptsidi, T. Papageorgiou, Geography, transportation, and endogenous trade costs, Econometrica 88 (2020) 657–691.

Management of railway stations exposed to a terrorist threat

Sakdirat Kaewunruen and Hamad Alawad
University of Birmingham, Birmingham, United Kingdom

5.1 Introduction

Stations form a critical point in the rail system where the passengers start and end their journeys. In addition to millions of passengers also many tons of goods are being transported daily across the globe and passing through the railway stations. It is an essential structural part of the railway network and provides a major part of the crucial public transport system as well as the backbone of an urban transportation system. At the same time, stations play a vital role in a country's economy, society, and daily life. For these reasons, stations need to be in continuous operation with a high level of health, safety, and security. Many of the stations are located at a susceptible point of the cities and possibly intersect with other transportation networks perhaps containing many levels and tunnels. Some stations nowadays have shops or malls and are linked with critical infrastructures such as airports, essential areas, and city centers. Owing to the high demand and the growth of travel and economic development as well as population growth, stations can become crowded and therefore face a range of challenges for safety and security such as terrorist threats, and other vulnerabilities and hazards. It has been noticed that such a critical structure is a target for terrorists due to the consequences an attack could have in crippling an economy, community, or nation [1] and owing to their level of accessibility [2,3]. Hence, railway stations and the network robustness, resilience, and effective preventative management to ensure the safe and secure operation are significant matters for research. The terrorist threat is a global risk with prior events often resulting in casualties, assets damage, and more such as passenger fear, outage of the entire network, and other indirect costs. The consequences of terrorist attacks can be severe involving deaths, injuries, damage to goods as well as economic and social impacts. Terrorism involves a wide variety of groups embracing violence to influence political and social discourse and it includes a wide range of groups and movements with religious, ethnic, right-wing, and issue-specific agendas with members who have been spread over many countries [4]. On a global scale from 1997 to 2000 it is shown that 40% of over 195 terrorist attacks targeted railways including trains, underground trains and stations (Fig. 5.1) [5].

Furthermore, throughout the world between 2004 and 2008 passenger rail systems were targeted by hundreds of terrorist acts, resulting in killing and injuring over 10,000 people [6]. A better comprehension of the actions of terrorists and the way they

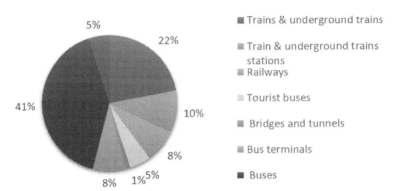

Fig. 5.1 Percentage distribution of the targets for terrorism.

select certain targets of attack can aid in making decisions and help security designers to allocate resources in the fight against terrorism [7]. To face the challenge of terrorism, new analytical methods and new institutional arrangements must be elaborate, as stated in the Making the Nation Safer, a report by the National Research Council [8]. Infrastructure of the station is not an independent system but rather is linked with other interdependent systems. A failure of the electrical power grid, for example, may affect not only the energy sector but also may result in the collapse of or severe disruption of transportation, telecommunications, public health, and banking and financial systems of the country [9]. The railway station infrastructure is typically complex, and many issues can arise for protecting such a system during the decision-making process. The management of railway stations is required to acknowledge all the parameters and uncertainties involved.

It has been shown that since 9/11 the threat of terrorist attacks has been increasing around the world and more specifically on the transportation systems. Thus, as a reaction to this, several countries have implemented rigorous security measures to defend and protect lives and assets and to moderate the security risk to the transportations system. For instance, air systems realized their vulnerabilities and implemented new countermeasures to cover the possible deficiencies. Also, in the marine industry security inspections were quite advanced already, but land transport security in general, and railway security in particular, had been afforded less consideration. In the railway industry, the concern was concentrated on safety rather than on security. However, ensuring security is a necessity for ensuring safety. The high density, overloading, and the strong relation between business systems in the railway station create an extraordinary challenge for ensuring security and safety [10]. This chapter will present some attack examples and then analyze the risk. After this, a review of the risk measures and then, the emergency and preincident management will be considered before finally, drawing some conclusions.

5.2 Previous studies

Risk management is essential for any critical infrastructure. Some studies have discussed the protection of critical infrastructures such as gas supply [11], power and water system supply [12,13], and transport [14,15]. Major risk categories, man-made or natural, include failures in the subsystems in the station, train delays, failure of the operation and maintenance, crowd management, technological challenges, collisions, sudden conditions, crimes, security failure, and severe weather conditions. Modeling risk and decision frameworks are appropriate for risk managers to derive the most suitable risk management measures [16,17]. Many methods have been devised in an attempt to model and assess the efficiency of counter-terrorism procedures. In general, it has been perceived that the risk analysis of terrorism is modernistic, and no particular method is destined to face this issue. From the literature, it has been observed that utilizing many tools such as probabilistic risk assessment (PRA), including any probabilistic approach with tools like event trees [18–20], logic trees [21], fault trees, and decision trees [22], event seniors [23] (Fig. 5.2) may not be a solution.

Additionally, other tools such as game-theoretic approaches and system dynamics may prove to be useful in dealing with an intelligent adversary [24]. This approach has been utilized to estimate the probabilities of terrorist attack inputs [25–28].

Some of these methods involve game-theoretic approaches which are used to model how intelligent attackers and defenders interact. Another set of methods is based on probabilistic risk assessment or PRA. PRA has been used to evaluate the risks associated with complex engineered entities and it recently started being used to assess terrorism risk [18]. Moreover, PRA aids analysis and helps decision-makers

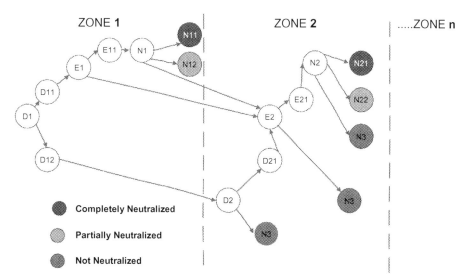

Fig. 5.2 Example of hypothetical asset with two security zones.

in understanding and describing the risks and in predicting the probable consequences [12].

Some countries have been carrying out trials on security procedures as part of a qualitative study, for example, the Department for Transport (DfT) in the United Kingdom carried out some transport security measures including passenger full-body scanner and luggage X-ray; passenger index finger swabbing, and sniffer dogs [29–31]. However, some of these methods cause delays and impact passenger attitude and privacy; thus, it has been concluded that technology should be developed for more effective security check methods as it is less intrusive. In fact, the framework of railway security is at a growth stage, compared with the security in air transport which is already mature, and which already has many specific existing measures. All of the incidents that have occurred so far in the railway industry provide increased pressure for the rail industry to develop a comprehensive and systematic framework for security [32].

5.3 Security risk analysis

In order to enhance the security of international and local utilities and infrastructure, stakeholders and decision-makers must take into account all the subtle and different ways a malicious attacker can cause human injury, fatalities, and structural damage. The main difficulty in the risk analysis of terrorism is the fact that terrorists, unlike nature or engineered systems, are intellectual challenges and may adjust to defensive countermeasures [24]. For this reason, limited resources need to be allocated accordingly across a variety of improvements that could mitigate the risk of an attack.

Reid and Reid [33] revealed that the terrorist principles and priorities for selecting targets generally include some of the following:

- High fatality possibility (crowd, confinement).
- General vulnerability (interconnected consequence, lack of security).
- High local visibility and media impact.
- Accessibility and high traffic population areas.
- Poor design for repelling or mitigating attack, dealing with perpetrator infiltration.
- Ability to damage or destroy the place that provides emergency care to victims.
- Difficulty evacuating victims and potential victims.

Therefore, a guide for designers and legislators in improving the infrastructure responses during an emergency is really needed [34].

The strategy of the attackers is to cause passenger fatalities as this will have a negative effect on the economy, and public transport industries. This in turn will have a media impact and affect public confidence, as evidenced by an immediate drop in revenue. An example of this is in 2004, the Madrid attack bombings on commuter trains (13 bombs and 10 explosions, 175 fatalities, and 626 injured) [35–37]. Some further examples are presented in Table 5.1. The terrorism led to a massive loss of life, assets damage, financial loss, reputational harm, and the breaking down of societal relations, which lead to impacts on cities' economic activities and the tourism [41].

Table 5.1 Examples of some previous terrorist attacks in train stations.

Date	Country	Deaths or injuries	Details
July 7, 2005	United Kingdom	Killed 52 passengers and injured more than 700 people	Four terrorists detonated suicide Bombs [38]
July 11, 2006	India	Killed at least 209 people and more than 700 others were injured	Bombings near the Suburban railway stations in Mumbai [39]
February 18, 2003	South Korea	198 fatalities, 147 injuries	A self-immolation in the Jungangno subway station [40]
February 6, 2004	Russia	39 fatalities and more than 100 injuries	A bomb in a crowded Moscow, Subway station [40]

Infrastructure is not an independent system but rather it is interconnected with other interdependent systems. A failure of the electrical power grid, for example, may affect not only the energy sector but also in a cascading effect that may result in the collapse or severe disruption of transportation, telecommunications, public health, and banking and financial systems of the country. More seriously, owing to the interdependence amidst distinct infrastructures, what happens in one infrastructure will possibly influence others, directly or indirectly. After the terrorist attacks of September 11th, 2001 in the United States, it was found that the communications' services were extensively damaged [9,16,42].

As infrastructures are typically and naturally complex, and many issues can arise simultaneously during the decision-making process, a risk management framework is required to acknowledge all the parameters and uncertainties involved. In practice, a local manager of a specific infrastructure is naturally responsible for risk management in a crisis; however, in an extreme event, the risk mitigation and monitoring may not be effective, causing irreversible severe consequences. Modeling risk and decision frameworks are therefore appropriate for risk managers to derive the most suitable risk management measures [16]. Hence, a guide for designers and legislators in advancing the infrastructure's responses during a crisis is necessary [23,43]. The risk analysis can be broken down into five main process components, as shown in Fig. 5.3. A separation of these main components describes and quantifies the risk provided by the terrorist attack and the critical parameters that add to the uncertainty and form a framework for the critical asset and the processes for risk analysis [44].

During this first step of the risk analysis, the possible threat scenarios are determined, where there is an agreement that the initial stage of risk management centers on the identification of risks. Since the 1990s transportation systems worldwide have

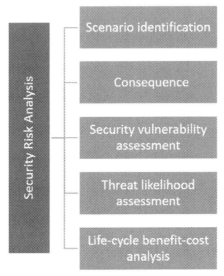

Fig. 5.3 Five main components of security risk analysis.

been predominantly targeted by terrorist attacks. This form of terrorist threat encompasses a wide range of potential attacks and more likely scenarios such as:

- Suicide attackers (the case in London, July 2005).
- Explosive devices (the Madrid bombings, 2004).
- A release of chemical, biological, radiological, or nuclear material (the sarin gas attack in Tokyo in several subway stations, March 1995) [37].

The main factors contributing to the plausibility of a threat scenario are the availability of resources for the terrorists, their ability to execute a particular type of attack, the asset's susceptibility to a threat, and possible outcomes of an attack.

By taking these factors into account, the array of threats can be reduced to a number of plausible and predominant threat scenarios with lesser consequences. It has been shown that by reducing the number of scenarios to a degree that is manageable, it will possibly avoid any additional bias and uncertainty in the risk analysis and will provide more accuracy [45]. Outlining the main scenarios has to be done with care because eliminating possible threat scenarios can introduce uncertainty and bias into the outcome of the risk analysis. For the Consequence and Criticality Assessment, the assessment provides the estimated losses of a successful attack or the severity of the losses that may be incurred. In the area of counter-terrorism, the term loss has various meanings; such as damage to the environment, human casualties, impact on society, as well as economic losses either direct or indirect due to physical damage, interruptions of business, and financial market insecurity. These different types of losses can be estimated using game-theory methods and modeling means, such as event trees, fault trees, and decision trees. For Security Vulnerability Assessment, Security is only one part of the risk, and the relationships among risk management, risk assessment, and vulnerability assessment are shown in Fig. 5.4.

Fig. 5.4 The relationships between risk management and vulnerability assessment.

To understand and measure the impact of threats such as terrorism, the risk analysis and evaluation of threats and vulnerabilities are an essential process and thus the output is precious information for decision-makers to adopt the optimal countermeasures to manage threats or propose effective improvements to the system. The vulnerability assessment is a system in which quantitative or qualitative methods are applied to predict, for example, the security system's components' effectiveness and overall performance, by identifying its exploitable weaknesses for a defined threat. Then, after finding the gaps, it is used for improvement. The vulnerability assessment process is part of the more significant risk assessment process. It is vital to differentiate security from safety when discussing a vulnerability assessment. Safety is defined as the measures (people, procedures, or equipment) used to prevent or detect an abnormal condition that can endanger people, property, or the enterprise. Security, on the other hand, includes the measures used to protect people, property, or the enterprise from malicious human threats. A proper security vulnerability assessment will consider safety controls because some safety measures aid in detection and response to security events [25]. The next step of this risk analysis reveals the probability that a terrorist is successful in attacking their target with the condition that they initiated the attack. By combining this probability with the estimates of possible losses of key assets conditioned on the success of the attack, the main conditional expected loss associated with a scenario is revealed. The successful attack is based on the terrorist's ability to defeat the security system. In traditional systems, the final step of risk management is that of control, funding, and monitoring which is designed to diminish the risk probability to occur and consequences of the losses, as well as the mechanisms for funding the losses [41].

5.4 Managing terrorism risk

The aim of the risk management process is controlling risk, but the foundation for making the decisions is knowing what the risks are faced. The science of risk assessment, mainly quantitative risk assessment, is an analytical process, and it has been proved to answer three basic questions:

1. What can go wrong?
2. How likely is that to happen?
3. What are the consequences if it does happen? [9,17].

Fig. 5.5 Risk is the connection of threat, vulnerability, and consequences.

The terrorism risk was described in the literature as a role or product of threat, vulnerability, and outcomes (consequences) [46–48]. The three components included in the definition play a significant part in the overall risk, and all three intersected to present the risk (Fig. 5.5).

From the three aspects, we can present the Risk as a Function of Threat, Vulnerability, and Consequences as probabilistic background follows [47]:

$$\text{Risk} = \text{Threat} * \text{Vulnerability} * \text{Consequence} \tag{5.1}$$

The next lines will show some details for each, and this approach of terrorism risk or formulation has two notable advantages:

- It provides an approach for comparing and aggregating terrorism risk.
- This definition of risk provides a clear mapping between risk and approaches to managing or reducing risk.

It is obvious that increasing preparedness and response will reduce the effects of consequences [48].

5.4.1 Threat

This includes two linked points, the intent, and capability. This act might be carried out by the individual people or organizations to represent a terrorist threat to impose damage to a target. For measuring the threat of attacks such as targeting the railway stations at specific times, it can be displayed by probability as a measure of the likelihood that an attack will occur. The equation is as follows:

$$\text{Threat} = P(\text{attack happens}) \tag{5.2}$$

This formal measure presents the probability (P) that a specific target will be attacked in a specific way during a specified time period, which is known as the attack probability. It is challenging to estimate and requires knowledge, and information about the motivations, plans, and capabilities of attackers [49,50].

5.4.2 Vulnerability

The vulnerability can be evaluated as being the capacity of a system to respond to the threat. Vulnerability is the manifestation of the inherent state of the system (e.g., physical, technical, organizational, cultural) that can result in damage if attacked by an adversary. In other words, not all threats of the same type are equally important. Thus, there is a need to measure vulnerability that will lead to damage when attacks occur. The damage may involve fatalities, injuries, infrastructure damage, and in this case, damage to the railway station at a specific time. Therefore, the station vulnerability can be explained as the likelihood of an attack happening assuming that (condition) the attack will be successful and produce specific types of damage (results) (i.e., there would be separate vulnerability assessments for deaths, injuries, and structure damage). It is calculated using the formula below:

$$\text{Vulnerability} = P(\text{attack results}|\text{attack happens}) \tag{5.3}$$

In this set, it has been noted that the specified magnitude of the damage is not measured and is not part of the definition of vulnerability. Also, it covers both a successful attack with damage or unsuccess with no damage which explains vulnerability as the probability of an attack's success in the case where it occurs.

5.4.3 Consequences

This component will measure the likely (expected (E)) magnitude of damage (e.g., deaths, injuries, direct and indirect economic impacts, structure damage, or other types of damage) given a specific attack type, at a specific time, that results in damage to a specific target. It covers the magnitude and kind of damage following successful terrorist attacks. The formula can be presented as:

$$\text{Consequence} = E(\text{damage}|\text{attack occurs and results in damage})$$

These consequences are the losses that occur given a successful attack [24,51]. Moreover, it has been noticed that risk management is based on established principles and involves the four actions shown in Fig. 5.6 [9].

5.5 The technology and terrorist threat

It has been noticed that some of the security measures as available security devices have been demanded or show a rise in requests, such as:

- Electronic check system for employee ID covering all entrance points and the vehicular gates at all station facilities and the programmable intrusion equipment.
- Closed-circuit TV and motion detection alarms for metro-rail permission areas covering all the perimeter fencing and shop facilities, including fiber-optic network and video recording devices.

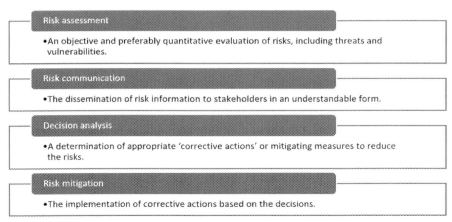

Fig. 5.6 The four risk management activities.

- Bomb-resistant containers at all rail stations, and teams and vehicles for explosive detection [17].
- Personal protection equipment, training, and satellite telephones for employees with chemical emergency sensor programs [52].

The latest technology using deep learning and CCTVs, drones, and other surveillance devices, will advance the detection resources. This would trigger an alarm in case of terrorist attacks, and detect attacks in many places like public transportation, schools, government offices, and hospitals where arms are entirely restricted [53]. Incredible improvements have been shown wherever AI, such as the deep learning and computer vision, has been employed. With the improvement in AI capabilities in computing, application of AI technologies to terrorist data can allow useful insights regarding the interaction of terrorists, governance, understanding patterns of terrorist behavior, and society [54]. Using images and videos, which can include object detection and classification, allows the video analytic software to differentiate between the objects such as trains, and human behaviors such as the trespassers and the flow and counting people to help aid emergency responses in case of an unwanted event [40]. Moreover, wireless sensors are being utilized to advance security. A sufficient level of security is required for all transportation systems including railway stations, which are intermodal transport hubs, where technologies such as WSN could be adopted to boost the security. Today, the infrastructure in a railway industry is equipped with various sensors and is closely coupled with information and communication technologies. However, intelligent monitoring, security, and risk management can be better realized via the usage of networked embedded devices [55]. Moreover, simulation of the training technology can offer training in a realistic, virtual situation, and allows employees to "encounter" a terrorist situation and then initiate immediate action [11,56].

5.6 Emergency and preincident management

The evolution and practice of emergency response plans are crucial for diminishing the impacts of the risk in emergencies in the station. The responders include the station operators, utility companies, ambulance services, fire department, medical service, police department, and others. Moreover, a preparedness plan for terrorism risk is vital for two critical reasons:

- Recently, world terrorism has increased and is occurring in unexpected places, at unexpected times, with unexpected losses.
- The consequences of not being prepared are too high [57].

Various incidents can happen in the station in separate locations such as tunnels, main entrance, parking, and on the platforms or on the trains. Consequently, the agencies responsible for operations should have emergency response plans for unwanted events and take into consideration the following points:

- The possibility of losing the main power of the station, structure failures, and means of communication.
- Evacuation of passengers from a train at any positions on the track within the station boundaries, including the tunnels.
- Managing the passenger flow, where it is to be expected passengers will panic and there will be disruption of services and possible police activities in the station.
- Dealing with hazardous materials in the station.
- Availability of First aid or medical care with adequate training in stations.
- The traffic conditions from and to the station and the locations of the emergency response agencies.
- Training of the employees must be considered, which is essential for the process of evacuation of the stations, trains, subway tunnels and helps fellow passengers during a rail disaster [52].

Lessons must be learned, and the events that have occurred in different locations around the world show the importance of preparedness and emergency planning as well as the importance of security in the railway stations. In fact, after all these incidents, the concern from the public and companies has led to some improvements in the security of the transport systems in addition to more understanding, and some points can be highlighted:

1. It has been shown that levels of difficulty are increased if the incidents occur in an underground railway system or in multifloor stations which have several exit routes. This leads to the importance of recognizing emergency events in the design stages of the railway stations. In such events like Madrid 2004, the authority's response centers could not properly predict the total number of sites or their places. Moreover, there was a need to hold some resources in reserve to respond to any possible routine incidents that may occur at the same time. The weakness of training and education of the emergency services to predict major incidents can lead to a lack of initial estimates and the forwarding of alarms [35,36,58].
2. It has been shown that there can be no perfect response to a significant incident ever, as in the major catastrophic events, situations of unusual pressure, uncertainty, and complexity, the decision-making process is under immense strain.

3. It is expected that for emergency events on the trains that have already left the station, even by a few minutes, there are uncertainties between the driver and passengers. The passengers generally do not know where they are, and they may also possibly be physically in the dark in tunnels. They do not know who anyone is, who has planned an evacuation of the train, or whether or not the current is still turned on and so there are difficulties in the communications between the driver and passengers, sometimes made worse when drivers are unable to communicate with their line control centers and there is no signal in the passenger phones' network coverage. It is possible that antennae are damaged as a consequence of emergency events. For example, in London on 7 July 2005 antennae were damaged by the explosion. The underground's radio systems were outdated, and they are expected sometimes to fail because of blind spots and temporary interruptions to the service [56,59].
4. For safety and security, effective decisions are required to ensure accuracy of the information, but this depends on the availability of the data which has been collected and analyzed, and conversely, any unavailability of this data will affect the level of effective decisions.

5.7 Conclusion

It has been clarified that the management and mitigation of terrorism risk are significant challenges due to the dynamic nature of terrorist risks. However, it is crucial to develop that understanding of the risks and learn from the accidents in the past for railway stations' future safety and security. We emphasize that new technology could be developed and implemented for the management of railway stations exposed to terrorist risk. Furthermore, the general aspects which play an essential role in any development in the security and safety system in the railway stations should be understood, so some main factors concluded are listed below:

- The security and safety in the railway station system are challenging by nature, as they are generally open structures with large volumes as well as public places for shopping and other activities rather than being an exclusive place for travelers.
- Safety and security are highly demanded from the passengers along with the convenience of mobility (from reservation to travel).
- The security and safety systems are a cost according to operations companies or station owners and managers of railway stations, but they wish to react to the customers' demands and meet the standards and regulations.
- The suppliers and scholars consider the security and safety in the railway stations to be a business and, on the other hand, a research field opportunity, utilizing the new technologies for a move toward smart railway stations which provide substantial raw data and enhance the security systems.
- From the previous incidents, the location, intelligence information, cost, and emergency process, as well as other factors, were key to the safety and security of the railway stations.
- The new technology is expected to play a significant role in the field to tackle these issues, so more research is demanded.
- Intelligence and other information are part of a key and active method, where many other measures are passive systems, and this information is the Vigor Point as we know the attackers are intelligent adversaries and may adapt to defensive measures.

Many approaches or control measures have limitations or constitute a complete solution, so modeling techniques should be constantly improved and should evolve in parallel with the increasing complexity of the systems and the intelligence of attackers. In the future, smart technology will play an essential role in security improvements and will aid decision-makers in managing risks. Finally, it should be remembered that security in railway stations is challenging by nature, as they are generally open structures as well as public places for shopping and other activities rather than being an exclusive place for travelers. They do, however, provide a rich environment for field research.

References

[1] J.G. Kappia, et al., The acceptability of counter-terrorism measures on urban mass transit in the UK, WIT Trans. Built Environ. 107 (2009) 627–636, https://doi.org/10.2495/UT090561.
[2] V.M. Bier, Risk in Extreme Environments Preparing, Avoiding, Mitigating, and Managing, 2018. (Accessed 25 January 2020).
[3] J.M. Pearce, et al., Encouraging public reporting of suspicious behaviour on rail networks, Polic. Soc. 30 (7) (2019) 1–19, https://doi.org/10.1080/10439463.2019.1607340.
[4] M.S. Townes, G.L. Blair, K. Hunter-zaworski, TCRP Synthesis 27 Emergency Preparedness for Transit Terrorism A Synthesis of Transit Practice, Chairman The City College of New York, Washington, DC, 1997.
[5] B.M. Jenkins, Protecting Surface Transportation Systems and Patrons From Terrorist Activities: Case Studies of Best Security Practices and a Chronology of Attacks, 1997.
[6] N. Barkakati, D. Maurer, Technology Assessment: Explosives Detection Technologies to Protect Passenger Rail, Government Accountability Office/DIANE Publishing Company, Washington, DC, 2010.
[7] G.L. Keeney, D. Von Winterfeldt, Identifying and structuring the objectives of terrorists, Risk Anal. 30 (12) (2010) 1803–1816, https://doi.org/10.1111/j.1539-6924.2010.01472.x.
[8] Committee on Counterterrorism Challenges for Russia and the United States, Terrorism: Reducing Vulnerabilities and Improving Responses: U.S-Russian Workshop Proceedings Committee, National Academies Press, 2006. Available from: *https://books.google.com/books?id=l5cvA11fgjkC&pgis=1*. (Accessed 25 January 2020).
[9] B.J. Garrick, et al., Confronting the risks of terrorism: making the right decisions, Reliab. Eng. Syst. Saf. 86 (2) (2004) 129–176, https://doi.org/10.1016/j.ress.2004.04.003.
[10] C. Guoqiang, J. Limin, Z. Liming, Research on rail safety security system, Int. J. Econ. Manag. Eng. 4 (8) (2010) 1938–1943.
[11] M. Ouyang, et al., A methodological approach to analyze vulnerability of interdependent infrastructures, Simul. Model. Pract. Theory 17 (5) (2009) 817–828, https://doi.org/10.1016/j.simpat.2009.02.001.
[12] J.W. Wang, L.L. Rong, Cascade-based attack vulnerability on the US power grid, Saf. Sci. 47 (10) (2009) 1332–1336, https://doi.org/10.1016/j.ssci.2009.02.002.
[13] S. Wang, L. Hong, X. Chen, Vulnerability analysis of interdependent infrastructure systems: a methodological framework, Physica A 391 (11) (2012) 3323–3335, https://doi.org/10.1016/j.physa.2011.12.043.
[14] C. von Ferber, T. Holovatch, Y. Holovatch, Attack vulnerability of public transport networks, in: Traffic and Granular Flow '07, Springer, 2009, pp. 721–731, https://doi.org/10.1007/978-3-540-77074-9_81.

[15] M.K. Jha, Dynamic bayesian network for predicting the likelihood of a terrorist attack at critical transportation infrastructure facilities, J. Infrastruct. Syst. 15 (1) (2009) 31–39, https://doi.org/10.1061/(ASCE)1076-0342(2009)15:1(31).

[16] T. Aven, O. Renn, The role of quantitative risk assessments for characterizing risk and uncertainty and delineating appropriate risk management options, with special emphasis on terrorism risk, Risk Anal. 29 (4) (2009) 587–600, https://doi.org/10.1111/j.1539-6924.2008.01175.x.

[17] S. Kaewunruen, H. Alawad, S. Cotruta, A decision framework for managing the risk of terrorist threats at rail stations interconnected with airports, Safety 4 (3) (2018), https://doi.org/10.3390/safety4030036.

[18] B.C. Ezell, Y.Y. Haimes, J.H. Lambert, Risks of cyber attack to water utility supervisory control and data acquisition systems, Mil. Oper. Res. 6 (2) (2011) 23–33, https://doi.org/10.5711/morj.6.2.23.

[19] G. Koller, Risk Modeling for Determining Value and Decision Making, Routledge, 2000, https://doi.org/10.1198/tech.2001.s633.

[20] W.K. Viscusi, R.J. Zeckhauser, Sacrificing civil liberties to reduce terrorism risks, J. Risk Uncertain. 26 (2–3) (2003) 99–120, https://doi.org/10.1023/A:1024111622266.

[21] J.G. Voeller, et al., Logic trees: fault, success, attack, event, probability, and decision trees, in: Wiley Handbook of Science and Technology for Homeland Security, John Wiley & Sons, 2009, https://doi.org/10.1002/9780470087923.hhs004.

[22] J.J. Bommer, et al., On the use of logic trees for ground-motion prediction equations in seismic-hazard analysis, Bull. Seismol. Soc. Am. 95 (2) (2005) 377–389, https://doi.org/10.1785/0120040073.

[23] H.A.H. Alawad, S. Codru, S. Kaewunruen, Complex-system decision framework for managing risks to rail stations at airports from terrorist threats, in: The 5th International Conference on Road and Rail Infrastructure, Zadar, Croatia, 17/05/18, University of Zagreb, 2018, pp. 855–861. *https://doi.org/10.5592/CO/CETRA.2018.665*.

[24] B.C. Ezell, S.P. Bennett, D. Von Winterfeldt, et al., Probabilistic risk analysis and terrorism risk, Risk Anal. 30 (4) (2010) 575–589, https://doi.org/10.1111/j.1539-6924.2010.01401.x.

[25] M. Garcia, Vulnerability Assessment of Physical Protection Systems, 2005 (Accessed 28 January 2020).

[26] W.L. McGill, B.M. Ayyub, M. Kaminskiy, Risk analysis for critical asset protection, Risk Anal. 27 (5) (2007) 1265–1281, https://doi.org/10.1111/j.1539-6924.2007.00955.x.

[27] E. Paté-Cornell, S. Guikema, Probabilistic modeling of terrorist threats: a systems analysis approach to setting priorities among countermeasures, Mil. Oper. Res. 7 (2002) 5–23.

[28] H. Rosoff, D. Von Winterfeldt, A risk and economic analysis of dirty bomb attacks on the ports of Los Angeles and Long Beach, Risk Anal. 27 (3) (2007) 533–546, https://doi.org/10.1111/j.1539-6924.2007.00908.x.

[29] C. Turley, Transport, V. S.-D. for and 2006, undefined, Sniffer Dogs Trials (London and Brighton), n.d.

[30] C. Turley, Transport, V. S.-D. for and 2006, U, Security Screening Trial at Heathrow Express, 2016.

[31] E. Carter, et al., Passenger acceptance of counter-terrorism security measures in stations, IET Intell. Transp. Syst. 10 (1) (2016) 2–9, https://doi.org/10.1049/iet-its.2015.0031.

[32] M. Finger, et al., Rail Passenger Security: Is It a Challenge for the Single European Railway Area? 2016, Available from: http://cadmus.eui.eu/bitstream/handle/1814/45148/Observer_2016_04_FSR_T.pdf?sequence=1.

[33] D.J. Reid, W.H. Reid, Managing facility risk: external threats and health care organizations, Behav. Sci. Law 32 (3) (2014) 366–376, https://doi.org/10.1002/bsl.2107.
[34] G.P. Cimellaro, S. Mahin, M. Domaneschi, Integrating a human behavior model within an agent-based approach for blasting evacuation, Comput. Aided Civ. Inf. Eng. 34 (1) (2019) 3–20, https://doi.org/10.1111/mice.12364.
[35] J. Peral-Gutierrez de Ceballos, et al., 11 march 2004: the terrorist bomb explosions in Madrid, Spain—an analysis of the logistics, injuries sustained and clinical management of casualties treated at the closest hospital, Crit. Care 9 (1) (2005) 104–111, https://doi.org/10.1186/cc2995.
[36] J. Peral Gutierrez De Ceballos, et al., Casualties treated at the closest hospital in the Madrid, March 11, terrorist bombings, Crit. Care Med. 33 (1 Suppl) (2005) 107–112, https://doi.org/10.1097/01.CCM.0000151072.17826.72.
[37] M.M. Sánchez, Security risk assessments in public transport networks, Proc. Inst. Mech. Eng. Part F J. Rail Rapid Transit 225 (4) (2011) 417–424, https://doi.org/10.1243/09544097JRRT409.
[38] W. Zhu, et al., Enhancing robustness of metro networks using strategic defense, Physica A 503 (August) (2018) 1081–1091, https://doi.org/10.1016/j.physa.2018.08.109.
[39] A. Rabasa, et al., The Lessons of Mumbai, Distribution, RAND Corporation, 2009. Available from: *www.rand.org*. (Accessed 29 January 2020).
[40] A.M. Rakoczy, S.T. Wilk, M.C. Jones, Security and safety of rail transit tunnels, Transp. Res. Rec. 2673 (1) (2019) 92–101, https://doi.org/10.1177/0361198118822819.
[41] D. McIlhatton, et al., Current considerations of counter terrorism in the risk management profession, J. Appl. Secur. Res. 14 (3) (2019) 350–368, https://doi.org/10.1080/19361610.2018.1545196.
[42] J.W. Seifert, The effects of September 11, 2001, terrorist attacks on public and private information infrastructures: a preliminary assessment of lessons learned, Gov. Inf. Q. 19 (3) (2002) 225–242, https://doi.org/10.1016/S0740-624X(02)00103-X.
[43] I. Jordaan, Decisions under uncertainty: probabilistic analysis for engineering decisions, in: Decisions Under Uncertainty: Probabilistic Analysis for Engineering Decisions, Cambridge University Press, 2005, pp. 1–672, https://doi.org/10.1017/CBO9780511804861. 9780521782.
[44] B.M. Ayyub, W.L. McGill, M. Kaminskiy, Critical asset and portfolio risk analysis: an all-hazards framework, Risk Anal. 27 (4) (2007) 789–801, https://doi.org/10.1111/j.1539-6924.2007.00911.x.
[45] A. Shafieezadeh, E.J. Cha, B.R. Ellingwood, A decision framework for managing risk to airports from terrorist attack, Risk Anal. 35 (2) (2015) 292–306, https://doi.org/10.1111/risa.12266.
[46] Y. Haimes, Risk Modeling, Assessment, and Management, 2005, John Wiley & Sons. (Accessed 29 January 2020).
[47] H.H. Willis, et al., Estimating Terrorism Risk, Security, RAND Corporation, Santa Monica, 2005.
[48] H.H. Willis, Guiding resource allocations based on terrorism risk, Risk Anal. 27 (3) (2007) 597–606, https://doi.org/10.1111/j.1539-6924.2007.00909.x.
[49] T. Bedford, R. Cooke Probabilistic Risk Analysis: Foundations and Methods, 2001 (Accessed 29 January 2020).
[50] S. Hora, et al., Advances in Decision Analysis: From Foundations to Applications, Cambridge University Press, 2007.
[51] R. Wilson, Combating Terrorism: An Event Tree Approach, books.google.com, 2003, pp. 122–143, https://doi.org/10.1142/9789812705150_0017.

[52] R. Tarr, V. McGurk, C. Jones, Intermodal transportation safety and security issues: training against terrorism, J. Public Transp. 8 (4) (2005) 87–102, https://doi.org/10.5038/2375-0901.8.4.6.

[53] G. Chandan, et al., Real time object detection and tracking using deep learning and OpenCV, in: Proceedings of the International Conference on Inventive Research in Computing Applications, ICIRCA 2018, IEEE, 2018, pp. 1305–1308, https://doi.org/10.1109/ICIRCA.2018.8597266.

[54] S.S. Gartner, D. Felmlee, Understanding Patterns of Terrorism in India Using AI Machine Learning: 2007-2017, Common Ground Research Networks, 2017, pp. 2007–2017.

[55] H. Alawad, S. Kaewunruen, Wireless sensor networks: toward smarter railway stations, Infrastructures 3 (3) (2018) 24, https://doi.org/10.3390/infrastructures3030024.

[56] L. Assembly, 7 July Review Committee, 2007.

[57] M. Harvey, et al., Engaging in duty of care: towards a terrorism preparedness plan, Int. J. Hum. Resour. Manag. 30 (11) (2019) 1683–1708, https://doi.org/10.1080/09585192.2017.1298651.

[58] E.R. Frykberg, Terrorist bombings in Madrid, Crit. Care 9 (1) (2005) 20–22, https://doi.org/10.1186/cc2997.

[59] K. Strom, J. Eyerman, Interagency coordination: lessons learned from the 2005 London train bombings, NIJ J. 261 (2008) 28–32. Available from: http://www.ncjrs.gov/pdffiles1/nij/224088.pdf.

Rail infrastructure systems and hazards

Chayut Ngamkhanong[a,b], Keiichi Goto[c], and Sakdirat Kaewunruen[a]
[a]School of Engineering, University of Birmingham, Birmingham, United Kingdom,
[b]Department of Civil Engineering, Faculty of Engineering, Chulalongkorn University, Bangkok, Thailand, [c]Railway Dynamic Division, Railway Technical Research Institute, Tokyo, Japan

6.1 Extreme temperature

6.1.1 Hot weather

At present, global temperatures have been increasing continuously by an average of 0.7°C per year. Due to the rise in temperature, this can cause damage and loss to the railway infrastructure around the globe. It is found that the failures of rail infrastructure are weather-related. This failure due to extreme weather tends to be more frequent. The future incidents of railway tracks can be predicted by the correlated data about the performance of rail infrastructure such as past failure with related weather conditions [1]. It is consequently seen that the extreme temperature tends to have the more significant impact on railway network in the future. The extreme heat not only induces loss to railway track but also other railway infrastructure assets in the whole network, as seen in Fig. 6.1.

The increase in global temperature has a direct effect on the steel rail which is naturally easy to expand due to the rise in temperature. It is noted that the rise of temperature of steel rail over neutral temperature (stress-free temperature) can build up the axial compression force in Continuous Welded Rail (CWR) due to such heat waves. Even if CWR can potentially provide the smooth ride of train, there are some drawbacks in which track tends to get buckled easily when the rail axial force reaches the limit level which depends on the track conditions. Note that track buckling is triggered by the small lateral misalignment which generates the lateral force in the form of follower force. It has been found in many literatures and reports that the number of hot days (over 20°C) has a proportional trend to the number of track buckles [3]. It is interesting that, in 2016–17, although there were fewer very hot days than previous years, track buckling was increased since the hottest day was observed in those years.

Note that different countries have different levels of rail neutral or stress-free temperature (SFT) depending on their weather conditions. Hot countries are likely to experience track buckling rather than cold countries and therefore the stress-free temperature of steel rail in hot countries is normally higher than that in cold countries. For example, in United Kingdom, the stress-free temperature is 27°C [4] while the stress-free temperatures vary from 35°C to 43°C in summer in United States [5].

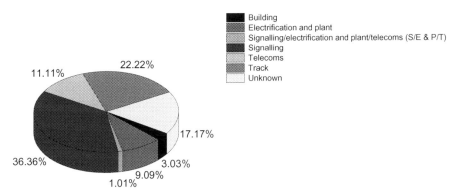

Fig. 6.1 Heat incidents by asset type from 15 June to 19 July 2015 [2].

Importantly, it has been found that track buckling has been one of the causes of train derailments which led to the huge loss of lives and assets as seen in many evidences [6–8]. Train derailments are frequently observed in hot season according to the records. The derailment risk of trains at railway turnout is studied [9] and it is found that the number of train derailments is relatively high in summer due to the irregularities and buckling of track.

Track buckling occurs in lateral plane so that the major factor undermining buckling resistance is lack of track lateral resistance. It is found that there are three main components that have a direct relationship to lateral resistance: bottom sleeper-ballast interface (bottom friction), side sleeper-ballast interface (side friction), and ballast shoulder at sleeper end (end restraint). These have significant effect on buckling temperature and shape of buckling. The buckling shapes of railway track depend on track conditions. Straight and curved tracks can be buckled in different shapes such as symmetrical, antisymmetrical, sinusoidal, etc. due to their lateral stiffness and weakness area. The typical buckling shape can be seen in Fig. 6.2. Moreover, the length of weakness zone significantly affects the number of buckling curves after buckling. For instance, larger area of lack of lateral resistance induces sinusoidal curves of buckling

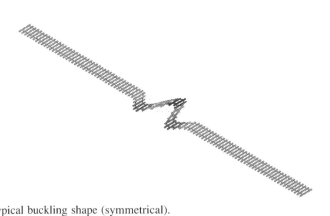

Fig. 6.2 Typical buckling shape (symmetrical).

Rail infrastructure systems and hazards

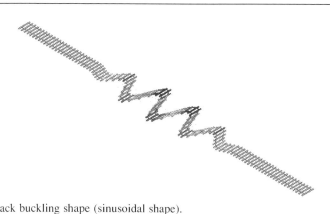

Fig. 6.3 Track buckling shape (sinusoidal shape).

as can be seen in Fig. 6.3. This is confirmed by the analytical solutions provided by Kerr [10]. It is also confirmed by the numerical simulations that the larger of unconstrained length or weaker zone, the lower buckling temperature [11].

Besides, maintenance activities, such as ballast tamping and cleaning, can undermine track lateral resistance due to the disturbance and uncompaction of railway ballast. The lateral resistance is significantly reduced after tamping and maintenance activities. The comparison of lateral resistance forces before and after tamping is presented by using Discrete Element Method (DEM) with the consideration of angularity index of ballast shape. It is found that tamping has a large influence on track lateral resistance especially when the ballast shape has high angularity index since it is reduced by 40%. While tamping has less impact on lateral resistance of track with low angularity index since its weak interlocking [12]. However, after train operating, the ballast is compacted and lateral resistance is increased again if the ballast is not degraded. Many more researchers have conducted the Single Sleeper (Tie) Push Tests (STPTs) to evaluate the lateral resistance of railway tracks [13]. Sleeper plays a significant role in lateral resistance. Heavier sleepers or textured/frictional sleepers can clearly improve the lateral resistance of ballasted tracks. It is clear that concrete sleepers can help improve the lateral resistance of track in comparison to the conventional timber sleepers resulting in higher buckling temperature [11].

Importantly, in the regions where the rail temperature tends to exceed the limit, the train speed needs to be restricted to prevent or reduce the likelihood of track buckling since the additional force can be applied by train. It is noted that speed restrictions are normally applied after the maintenance activities until the track stabilization is achieved. This may cause train delays and disruptions as can be seen in the news. The waterloo station in United Kingdom is packed with passengers due to the train delays and cancelations on the hottest day in 2018 (Fig. 6.4). There are train speed restriction policies provided over the network in different countries based on the air temperature, expected rail temperature, sleeper types, and maintenance activities [14–17]. For example, Network Rail proposes the speed restrictions that need to be applied when the air temperature reaches 36°C and the expected rail temperature is 53°C. Note that, for this case, the speed on timber and concrete sleeper tracks are 45 and 90 mph, respectively [14].

Fig. 6.4 The passengers at Waterloo station on the hottest day of the year in United Kingdom (www.thesun.co.uk).

Apart from speed restriction, rail stress must be controlled and managed well [18]. It is known that SFT is likely to decrease over time and this may consequently increase the likelihood of track buckling if the temperature is constant. SFT can be determined immediately on-site using VERSE nondestructive SFT testing. This method eliminates the need of cutting rail to measure rail stress. It is recommended that SFT must be measured after realigning or maintenance activities. In addition, the real-time monitoring of rail temperature using thermometer sensors should be undertaken by placing the sensors on rail web. It should be noted that if the temperature tends to exceed the limit, the maintenance work including ballast tamping, track realignment, opening out ballast crib and shoulder, rail fastening removal, must be stopped and restored to the normal condition [19]. Furthermore, track inspection should be carried out on critical track section, where maintenance taking place or misalignment occurred, and the area where track buckling has been observed.

6.1.2 Low temperature

Even though the global temperature tends to increase and the duration of winter or ice seasons decreases, many areas still suffer from extreme low temperatures and snow [1]. Obviously, in winter, snow can widely cause significant issues related to railway systems. Furthermore, the snowfall data are correlated to the incident appearance at the selected location. It is important to note that a majority of the incidents usually happen at the beginning of winter, or when the first snow falls, as the systems are not well prepared for winter in some areas. There are several direct potential impacts associated with snowfall, permafrost, and ice. Snow and ice can build upon a railway track. The electrical system may fail, as ice can easily coat the third rail or the overhead line equipment, which is a conductor; therefore, an electric train may not be operated because of the failure to supply power to the train. In order to keep the train running regularly, several types of equipment are used to remove snow from tracks, such as snowplows, steam jets, brushes, scrapers, hot air blowers, and antifreeze [4]. Before normal operation, a locomotive train fitted with a snowplow, Multi-Purpose Vehicle (MPV), and Snow and Ice Treatment Train (SITT), is run on the risky tracks to clear the snow and ice from the route before the passenger and freight trains run. MPV and SITT can also spray the rails with a hot liquid antiicer that prevents ice from sticking to the conductor railhead (Fig. 6.5).

In terms of soil and subgrade, in winter, there is a reduction in soil stability and the subsidence of the substructure because of the thawing of permafrost. These will increase the long-term settlement if the monitoring and protection regimes are not selected properly. The conditions of the railway substructure and the embankment under extreme cold weather have been studied by monitoring the frost depth, frost heave, thaw settlement, and moisture in the track embankments. The results have indicated that the railway tracks, which do not meet the requirement of protection, have a possibility of embankment failure because of the frost heave, which leads to permanent deformation. The data indicate that the construction of the railway embankment has enhanced the interannual variability of snow depth in the adjacent cleared and undisturbed areas by modifying the microclimatic conditions. Materials that can used for preventing the thaw degradation of the ice-rich foundation soils and the railway subgrade have been proposed [20]. Thermosyphon devices, penoplex insulation, and snow/solar sheds have been extensively used on the berms in many countries located in cold regions, such as China and Russia. They have shown a cooling effect on a subgrade soil embankment and a raising effect on a permafrost table, resulting in improving the stability of railway embankments.

Fig. 6.5 Trains cannot be operated because of the extreme snow. Picture taken at the University of Birmingham station.

Moreover, the low temperature can affect the concrete, which is normally used for the sleepers. Freeze-thaw can occur in concrete sleepers. This undermines the strength properties, i.e., the compressive, tensile, and flexural properties, of concrete [21]. Furthermore, the low temperature could lead to the formation of ice inside the concrete because the water surface can transmit through pores and the temperature freezes the water. The concrete is expanded and creates tensile stress itself upon the increase in the frost depth inside. Hence, cracks can occur if the stress that occurs is greater than the concrete tensile strength [22].

Over time, small defects in rails can grow dramatically and change their orientation, resulting in rail cracks. Most of the broken rails occur in cold weather because of the contraction of the defected rails. Moreover, broken rail is known to be one of the major causes of train derailment. Consequently, the rail condition should be inspected regularly. Interestingly, the rail breaking rate can be reduced by using risk-based test scheduling techniques to determine the optimum rail test intervals based on the level of risk for each rail segment [23]. Presently, broken rails can be prevented by rail monitoring using an ultrasonic nondestructive test (NDT). Several researchers have reported that the use of NDT can significantly reduce the risk of broken rails by capturing a rail defect before it propagates and breaks the rail. A summary of the extreme temperature-related incidents is presented in Table 6.1.

6.2 Earthquakes

In the past, there have been records of earthquakes around the world having high magnitudes, such as the Northridge, El Centro, Kobe, Niigata, Wenchuan, and Kaohsiung earthquakes that caused huge losses of rail infrastructure systems, such as train derailments, track misalignment, and loss of track geometry [25]. Note that the lateral seismic waves have a considerable influence on the lateral vibrations of trains and tracks but slight impacts on the vertical vibrations [26]. This has led to the great concern that the train safety criterion under earthquakes should be the displacement response of the wheel-rail. It is important to note that the wheel is likely to be considerably larger than the rail, leading to higher relative displacement. This indicates that the rail is more reliable in the case of an earthquake. However, small rail irregularities may later cause higher dynamic responses in the train-track system and an increasing risk of track buckling. Lai et al. [25] also mentioned that the longer duration of earthquake ground motion results in higher alignment irregularities. Moreover, the increasing earthquake intensity significantly increases the alignment irregularities, making the damage more evident.

Moreover, during an earthquake, the wheel may climb the rail or jump on the rail [27]. In case of jumping on the rail, the train may overturn causing fatal or severe injuries to the passengers. The consequences of derailments are not only the temporary interruption of railway lines but also the varying severity of personnel and property losses. If the trains climb off the rail, it will cause an extreme impact load on the track slab. This causes serious damage and cracks in the railway tracks, as observed in the postderailment behavior. For instance, the conventional modular slab track was

Table 6.1 Potential events caused by extreme temperatures [24].

Phenomenon	Track	Catenary	Substations	Signaling	Telecommunications	Rolling stock
High temperature	Rail buckling	Catenary dilatation	Component heating			Component heating
						Air-conditioning shutdown
Low temperature	Rail breaking	Catenary freezing	Component freezing			Pantograph failure
	Switch malfunction					Brake malfunction (hydraulic system)
	Ballast stones thrown					Component damage by thrown ballast stones
						Doors freezing

loaded by the derailment actions to study the structural capacity under the accidental derailment events [6]. The failures of the components of a modular slab track were observed in the critical zone, e.g., slab, shear stud, and bridge stringer. This implied that the design standards and calculations relevant to the composite rail track slabs should be improved for a better performance and capacity to prevent damage from a dynamic load caused by train derailment.

Fault rupture because of earthquakes is also a main cause that can extremely collapse the rail infrastructure if a near-field earthquake occurs. For instance, insulated rail joints were pulled apart because of a large offset in the track alignment, as observed in 1999 in Turkey during the Kocaeli earthquake [28]. The impacts of the earthquakes on the railway infrastructure could also be indirect. For instance, earthquakes can knock down the trees or any structure along the railway line, and thus, the trains could be disrupted by the blocking trees, rockfalls, or any electrical mast structures, and cause track damage [29].

On this ground, it is necessary to enhance the vehicle running safety on structures during an earthquake [30]. Note that the angular rotation of two rails can be one of the indicators to identify the risk of derailment and running safety on the track structure. It is important to reinforce the existing structures in an effective and efficient manner. It is also important to identify the potential critical structures for the running safety in continuous railway structures and understand the dominant factor that contributes to causing derailment and providing an appropriate reinforcement to the existing structures. The study proposes a novel derailment index based on the safety limits of the vibration displacement and differential displacement. Two safety limits are simply summed up to represent the derailment index. The results showed that the safety limit tended to be larger when a lower-speed train was passing. The paper also reports that derailments occur when the angular rotation is large, indicating that the differential displacement (angular rotation) exhibits a higher impact than that exhibited by the vibration displacement in this line while determining the vehicle running safety during an earthquake.

6.3 Flooding

The most devastating natural disasters experienced are floods and their consequential landslides. Among all the natural disasters, flooding is the most frequent disaster that has been observed, particularly in Asian countries [31]. The southern Indian state of Kerala just suffered from the worst flood in 100 years [32,33]. Binti Sa'adin et al. [34] pointed out that the accompanying extreme weather events have grown and are expected to increase in the future. Climate change can cause the existing structures to exhibit different performance characteristics than normal [35,36], as the components of the railway track are placed under flood conditions. As the main component of a ballasted railway system, the railway ballast is frequently used by the railway industry to enhance constructability and practicality. Numerous studies on train-track interactions have focused on ballast modeling and idealization in completely dry environments, but recent studies have found that, under extreme weather conditions such as floods, water can clog natural ballast beds and change the initial state of their properties. It has been observed that climate uncertainty has a significant influence on railway networks that affects the serviceability and performance of railway tracks [37]. Note that the train speed must be reduced to prevent damage to the train when the floodwater rises above the rails. However, for safety reasons, trains do not usually run on flooded railway tracks because of the high risk of short circuits and power cuts if the tracks use the conductor rail system. Another reason is the lack of information, either about the dynamic characteristics of the railway ballast under variable flooding conditions or about the dynamic train-track modeling to capture the flooding conditions.

It has been found that only considering the dynamic performance under dry conditions is insufficient, as an abnormal climate can occur frequently. Hence, for the engineering structures, there is a compelling need to consider not only safety and applicability but also reliability, resilience, and durability [38]. Thus far, most of the investigations have scarcely involved the area of the flooded ballast. It has been pointed out that it is necessary to determine their railway track exposure and the modal parameters of the railway tracks under flood conditions.

Numerous studies on train-track interactions have focused on ballast modeling and idealization in completely dry environments, but recent studies have found that, under extreme weather conditions such as floods, water can clog natural ballast beds and change the initial state of their properties. Ballast models used in multibody simulations have been mostly developed on the basis of the instrumented impact hammering method considering the ballast as a spring/dashpot. The effect of flooding on the dynamic behavior of a railway ballast has been investigated using the impact excitation technique [39]. The dynamic modal identification of a railway ballast under flooding conditions has been investigated. It also highlights the experimental results obtained as part of the railway engineering research activities at the University of Birmingham (UoB), aimed at improving the dynamic performance and modeling of railway tracks globally. The proposed relationships could be incorporated into track analysis and design tools for a more realistic representation of the dynamic train-track interaction and load transfer mechanisms under extreme events (Figs. 6.6 and 6.7).

For a better understanding of mechanical behaviors involving the railway ballast, we need reasonable test methods and appropriate numerical models. Therefore, the study of ballasts under flood conditions by identifying the changing law and establishing an ideal dynamic model is particularly important. The previous study embarked on the full-scale experiments and formulated six models of ballast idealization concepts. The modal identification results revealed that the fundamental model with spring and dashpot (Fig. 6.8) prevails over the other models in terms of accuracy and precision. The more complicated models with more springs and dashpots are not suitable for ballast idealization, particularly under a flood condition. The results based on the impact excitation technique show that the increase in the flooding level will change the natural frequency and decrease the frequency response function. The system stiffness will decrease with a decrease in the natural frequency. The novel insights are of considerable significance for exploring the nonlinear dynamic traits of ballasts in extreme environments, which can be integrated into the coupled train-track analysis

Fig. 6.6 Typical ballasted track (left). The capping layer called "subballast" is prepared by using compacting roller prior to laying ballast. In reality, the capping layer and subgrade are nonhomogenous and cannot be accurately modeled by a simple constant elastic half-space or a continuous layer. These layers are thus designed under higher safety margin or higher factor of safety by using the allowable stress design concept. The design takes into account the accumulated strains of these layers, which are often limited over a target design period (e.g., 15–25 years). Track maintenance cost function of deteriorated ballasted tracks will increase over time.

Fig. 6.7 Train standing on a flooded track.

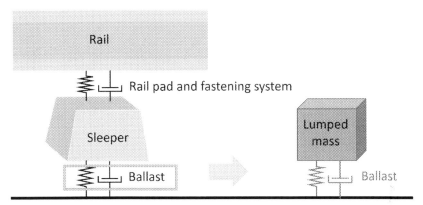

Fig. 6.8 Ballast idealization model for train-track simulation.

that can better express more realistically the dynamic train-track interaction and the load transfer mechanism of flooded railway tracks.

Importantly, it has been reported that when the water drains away, railway tracks usually lose their support as the ballast is washed away. In a track in a poor condition, large voids and gaps can easily be observed between sleepers and the ballast, usually caused by the wet track beds (highly moist ground) from natural water springs or poor drainage [40]. The strength and drainage aspects of ballasted tracks are compromised by the increasing level of ballast fouling. Different support conditions of tracks due to the loss of the sleeper-ballast contact can create different vibration characteristics of the superstructure [41]. This leads to larger particle movement, resulting in a more severe loss of support conditions. Thus, it is very important that the ballast be re-laid to make it safe again.

6.4 Summary

This chapter presented a systems review of weather-related hazards on railway infrastructure systems. The effects of extreme temperatures, earthquakes, and flooding on the failure of railway infrastructure are presented. The weather-related failures due to extreme weather conditions tend to occur more frequently and extremely. It is a challenge to improve the future capacity and adaptability of railway infrastructure systems to climate change and extreme events due to natural hazards. It is very essential that researchers, engineers, and scientists meet the challenge of providing sustainable, smart, and resilient transportation infrastructure systems critical for maintaining regional competitiveness and tackling the future natural hazards.

References

[1] I.S. Oslakovic, H. ter Maat, A. Hartmann, G. Dewulf, Climate change and infrastructure performance: should we worry about? Procedia Soc. Behav. Sci. 48 (2012) 1775–1784.

[2] E. Ferranti, L. Chapman, S. Lee, D. Jaroszweski, C. Lowe, S. McCulloch, A. Quinn, The hottest July day on the railway network: insights and thoughts for the future, Meteorol. Appl. 25 (2018) 195–208.
[3] J. Edgley, Summer Track Engineering Conference, 2018.
[4] Network Rail, Specialist machines help our engineers to keep the railway running whatever the weather, 2019. https://www.networkrail.co.uk/running-the-railway/looking-after-the-railway/our-fleet-machines-and-vehicles/seasonal-track-treatment-and-weather-support-fleet.
[5] United States Department of Transportation, Track Buckling Research, John A. Volpe National Transportation Systems Center, United States, 2003.
[6] S. Kaewunruen, Y. Wang, C. Ngamkhanong, Derailment-resistant performance of modular composite rail track slabs, Eng. Struct. 160 (2018) 1–11.
[7] L. Ling, X.B. Xiao, X.S. Jin, Development of a simulation model for dynamic derailment analysis of high-speed trains, Acta Mech. Sinica 30 (2014) 860–875.
[8] C. Ngamkhanong, S. Kaewunruen, B.J.A. Costa, State-of-the-art review of railway track resilience monitoring, Infrastructures 3 (2018) 3.
[9] S. Dindar, S. Kaewunruen, M. An, O. Mohd, Natural hazard risks on railway turnout systems, Procedia Eng. 161 (2016) 1254–1259.
[10] A.D. Kerr, Analysis of thermal track buckling in the lateral plane, Acta Mech. 30 (1978) 17–50.
[11] C. Ngamkhanong, C.M. Wey, S. Kaewunruen, Buckling analysis of interspersed railway tracks, Appl. Sci. 10 (2020) 3091.
[12] E. Tutumluer, H. Huang, Y.M.A. Hashash, J. Ghaboussi, Aggregate shape effects on ballast tamping and railroad track lateral stability, in: AREMA Conference, USA, 2006.
[13] G. Jing, P. Aela, Review of the lateral resistance of ballasted tracks, Proc. Inst. Mech. Eng. F J. Rail Rapid Transit 234 (2020) 807–820.
[14] Network Rail, Continuous Welded Rail (CWR) Track, Network Rail, London, 2006.
[15] Network Rail, Why rails buckle in Britain, 2008. https://www.networkrail.co.uk/stories/why-rails-buckle-in-britain/.
[16] Queensland Railways, in: Committee, R. O. A. T. D. A. A. (Ed.), Track Buckling, Railways of Australia, 1988.
[17] RailCorp, TMC211 track geometry & stability, in: Engineering Manual, RailCorp., Australia, 2013.
[18] CRC for Rail Innovation, Track Stability Management – Literature Review: Theories and Practices, CRC for Rail Innovation, Brisbane, Australia, 2009.
[19] TrainsAdelaide, Track and civil infrastructure code of practice volume two-train system [CP2] Rail Stress Control, Government of South Australia, 2008.
[20] S.P. Varlamov, Thermal monitoring of railway subgrade in a region of ice-rich permafrost, Yakutia, Russia, Cold Reg. Sci. Technol. 155 (2018) 184–192.
[21] E. Solatiyan, M. Asadi, M. Bozorgmehrasl, Investigating the effect of freeze-thaw cycles on strength properties of concrete pavements in cold climates, Indian J. Fundam. Appl. Life Sci. 5 (2015) 2231–6345.
[22] J. Bijen, Durability of Engineering Structures: Design, Repair and Maintenance, Elsevier, 2003.
[23] A.M. Zarembski, J.W. Palese, Characterization of broken rail risk for freight and passenger railway operations, Proceedings of the AREMA 2005 Annual Conference, AREMA, Chicago, IL, 2005.
[24] A.D. Quinn, A. Jack, S. Hodgkinson, E.J.S. Ferranti, J. Beckford, J. Dora, Rail adapt: adapting the railway for the future, 2017. A report for the International Union of Railways (UIC).

[25] Z. Lai, X. Kang, L. Jiang, W. Zhou, Y. Feng, Y. Zhang, J. Yu, L. Nie, Earthquake influence on the rail irregularity on high-speed railway bridge, Shock. Vib. 2020 (2020), 4315304.
[26] L. Xu, W. Zhai, Stochastic analysis model for vehicle-track coupled systems subject to earthquakes and track random irregularities, J. Sound Vib. 407 (2017) 209–225.
[27] L. Cao, C. Yang, J. Zhang, P. Zeng, S. Li, Earthquake response of the train–slab ballastless track–subgrade system: a shaking table test study, J. Vib. Control 27 (17–18) (2020) 1979–1990.
[28] W.G. Byers, C. Edwards, A. Tang, J. Eidinger, C. Roblee, M. Yashinsky, J.-P. Bardet, J. Swift, Performance of transportation systems after the 1999 Kocaeli earthquake, Earthquake Spectra 16 (2000) 403–410.
[29] P. Leviäkangas, A. Tuominen, R. Molarius, J. Schabel, S. Toivonen, J. Keränen, J. Törnqvist, L. Makkonen, A. Vajda, H. Tuomenvirta, Extreme Weather Impacts on Transport Systems, VTT Technical Research Centre of Finland, Finland, 2011.
[30] K. Goto, M. Sogabe, M. Tokunaga, Evaluation of vehicle running safety on railway structures during earthquake, Int. J. Transp. Dev. Integ. 4 (2020) 113–128.
[31] S.L. Binti Sa'adin, S. Kaewunruen, D. Jaroszweski, Heavy rainfall and flood vulnerability of Singapore-Malaysia high speed rail system, Aust. J. Civ. Eng. 14 (2016) 123–131.
[32] P. Kellermann, et al., Frequency analysis of critical meteorological conditions in a changing climate—assessing future implications for railway transportation in Austria, Climate 4 (2) (2016), 25.
[33] BCC News, Kerala Floods: Monsoon Waters Kill Hundreds in Indian State, 2018. *https://www.bbc.com/news/world-asia-india-45216671#:~:text=At%20least%20324%20people%20have,in%20the%20past%2024%20hours.*
[34] S.L. Binti Sa'adin, S. Kaewunruen, D. Jaroszweski, Climate change vulnerability and adaptation for the Singapore-Malaysia high-speed rail system, Ingenieur 66 (2016) 44–54.
[35] S. Kaewunruen, R. Janeliukstis, C. Ngamkhanong, Dynamic properties of fibre reinforced foamed urethane composites in wet and dry conditions, Mater. Today Proc. 29 (2020) 7–10.
[36] S. Kaewunruen, C. Ngamkhanong, M. Papaelias, C. Roberts, Wet/dry influence on behaviors of closed-cell polymeric cross-linked foams under static, dynamic and impact loads, Constr. Build. Mater. 187 (2018) 1092–1102.
[37] S. Kaewunruen, A.M. Remennikov, Current state of practice in railway track vibration isolation: an Australian overview, Aust. J. Civ. Eng. 14 (2016) 63–71.
[38] A. Freimanis, S. Kaewunruen, Peridynamic analysis of rail squats, Appl. Sci. 8 (2018) 2299.
[39] S. Kaewunruen, T. Tang, Idealisations of dynamic modelling for railway ballast in flood conditions, Appl. Sci. 9 (2019) 1785.
[40] S. Kaewunruen, A.M. Remennikov, Investigation of free vibrations of voided concrete sleepers in railway track system, Proc. Inst. Mech. Eng. Part F J. Rail. Rapid. Transit. 221 (4) (2007) 495–507.
[41] B. Feng, W. Hou, E. Tutumluer, Implications of field loading patterns on different tie support conditions using discrete element modeling: dynamic responses, Transp. Res. Rec. 2673 (2019) 509–520.

Wheel-rail dynamic interaction

Zhen Yang and Zili Li
Delft University of Technology, Delft, The Netherlands

7.1 Introduction

7.1.1 Wheel-rail contact vs wheel-rail dynamic interaction

When a railway wheel rolls along a straight rail without significant geometric or stiffness irregularity, wheel-rail contact can be modeled analytically [1–4] or with a boundary element method (BEM) [5] based on the assumptions of half-space and linear elasticity. The analyzing scope of the contact problem is then concentrated on the potential contact region, as indicated by the red frame in Fig. 7.1A. When a wheel passes a more complex section of the track, e.g., a sharp curve, rail joint, or turnout, arbitrary contact geometry, and plastic deformation may occur, and the wheel-rail contact is generally better to be modeled with a finite element method (FEM) [6–19], in which arbitrary contact geometries and material nonlinearities can be handled. By discretizing the structures of interest in the spatial domain, the stresses/strains, deformation, and motion of the whole wheel and rail structures can be analyzed, as shown in Fig. 7.1B. In contrast to "wheel-rail contact" where only a local contact region is of concern, the term "wheel-rail interaction" denotes that the behavior of wheel and rail structures with certain geometric and material properties contribute to and are influenced by wheel-rail contact.

Early finite element (FE) studies on the wheel-rail interaction generally assumed quasistatic-state contact [7,8,11]; however, when a wheel passes complex sections of the track, dynamic effects often play a significant role in the wheel-rail interaction. Here the dynamic effects denote that the inertia of wheel/rail material influences the stress field in and surrounding the contact area, as well in the structure; the contact-induced wheel/rail structural vibration may significantly affect the wheel-rail contact [10,14,16–18]. Wheel-rail dynamic interaction should thus be considered, as schematically indicated in Fig. 7.1C. The explicit FEM is an ideal approach to deal with wheel-rail dynamic interaction, owing to its capabilities of handling nonlinear material properties and arbitrary contact geometries and considering dynamic effects [18] (to be elaborated and demonstrated in Section 7.2).

7.1.2 Weak spots of the track

The aforementioned complex track sections where wheel-rail dynamic interaction often occurs can be roughly categorized as the impact-inducing sections and the large-friction-inducing sections. For both, large wheel-rail contact force in the normal or/and tangential directions can be expected, consequently causing high-amplitude

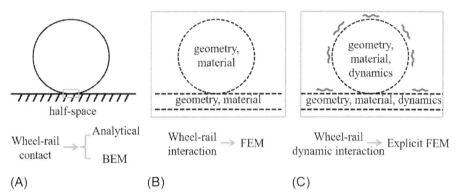

Fig. 7.1 From wheel-rail contact to the dynamic interaction (A) wheel-rail contact; (B) wheel-rail interaction; (C) wheel-rail dynamic interaction.
Reproduced from Z. Yang, Numerical Modelling of Wheel-Rail Dynamic Interactions With an Explicit Finite Element Method (PhD thesis), Delft University of Technology, 2018, https://doi.org/10.4233/uuid:8acb9b48-bf77-45b2-a0d6-1cf6658f749e.

structural vibration and noise, accelerating track degradation. These complex sections can thus be regarded as weak spots of the track.

Typical impact-inducing track sections include rail joints and turnouts. When trains pass over these sections, wheel-rail impacts are generated due to the track stiffness and geometric irregularities. The impact, besides leading to vibration and noise, initiates and accelerates track deterioration, e.g., squat, plastic deformation, corrugation, fastening failure, hanging sleepers, and uneven track settlement. The impact-inducing track sections contribute to a large portion of the maintenance and renewal costs of the track infrastructure. For example, the impact load at a turnout can be more than two times the static wheel-rail contact load in amplitude, resulting in eight unexpectedly broken frog points per week in the Netherlands.

The large-friction-inducing sections include curves and places where acceleration and braking frequently take place. When a train negotiates a tight curve, large lateral friction force occurs at the contact between the tread of the inner wheel and the top surface of the low rail, and between the flange of the outer wheel and the gauge corner of the high rail, if a wheelset fails to align itself tangentially to the rail. The large friction force may induce full slip and stick-slip contact behavior and, consequently, severe wear, rolling contact fatigue, corrugation, and squeal noise. The flange contact introduces large spin creepage, causing severe wear and sometimes flange squeal. Large longitudinal friction force is generated in the acceleration and braking sections, as well as on slopes, where local wear of wheel tread (wheel flat) and railhead (wheel burn in a severe condition) may take place.

7.1.3 Consequences of the wheel-rail dynamic interaction

Based on the commonly-observed problems at the weak spots of the track, the consequences of the wheel-rail dynamic interaction can be summarized as the increased wheel/rail structure degradation and the contact-induced vibration and noise.

7.1.3.1 Wheel/rail structure degradation

Severe wear
Shown as the removal of material from the contacting surfaces of structures, wear is the most common failure mode of wheel/rail interface [20]. Severe wear is often observed at the gauge corner of the high rail on curves where flange contact takes place, as well as at the turnout noses and rail joints where impacts are generated. The worn profiles of wheel and rail further increase wheel-rail contact force and accelerate structure degradation.

Wheel polygonal wear, shown as periodic wave-like wear along the circumference of the wheel tread, commonly occurs in locomotives, metros, and high-speed trains. The dynamic effects play important roles in the generation of wheel polygon: the resonances of the train/track system are closely related to the wavelengths of the polygon [21–23].

Rolling contact fatigue
Rolling contact fatigue in the rail surface mainly includes squats and head checks. Squats mainly occur at welds, joints, turnouts, indentations, wheel burns, and corrugation, where wheel-rail impacts and excessive dynamic contact force are generated [10,14,19,24]. Squats in turn induce wheel-rail impacts. Head checks mainly occur on the curves with radii between 500 m and 3000 [25,26]. Rolling contact fatigue threatens the safety of railway traffic because the propagation of cracks may result in sudden rail fracture and derailment.

Corrugation
Corrugation comprises the formation of a periodic wave-shaped irregularity of the rail surface, and appears in almost all types of tracks, from ballast tracks to slab and embedded tracks, from heavy-haul to light rail and rapid transit [27,28]. The generation and development of corrugation can be explained by a short-term dynamic process and a long-term damage mechanism. The short-term dynamic process is believed to fix the wavelength of the periodic defect pattern to the dynamics of the train/track system, which is also the origin of the dynamic effects involved in the wheel-rail dynamic interaction. As to the long-term damage mechanism, the corrugation is considered to be formed by wear and plastic deformation.

7.1.3.2 Contact-induced vibration and noise

Impact vibration and noise
Impact vibration and noise are generated at the impact-inducing track sections. The vertical contact irregularities in the running surface generate high wheel-rail contact force and high-amplitude impact vibration and noise. The impact noise has discrete nature of the event, which can, however, be quite loud [29].

Friction induced vibration and squeal
Frictional induced unstable vibration of a railway wheel and the consequent squeal noise radiation often occur in sharp curves. The tonal squeal is one of the loudest and most disturbing railway noise sources. The wheel-rail dynamic interaction is

considered to be directly related to squeal generation [30,31]. Unsteady lateral creepage, particularly between the leading inner wheel and low rail, is thought to be the main cause of the unstable wheel vibration and curve squeal noise [29]. Flange contact has been found to reduce the likelihood of curve squeal due to lateral creepage under certain conditions [29], but may generate a different form of noise referred to as "flanging noise" or "flange squeal," which is much more broad-band in nature and also a source of considerable annoyance.

7.2 Modeling of wheel-rail dynamic interaction

Reliable modeling of wheel-rail dynamic interaction contributes to the mitigation and elimination of the aforementioned degradation and environmental impacts (i.e., vibration and noise), especially at the weak spots of the track. In addition, because the dynamic effects involved in wheel-rail interaction increase with the train speed and axle load, a better understanding of the wheel-rail dynamic interaction is necessary for the capacity increase of railway transportation.

7.2.1 Methodology

Owing to the capabilities of handling nonlinear material properties and arbitrary contact geometries and considering dynamic effects, the explicit FEM has been broadly employed to model wheel-rail dynamic interaction [10,13–18,28,31–35]. In comparison with the implicit FEM, the explicit integration scheme is more robust in handling difficult contact problems because it avoids the convergence difficulties caused by demanding contact conditions [36] and the regularization of the friction law required to treat the no-slip condition in the adhesion area [37]. Moreover, by avoiding the need for matrix evaluation, assembly, and decomposition as required by the implicit integration algorithms, the explicit procedure is computationally attractive for analyzing high-frequency dynamic problems of short-duration, especially when the total dynamic response time that must be modeled is only a few orders of magnitude longer than the stability critical time step, which is frequently the case in wave propagation analyses [38].

The core algorithms employed in the explicit FE wheel-rail dynamic interaction were systematically presented in [18]. Table 7.1 briefly presents the solution procedure of the explicit FEM that contains two loops. The outer loop is constructed mainly by formulating the equation of motion and solving the equation with the central difference scheme; and the inner loop calculates the wheel-rail contact, which is called as a subroutine at each time step prior to updating the structural dynamic responses. This solution procedure indicates that the calculation of wheel/rail dynamics and the calculation of wheel-rail frictional contact can be coupled in the numerical algorithm, and thus demonstrates that the explicit FEM is an ideal approach for the modeling of wheel-rail dynamic interaction.

Wheel-rail dynamic interaction 115

Table 7.1 The solution procedure for the explicit FE wheel-rail dynamic interaction analysis.

Initialize algorithm: apply initial conditions; define contact pairs; construct the lumped mass matrix, and set the termination time
Loop 1: time step = 0, 1, 2, ...
(I) Apply load conditions to construct the external force vector
(II) Process elements to construct the internal force vector
(III) Construct the wheel-rail contact force vector using the penalty contact method
    Loop 2: number of the node on the slave (wheel) surface = 1, 2, ...
      (i) Locate the master (rail) segment for each slave (wheel) node
      (ii) Locate the wheel-rail contact points (projection of the slave node on the master segment)
      (iii) Calculate the contact force at the contact points
    END Loop 2
(IV) Update the nodal accelerations, velocities, and displacements using the central difference method
(V) Check for termination
END Loop 1
Output: wheel/rail nodal force and nodal motion (i.e., acceleration, velocity, and displacement)

7.2.2 Case studies

Three typical examples of using the explicit FEM to deal with wheel-rail dynamic interaction are presented. The first example models the wheel-rail impact at a rail joint [39]; the second and third examples deal with wheel-rail dynamic interaction at curves, i.e., frictional instability of the inter wheel [31] and flange contact of the outer wheel [40], respectively.

7.2.2.1 Wheel-rail impacts at a rail joint

This example demonstrates the reliability of the explicit FEM for the modeling of wheel-rail dynamic interaction. Because the explicit FEM couples the calculation of wheel/rail dynamic responses with the calculation of wheel-rail contact (see Table 7.1), the reliability of the wheel-rail dynamic interaction solutions are confirmed by separately validating the wheel/rail dynamic responses and contact solutions.

A typical Dutch insulated rail joint (IRJ) without visible damage was selected as the study target in the trunk line Amsterdam-Utrecht of the Dutch railway network. A 3D explicit FE wheel-IRJ dynamic interaction model was built up, as shown in Fig. 7.2. The model consists of a 10-m-long half-track with an IRJ in the middle and a half-wheelset with the sprung mass of the car body and bogie. The ballast was simplified as vertical spring and damper elements, with the displacements constrained in the lateral and longitudinal directions. Since the stiffness and damping parameters used to

Fig. 7.2 A finite element wheel-IRJ dynamic interaction model.
Modified from Z. Yang, A. Boogaard, R. Chen, R. Dollevoet, Z. Li, Numerical and experimental study of wheel-rail impact vibration and noise generated at an insulated rail joint, Int. J. Impact Eng. 113 (2018) 29–39, https://doi.org/10.1016/j.ijimpeng.2017.11.008.

model the fastenings and ballast can hardly be measured directly in the field, these parameters were calibrated by fitting the simulated frequency response functions to the measurement results, which will be illustrated in detail in Section 7.3.1.

The simulated wheel-rail impact vibration response was validated against a trackside pass-by measurement. Fig. 7.3 compares the wavelet power spectra of the simulated (Fig. 7.3A) and measured (Fig. 7.3B–E) impact vibration up to 10 kHz. The results of the four pass-bys shown in Fig. 7.3 were successively measured when the four wheelsets of a coach passed the target IRJ. The simulated impact vibration agrees well with the pass-by measurement results in both the time domain and frequency domain. More details about the measurement campaign and model validation can be found in [39].

After validating the dynamic response, the simulated impact contact solutions are analyzed including the contact patch area, stress magnitude and direction, micro-slip, and adhesion-slip distributions. Detailed analyses and validation of the wheel-IRJ impact contact solutions can be found in [17]. By plotting a trail of transient contact areas, the "footprints" of the simulated contact patch are presented in Fig. 7.4A. Good correspondence can be obtained by comparing the simulated "footprints" to the in situ running band of the target IRJ shown in Fig. 7.4B, which implies that the simulation accurately reproduced the transient impact contact solutions.

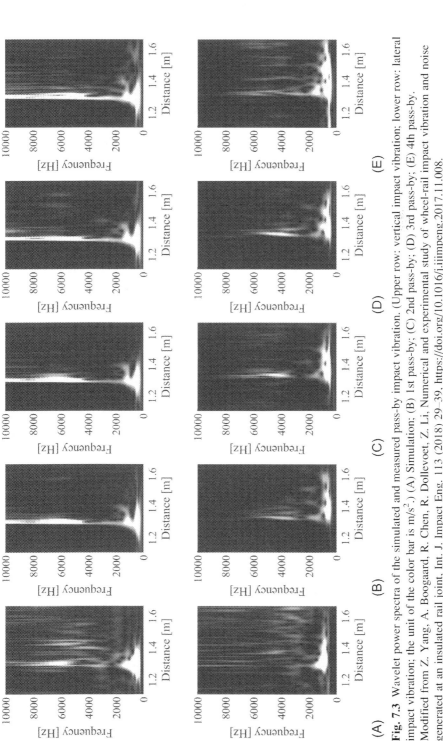

Fig. 7.3 Wavelet power spectra of the simulated and measured pass-by impact vibration. (Upper row: vertical impact vibration; lower row: lateral impact vibration; the unit of the color bar is m/s^2.) (A) Simulation; (B) 1st pass-by; (C) 2nd pass-by; (D) 3rd pass-by; (E) 4th pass-by. Modified from Z. Yang, A. Boogaard, R. Chen, R. Dollevoet, Z. Li, Numerical and experimental study of wheel-rail impact vibration and noise generated at an insulated rail joint, Int. J. Impact Eng. 113 (2018) 29–39, https://doi.org/10.1016/j.ijimpeng.2017.11.008.

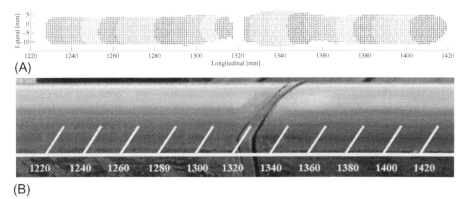

Fig. 7.4 Comparison of the simulated contact patch "footprints" and the in situ running band (A) "Footprints" of the simulated contact patch (the color differentiates an individual contact patch from the others); (B) in situ condition of the running band.
Reproduced with permission from Z. Yang, A. Boogaard, Z. Wei, J. Liu, R. Dollevoet, Z. Li, Numerical study of wheel-rail impact contact solutions at an insulated rail joint, Int. J. Mech. Sci. 138–139 (2018) 310–322, https://doi.org/10.1016/j.ijmecsci.2018.02.025.

7.2.2.2 Frictional instability at curves

The second example simulates the curving behavior of the inner wheel to deal with the frictional instability, i.e., the squeal-exciting wheel-rail contact. The contact modeling accounting for the frictional instability of a vibrating wheel is considered the central part of the squeal prediction model [41]. Fig. 7.5 shows the employed 3D explicit FE wheel-rail, dynamic interaction model, with wheel lateral motion during curving. The wheel-rail contact was defined with nominal geometry and with the wheel flange being included. More detailed descriptions of the model can be found in [31].

In the simulation, the wheel rolled from its initial position at time t_0 to the solution zone along the rail, as shown in Fig. 7.5B. The dynamics arising from the wheel/rail initial kinematic and potential energy due to imperfect static equilibrium [13] have relaxed at time t_1. Wheel lateral motion was subsequently simulated from time t_2 by applying the prescribed displacement boundary conditions listed in Table 7.2 to both ends of the wheel half-axle, except for simulation case 1. The wheel entered the solution zone at time t_3 and exited at t_4. The dynamic evolution of the contact solutions was captured between t_3 and t_4.

Unsteady lateral creepage, particularly between the leading inner wheel and the low rail, is thought to be the main cause of squeal [29]. When simulating wheel lateral motion, the resulting wheel-rail lateral creepage causes asymmetric distributions of contact stress and micro-slip within the contact patch, as shown in Fig. 7.6. The asymmetry of the distribution of the contact patch is characterized by an orientation angle θ, as shown in Fig. 7.6C. The orientation angle increases with the lateral creepage from Fig. 7.6A–F. The orientation angles obtained with CONTACT [5] (4th row), are consistent with those obtained with the explicit FEM. The distributions of the adhesion-slip regions determined by the simulated explicit FE contact stresses in the 1st row and

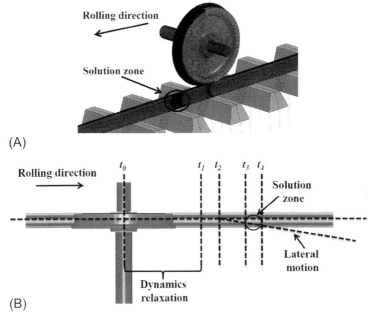

Fig. 7.5 The wheel-rail dynamic interaction model to simulate curving behavior (A) 3D FE model; (B) simulated frictional rolling with wheel lateral motion.
Reproduced with permission from Z. Yang, Z. Li, Numerical modeling of wheel-rail squeal-exciting contact, Int. J. Mech. Sci. 153–154 (2019) 490–499, https://doi.org/10.1016/j.ijmecsci.2019.02.012.

Table 7.2 Prescribed displacement boundary conditions applied to different simulation cases.

Prescribed displacement boundary conditions	Simulation cases
No lateral motion of the wheel	Case 1
Small lateral motion of the wheel	Case 2
Medium lateral motion of the wheel	Case 3
Large lateral motion of the wheel	Case 4

micro-slips in the 3rd row are in line with each other. They are also in reasonable agreement with the distributions of adhesion-slip regions determined by CONTACT in the 4th row. The main difference of the FE and boundary element (obtained with CONTACT) contact solutions is that the waves are observed in the micro-slip distributions calculated with the explicit FEM in Fig. 7.6E and F. The waves have been identified as the Rayleigh wave [42]. This confirms the transient/dynamic effects are included in the explicit FE contact modeling, which is needed for a correct representation of the squeal mechanisms [43].

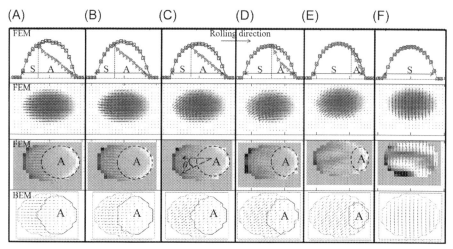

Fig. 7.6 Simulated contact solutions with lateral creepage. (1st row: stress distributions along the longitudinal centerline; 2nd row: stress distributions within the contact patch; 3rd row: micro-slip distributions within the contact patch; and 4th row: shear stress and adhesion-slip region distributions calculated with Kalker's CONTACT [5]. η and ξ are lateral and longitudinal creepage, respectively.) (A) simulation case 1 at 17.58 ms; (B) case 2 at 17.58 ms; (C) case 3 at 17.58 ms; (D) case 4 at 17.07 ms; (E) case 3 at 18.54 ms; and (F) case 4 at 17.94 ms. Reproduced with permission from Z. Yang, Z. Li, Numerical modeling of wheel-rail squeal-exciting contact, Int. J. Mech. Sci. 153–154 (2019) 490–499, https://doi.org/10.1016/j.ijmecsci.2019.02.012.

The time histories and power spectrum densities (PSDs) of the wheel dynamic lateral vibration calculated in simulation cases 1–4 are presented in Fig. 7.7A and B, respectively. The squeal-like vibration signals represented by large amplitude limit-cycles were produced when applying the wheel lateral motion. The amplitudes of the time histories increased with increasing amplitude of the lateral motion applied to the wheel model. Comparing the dominant frequencies of the vibration signals to the identified wheel modal frequencies, the wheel modes excited in the simulations can be determined: the mode with zero nodal circle and zero nodal diameter (0,0), the mode with zero nodal circle and three nodal diameters (0,3), and the mode with zero nodal circle and four nodal diameters (0,4), as detonated in Fig. 7.7B. The results correspond well to the conclusion that zero nodal circle modes tend to be excited in curve squeal [29].

7.2.2.3 Flanging and contact transition

The third example models nonsteady-state transition from single-point to two-point contact. The explicit FE wheel-rail dynamic interaction model presented in Section 7.2.2.2 (see Fig. 7.5A) was employed again but different boundary conditions were applied to the wheel in the manner shown in Fig. 7.8. The lateral displacement was constrained at the inner side of the wheel half-axle, and the outer end of the axle

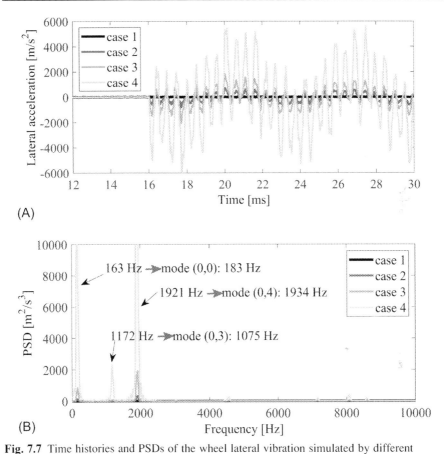

Fig. 7.7 Time histories and PSDs of the wheel lateral vibration simulated by different simulation cases. (A) Time histories; (B) PSDs.
Reproduced with permission from Z. Yang, Z. Li, Numerical modeling of wheel-rail squeal-exciting contact, Int. J. Mech. Sci. 153–154 (2019) 490–499, https://doi.org/10.1016/j.ijmecsci.2019.02.012.

was free. In the transient dynamic simulation, the wheel rolled from its initial position to the solution zone, forming an angle of attack (AoA) with respect to the rail longitudinal direction.

The contact transition is indicated by the simulated wheel-rail contact positions shown in Fig. 7.9. Initially, the wheel-rail contact occurred only between the rail top and wheel tread. A trail of blue patches represents the "footprints" of the single-point contact (Patch 1) in a series of time steps. Due to the presence of AoA, the wheel flange moved toward the gauge corner as the time step goes on and the 2nd contact patch (Patch 2) starts to appear. The "footprints" of the contact patches in the case of two-point contact are shown as two trails of red patches on the rail top and gauge corner.

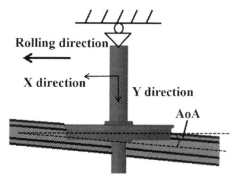

Fig. 7.8 Wheel-rail interaction model and the boundary conditions.
Reproduced with permission from Z. Yang, Z.L. Li, R. Dollevoet, Modelling of non-steady-state transition from single-point to two-point rolling contact, Tribol. Int. 101 (2016) 152–163, https://doi.org/10.1016/j.triboint.2016.04.023.

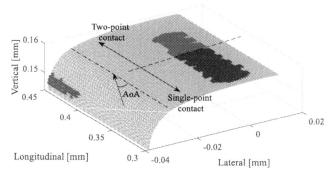

Fig. 7.9 Contact patches "footprints" (*blue* (*dark gray* in print version) for single-point contact and *red* (*light gray* in print version) for two-point contact).
Reproduced with permission from Z. Yang, Z.L. Li, R. Dollevoet, Modelling of non-steady-state transition from single-point to two-point rolling contact, Tribol. Int. 101 (2016) 152–163, https://doi.org/10.1016/j.triboint.2016.04.023.

The evolution of the contact pressure and surface shear stress is plotted in the contour/vector diagrams in Figs. 7.10 and 7.11 for Patches 1 and 2, respectively. The pressure magnitude corresponds to the depth of color within the contact patch, as indicated by the color bar. It shows that the contact pressure and patch area decreases in Patch 1 but increases in Patch 2 once Patch 2 has arisen. The surface shear stresses in single-point and two-point contact are indicated by the blue and red arrows, respectively. The arrows point in the direction of the shear stress, and their lengths are proportional to the magnitudes. To characterize the direction of the surface shear stress, the orientation angle between the stress vector and the negative direction of the longitudinal axis is denoted as $\theta 1$ and $\theta 2$ in Patches 1 and 2, respectively. $\theta 1$ is typically small under a single-point contact, but it increases rapidly to approximately 90 degrees as the

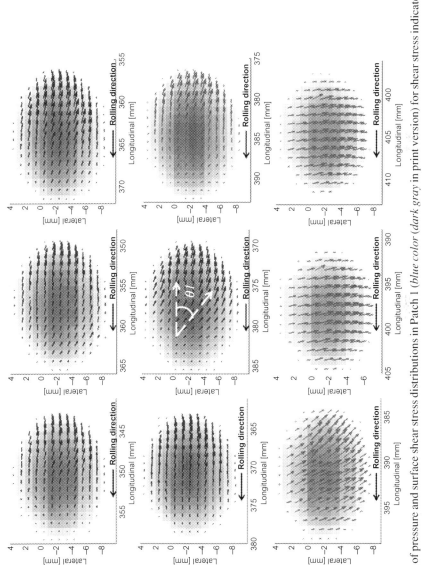

Fig. 7.10 Evolution of pressure and surface shear stress distributions in Patch 1 (*blue color* (*dark gray* in print version) for shear stress indicates single-point contact; *red* (*light gray* in print version) indicates two-point contact). Reproduced with permission from Z. Yang, Z.L. Li, R. Dollevoet, Modelling of non-steady-state transition from single-point to two-point rolling contact, Tribol. Int. 101 (2016) 152–163, https://doi.org/10.1016/j.triboint.2016.04.023.

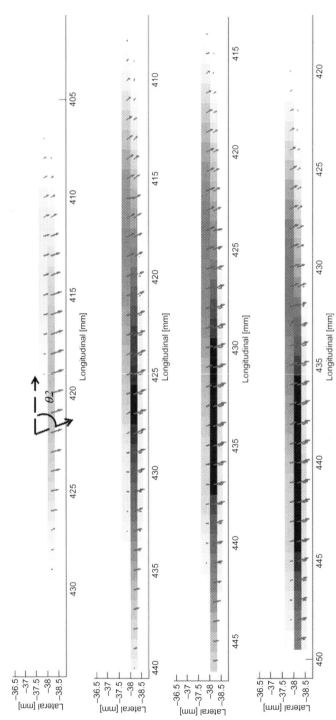

Fig. 7.11 Evolution of pressure and surface shear stress distributions in Patch 2 (*red color* (*dark gray* in print version) for shear stress indicates two-point contact).
Reproduced with permission from Z. Yang, Z.L. Li, R. Dollevoet, Modelling of non-steady-state transition from single-point to two-point rolling contact, Tribol. Int. 101 (2016) 152–163, https://doi.org/10.1016/j.triboint.2016.04.023.

participation of the 2nd contact patch increases. $\theta 2$ changes only slightly from the first occurrence of Patch 2 onwards. Detailed analyses of contact solutions during the nonsteady-state transition, including the contact-induced waves, can be found in [40].

7.2.3 Waves induced by wheel-rail dynamic interaction

Contact-induced waves inherently exist in the wheel-rail dynamic interaction. With the explicit FE modeling of dynamic interaction, the wheel-rail contact-induced waves were reproduced by the three aforementioned examples. According to the generation mechanism, the waves are categorized as the impact-induced wave, large-creepage-induced wave, and perturbation-induced wave. See [42] for the descriptions of the three types of waves. The generation mechanisms of the former two are evident: the significant dynamic effect or kinetic energy [44] induced by the wheel-rail impact or large creepage results in large oscillations of the wheel/rail surface material particles in the vicinity of the contact patch; the large local oscillations then propagate and form regular wave patterns.

The perturbation arises during seemingly steady-state rolling. The initiation of the perturbation has been found either close to the leading edge of the contact patch or at the juncture of the adhesion-slip regions, where the surface shear stress is close, but not equal, to the traction bound; therefore, the contacting particles originally in adhesion at these locations are more likely to slip than those elsewhere with an increase of the surface shear stress or a decrease of the contact pressure. Because the variation in the stress distribution is caused by vibration [45], the perturbation-induced waves are intrinsically generated by dynamic effects similar to the impact-induced wave and the creepage-induced wave. In addition, the perturbation-induced waves are found to generally initiate at the leading edge of the contact patch in the wheel braking condition, whereas the waves are initiated more often at the juncture of the adhesion-slip regions in wheel traction conditions.

The physical characteristics of the wheel-rail contact-induced waves can be revealed by analyzing the nodal motion on the contact surface that forms the wave. A typical large-creepage-induced wave is analyzed in [42]. The nodal velocities are narrow-banded signals and have a dominant frequency around 0.495 MHz. By multiplying the wavelength and the frequency of nodal velocity, the wave speed can be calculated, which is about 3 km/s. In addition, the phases out of the tangential and normal nodal velocities are approximately $\pi/2$. These characteristics correspond to the simulated large-creepage-induced wave to Rayleigh wave.

Experimental investigations [46,47] have suggested that Rayleigh waves can be used to detect rail cracks. The successful reproduction of the Rayleigh wave qualifies the explicit FEM for the simulation of crack-induced waves, which may contribute to the early-stage detection of cracks in the rail surface. Fig. 7.12A and B shows the distribution of the rail surface nodal velocities within the whole solution zone when the wheel was approaching and rolling over the crack in the simulation, respectively. The contact patch is indicated by the dashed black oval; the crack is denoted by the thick black line. The wave patterns observed in Fig. 7.12 were generated at the location of the crack and propagated radially. The wave generated when the wheel was rolling

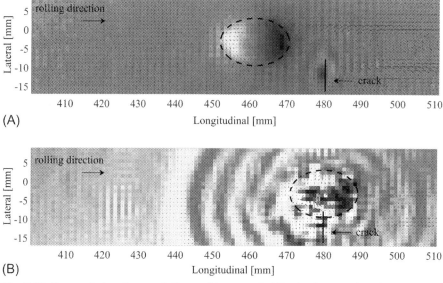

Fig. 7.12 Contact-induced waves influenced by a crack (A) wheel approaches the crack; (B) wheel rolls over the crack.
Reproduced with permission from Z. Yang, Z. Li, A numerical study on waves induced by wheel-rail contact, Int. J. Mech. Sci. 161–162 (2019) 105069, https://doi.org/10.1016/j.ijmecsci.2019.105069.

over the crack is more significant than the wave generated when the wheel was approaching the crack. See [42] for the simulation model and more detailed results analyses.

7.3 Detection and maintenance

7.3.1 Detection of wheel-rail dynamic interaction-related problems

Because wheel-rail dynamic interaction generates high-amplitude wheel-rail contact force and structural vibration, the related problems, i.e., wheel/track structural degradation, vibration, and noise radiation can be detected with the measurements of wheel-rail contact force, wheel, and track vibration. Hardly can the vibration of rolling wheels be directly measured. Alternatively, axle box acceleration (ABA) measurement has been used extensively to quantify the wheel axle responses and to identify the wheel-rail dynamic interaction-related problems [48–53]. Note that here the discussion is about the detection techniques based on the responses of wheel-rail dynamic interaction; therefore, the detection methods such as ultrasonic measurements, eddy current testing, image recognition are not included.

7.3.1.1 Detection with the contact force measurement

Wheel-rail contact force can be measured with track-side or/and wheelset instrumentation of strain gauge. The wheel-rail contact force up to 2 kHz can be obtained with the strain gauge measurement [54]. Track-side measurements are generally employed when the condition of a certain section of the track, usually a weak spot, needs to be monitored. Fig. 7.13 shows an example of the strain gauge instrumentation.

The large contact force can be correlated to several types of track/wheel degradations such as bad rail joints, wheel flats, corrugations, and fastening failure. The dynamic load factor, i.e., the ratio between the dynamic force and the static/nominal load, has been widely used to estimate the track [55] and wheel conditions [56,57]. In addition, the adhesion condition (i.e., traction of wheels to accelerate and brake) and climb index (i.e., L/V ratio) at curves can be evaluated based on the measured force signals. The strain gauge instrumented wheelset can efficiently measure the wheel-rail contact force along a railway line, which is suitable for wide-scale track condition monitoring and detection of different classes of rail irregularities [54].

7.3.1.2 Detection with the ABA measurement

ABA measurements have been employed to detect track defects such as corrugation, welds, squats, poor-quality joints, and turnouts [48–53]. A train-borne ABA measurement system is schematically shown in Fig. 7.14, which offers the following advantages [50]:

- ABA is a low-cost measurement system compared to other types of detection methods.
- The sensors are easy to maintain.
- ABA can be implemented on in-service operational trains.
- Early rail defects are possibly detected with no need for expensive and complex instrumentation.
- The ABA signals can reflect the severity level of the dynamic contact force.

Fig. 7.13 Strain gauges instrumentation.

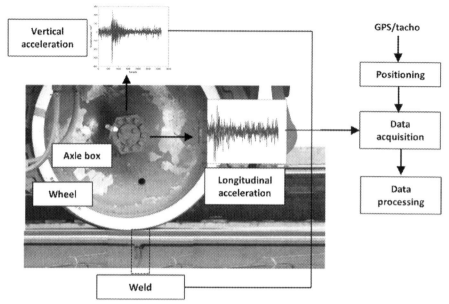

Fig. 7.14 Train-borne ABA measurement system.
Reproduced from A. Nunez, A. Jamshidi, H. Wang, Pareto-based maintenance decisions for regional railways with uncertain weld conditions using the Hilbert spectrum of axle box acceleration, IEEE Trans. Ind. Inf. 15 (2019) 1496–1507, https://doi.org/10.1109/tii.2018.2847736, open source.

The ABA measurements can have some drawbacks [43]: First, when the wheel is damaged, the ABA signals and the assessment of rail irregularities are affected by vibration originated from the wheel; this can be overcome with proper instrumentation and signal processing [48,58]. Second, ABA measurements are speed-dependent. To eliminate the influence of the train speed, the measurements should be performed at a constant speed, and were not possible, a mapping between ABA measurements and speed is needed. Thirdly, the frequencies of track vibration are dependent on the stiffness and damping properties of the track. This needs to be considered especially when the detected railway line consists of different tracks (e.g., ballast and nonballast).

7.3.1.3 Detection with the track vibration measurement

Track vibration is often measured with controlled excitations, e.g., by hammers [59,60] or falling weights [61,62], or under dynamic train, loading using specialized measurement vehicles [63] or operational trains [39]. Hammer tests are broadly used to measure the frequency response functions of track structures. The frequency response functions, on one hand, reflect the track structural dynamic behavior (i.e., the modal frequency, modal shape, and damping ratio) and thus healthy conditions in terms of dynamics, and on the other, can be used to derive the parameters of the track components, e.g., the support stiffness and damping [59]. Since the support stiffness and damping of the track needs to be identified in the ABA and train pass-by measurements, hammer tests are sometimes conducted as supplements to the ABA and train pass-by detection.

Track vibration in the low-frequency range (below 150 Hz) corresponds mainly to the dynamics of the substructure components (i.e., subgrade and ballast) [64]. To fully excite these components, the falling weight test is a good option to provide sufficient energy. The track vibration at 150–800 Hz is closely related to the rails and track superstructure components (i.e., fastening, railpad, and sleeper), the vibration in this frequency range is best tested using a sledgehammer with a soft tip [65]. At higher frequencies (800–5000 Hz), track vibration is dominated by the rails, and a light hammer with a hard tip should be used [39].

Track vibration induced by passing trains can be measured to assess the condition of a certain section of track, usually a weak spot [61]. Passing trains can excite tracks to vibrate in a wide frequency range, e.g., up to 10 kHz when impacts occur at an IRJ [39]. In combination with a hammer test, the resonances of the track structure contributing greatly to the impact vibration at the IRJ can be identified: the rail resonance (rail mass on the fastening stiffness [66]) and the pinned-pinned resonance.

7.3.2 Maintenance of weak spots of the track

Timely maintenance can prevent failures of track components and traffic disruptions and minimize the life cycle cost of the railway infrastructure. The failures of track components caused by wheel-rail dynamic interaction occur mainly at weak spots of the track. This section briefly discusses the maintenance of the track weak spots, i.e., the impact-inducing track sections and large-friction-inducing sections.

7.3.2.1 Maintenance of the impact-inducing track sections

Because wheel-rail impacts are caused by stiffness or/and geometric irregularities of railway tracks, the countermeasures against impact may be divided into stiffness control and geometric control. Taking the joints as an example, the dynamic stiffness at joints can be controlled by ballast maintenance and optimizing the sleeper set-up and distribution. Ballast tamping and renewal mitigates the track irregularities in both stiffness and geometry. To reduce the deflection of the joint, two commonly used sleeper set-ups are the suspended rail joint (with the joint between two adjacent sleepers) and the supported rail joint (one sleeper right beneath the joint). A conceptual design, termed as embedded IRJ was presented in [67], in which the rails were embedded into the concrete of the posttensioned sleepers. The numerical simulations indicate that the performance of the embedded IRJ is superior under wheel-rail impacts. An experimental study [68] conducted in the Dutch railway indicates that the amplitude of the rail resonance decreases with the 1st sleeper span after the joint, whereas the amplitude of the pinned-pinned resonance decreases with the difference between the 1st and 2nd sleeper spans. The nonuniform sleeper distribution at the IRJ is thus proposed to reduce wheel-rail impacts at IRJs.

As to the geometric control, timely grinding is an effective way to reduce wheel-rail impact force. As reported in [17], the numerically simulated wheel-rail contact force increases from 130 kN (quasistatic load) to 160 kN at a new joint (with nominal rail profile), and to 230 kN at a worn joint (with measured rail profile). In addition, antiimpact rail profiles [69–71] have been proposed for the rail joint and turnout sections.

At turnouts, the grinding-welding-grinding method can be performed to control the wheel-rail contact geometry and to extend the service life. First, the crossing rails were ground to remove surface damage (e.g., cracks, uneven wear, and plastic deformation); second, the hollow regions caused by the grinding were filled by welding; finally, the crossing rails were ground again to the desired profiles [53]. The repair was performed manually because the train-borne grinding machine is incapable of operating at the turnout section.

7.3.2.2 Maintenance of the large-friction-inducing sections

A straightforward way to maintain the large-friction-inducing sections is to manage wheel-rail friction. A low coefficient of friction (COF) is desired to reduce rail wear (especially at shape curves [72]), head checks (HCs) [73], noise [74], and energy consumption of railway transportation [75]. However, low COFs may result in low adhesion, which then generates severe rail burn and wheel flat, affects punctuality in traction operation, and threaten safety during braking. To guarantee the safe acceleration and braking which has maximum demand for COF, the required COFs are 0.25 in traction and 0.14 in braking in the Dutch railway [76]. The COF required in traction is larger because a considerably lower number of wheel axles is involved in the acceleration than in braking.

Friction modifiers (FMs) can be applied to the wheel-rail interface either by trackside or train-borne installations. However, providing a proper dosage of the FMs is challenging because it varies case by case, influenced by weather and characteristics of the FMs, etc. Smaller quantities of FMs do not provide a stable level of adhesion and the lasting effects are limited, whereas bigger quantities have a longer-lasting effect, but critically low adhesion levels may be caused after the application of FMs.

Grinding is a commonly used approach for the maintenance of the large-friction-inducing sections to restore the rail profile and to remove RCF cracks. The best treatment available for HCs and corrugation would be early detection and removal by grinding. In addition, rail profile optimization can effectively prevent or retard HC initiation. The anti-HC profile presented in [77] has been normalized as a standard European rail profile. By avoiding contact in the HC-prone part of the rail, the maximum surface shear stress is greatly reduced, mainly owing to the decrease of spin in the contact. Large-scale application of the anti-HC profile on the Dutch railway network shows that HCs in 2008 were reduced by about 70% with respect to 2004 when HC was the most widespread.

References

[1] H. Hertz, Ueber die Berührung fester elastischer Körper, J. Reine Angew. Math. 1882 (1882), https://doi.org/10.1515/crll.1882.92.156.
[2] R.D. Mindlin, Compliance of elastic bodies in contact, J. Appl. Mech. 16 (3) (1949) 259–268.
[3] F.W. Carter, On the action of a locomotive driving wheel, P. R. Soc. Lond. A 112 (1926) 151–157, https://doi.org/10.1098/rspa.1926.0100.
[4] P.J. Vermeulen, K.L. Johnson, Contact of nonspherical elastic bodies transmitting tangential forces, J. Appl. Mech. 31 (1964) 338, https://doi.org/10.1115/1.3629610.
[5] J.J. Kalker, Three-Dimensional Elastic Bodies in Rolling Contact, first ed., Springer, The Netherlands, 1990.
[6] J. Ringsberg, Rolling contact fatigue analysis of rails inculding numerical simulations of the rail manufacturing process and repeated wheel-rail contact loads, Int. J. Fatigue 25 (2003) 547–558, https://doi.org/10.1016/s0142-1123(02)00147-0.
[7] T. Telliskivi, U. Olofsson, Wheel–rail wear simulation, Wear 257 (2004) 1145–1153, https://doi.org/10.1016/j.wear.2004.07.017.
[8] A. Sladkowski, M. Sitarz, Analysis of wheel–rail interaction using FE software, Wear 258 (2005) 1217–1223, https://doi.org/10.1016/j.wear.2004.03.032.
[9] M. Wiest, E. Kassa, W. Daves, J.C.O. Nielsen, H. Ossberger, Assessment of methods for calculating contact pressure in wheel-rail/switch contact, Wear 265 (2008) 1439–1445, https://doi.org/10.1016/j.wear.2008.02.039.
[10] Z. Li, X. Zhao, C. Esveld, R. Dollevoet, M. Molodova, An investigation into the causes of squats—correlation analysis and numerical modeling, Wear 265 (2008) 1349–1355, https://doi.org/10.1016/j.wear.2008.02.037.
[11] A. Ekberg, J. Sandström, Numerical study of the mechanical deterioration of insulated rail joints, Proc. Inst. Mech. Eng. F J. Rail Rapid Transit 223 (2009) 265–273, https://doi.org/10.1243/09544097jrrt243.
[12] A. Johansson, B. Pålsson, M. Ekh, J.C.O. Nielsen, M.K.A. Ander, J. Brouzoulis, E. Kassa, Simulation of wheel–rail contact and damage in switches & crossings, Wear 271 (2011) 472–481, https://doi.org/10.1016/j.wear.2010.10.014.
[13] X. Zhao, Z.L. Li, The solution of frictional wheel-rail rolling contact with a 3D transient finite element model: validation and error analysis, Wear 271 (2011) 444–452, https://doi.org/10.1016/j.wear.2010.10.007.
[14] Z. Li, R. Dollevoet, M. Molodova, X. Zhao, Squat growth—some observations and the validation of numerical predictions, Wear 271 (2011) 148–157, https://doi.org/10.1016/j.wear.2010.10.051.
[15] X. Zhao, Z.L. Li, A three-dimensional finite element solution of frictional wheel-rail rolling contact in elasto-plasticity, Proc. Inst. Mech. Eng. B J. Eng. 229 (2015) 86–100, https://doi.org/10.1177/1350650114543717.
[16] Z. Wei, C. Shen, Z. Li, R. Dollevoet, Wheel–rail impact at crossings: relating dynamic frictional contact to degradation, J. Comput. Nonlinear Dyn. 12 (2017), https://doi.org/10.1115/1.4035823, 041016.
[17] Z. Yang, A. Boogaard, Z. Wei, J. Liu, R. Dollevoet, Z. Li, Numerical study of wheel-rail impact contact solutions at an insulated rail joint, Int. J. Mech. Sci. 138–139 (2018) 310–322, https://doi.org/10.1016/j.ijmecsci.2018.02.025.
[18] Z. Yang, X. Deng, Z. Li, Numerical modeling of dynamic frictional rolling contact with an explicit finite element method, Tribol. Int. 129 (2019) 214–231, https://doi.org/10.1016/j.triboint.2018.08.028.

[19] X. Deng, Z. Li, Z. Qian, W. Zhai, Q. Xiao, R. Dollevoet, Pre-cracking development of weld-induced squats due to plastic deformation: five-year field monitoring and numerical analysis, Int. J. Fatigue 127 (2019) 431–444, https://doi.org/10.1016/j.ijfatigue.2019.06.013.

[20] Z. Li, J.J. Kalker, Simulation of severe wheel-rail wear, in: Proceedings of the 6th International Conference on Computer Aided Design, Manufacture and Operation in the Railway and Other Advanced Mass Transit Systems, Lisbon, Portugal, 1998, pp. 393–402, https://doi.org/10.2495/CR980381.

[21] Y. Wu, X. Du, H.-J. Zhang, Z.-F. Wen, X.-S. Jin, Experimental analysis of the mechanism of high-order polygonal wear of wheels of a high-speed train, J. Zheijang Univ. Sci. A 18 (2017) 579–592, https://doi.org/10.1631/jzus.A1600741.

[22] G. Tao, Z. Wen, X. Liang, D. Ren, X. Jin, An investigation into the mechanism of the out-of-round wheels of metro train and its mitigation measures, Veh. Syst. Dyn. 57 (2018) 1–16, https://doi.org/10.1080/00423114.2018.1445269.

[23] W. Cai, M. Chi, X. Wu, F. Li, Z. Wen, S. Liang, X. Jin, Experimental and numerical analysis of the polygonal wear of high-speed trains, Wear 440–441 (2019), https://doi.org/10.1016/j.wear.2019.203079.

[24] X. Deng, Z. Qian, Z. Li, R. Dollevoet, Investigation of the formation of corrugation-induced rail squats based on extensive field monitoring, Int. J. Fatigue 112 (2018) 94–105, https://doi.org/10.1016/j.ijfatigue.2018.03.002.

[25] R. Dollevoet, Z. Li, O. Arias-Cuevas, A method for the prediction of head checking initiation location and orientation under operational loading conditions, Proc. Inst. Mech. Eng. F J. Rail Rapid Transit 224 (2010) 369–374, https://doi.org/10.1243/09544097jrrt368.

[26] A. Zoeteman, R. Dollevoet, Z. Li, Dutch research results on wheel/rail interface management: 2001–2013 and beyond, Proc. Inst. Mech. Eng. F J. Rail Rapid Transit 228 (2014) 642–651, https://doi.org/10.1177/0954409714524379.

[27] S.L. Grassie, Rail corrugation: characteristics, causes, and treatments, Proc. Inst. Mech. Eng. F J. Rail Rapid Transit 223 (2009) 581–596, https://doi.org/10.1243/09544097jrrt264.

[28] S. Li, Z. Li, A. Núñez, R. Dollevoet, New insights into the short pitch corrugation enigma based on 3D-FE coupled dynamic vehicle-track modeling of frictional rolling contact, Appl. Sci. 7 (2017) 807, https://doi.org/10.3390/app7080807.

[29] D.J. Thompson, Railway Noise and Vibration: Mechanisms, Modelling and Means of Control, Elsevier, 2009.

[30] Z.Y. Huang, D.J. Thompson, C.J.C. Jones, Squeal prediction for a bogied vehicle in a curve, Notes Numer. Fluid Mech. 99 (2008) 313–319, https://doi.org/10.1007/978-3-540-74893-9_44.

[31] Z. Yang, Z. Li, Numerical modeling of wheel-rail squeal-exciting contact, Int. J. Mech. Sci. 153–154 (2019) 490–499, https://doi.org/10.1016/j.ijmecsci.2019.02.012.

[32] Z.L. Li, X. Zhao, R. Dollevoet, M. Molodova, Differential wear and plastic deformation as causes of squat at track local stiffness change combined with other track short defects, Veh. Syst. Dyn. 46 (2008) 237–246, https://doi.org/10.1080/00423110801935855.

[33] X. Zhao, Z. Li, A solution of transient rolling contact with velocity dependent friction by the explicit finite element method, Eng. Comput. 33 (2016) 1033–1050, https://doi.org/10.1108/ec-09-2014-0180.

[34] X. Zhao, X.G. Zhao, C. Liu, Z.F. Wen, X.S. Jin, A study on dynamic stress intensity factors of rail cracks at high speeds by a 3D explicit finite element model of rolling contact, Wear 366 (2016) 60–70, https://doi.org/10.1016/j.wear.2016.06.001.

[35] M. Pletz, W. Daves, H. Ossberger, A wheel set/crossing model regarding impact, sliding and deformation-explicit finite element approach, Wear 294 (2012) 446–456, https://doi.org/10.1016/j.wear.2012.07.033.

[36] D.M. Mulvihill, M.E. Kartal, D. Nowell, D.A. Hills, An elastic–plastic asperity interaction model for sliding friction, Tribol. Int. 44 (2011) 1679–1694, https://doi.org/10.1016/j.triboint.2011.06.018.
[37] P. Wriggers, Computational Contact Mechanics, Springer, Berlin Heidelberg, 2006.
[38] G. Noh, K.-J. Bathe, An explicit time integration scheme for the analysis of wave propagations, Comput. Struct. 129 (2013) 178–193, https://doi.org/10.1016/j.compstruc.2013.06.007.
[39] Z. Yang, A. Boogaard, R. Chen, R. Dollevoet, Z. Li, Numerical and experimental study of wheel-rail impact vibration and noise generated at an insulated rail joint, Int. J. Impact Eng. 113 (2018) 29–39, https://doi.org/10.1016/j.ijimpeng.2017.11.008.
[40] Z. Yang, Z.L. Li, R. Dollevoet, Modelling of non-steady-state transition from single-point to two-point rolling contact, Tribol. Int. 101 (2016) 152–163, https://doi.org/10.1016/j.triboint.2016.04.023.
[41] I. Zenzerovic, W. Kropp, A. Pieringer, An engineering time-domain model for curve squeal: tangential point-contact model and Green's functions approach, J. Sound Vib. 376 (2016) 149–165, https://doi.org/10.1016/j.jsv.2016.04.037.
[42] Z. Yang, Z. Li, A numerical study on waves induced by wheel-rail contact, Int. J. Mech. Sci. 161–162 (2019), https://doi.org/10.1016/j.ijmecsci.2019.105069, 105069.
[43] D.J. Thompson, G. Squicciarini, B. Ding, L. Baeza, A state-of-the-art review of curve squeal noise: phenomena, mechanisms, modelling and mitigation, in: Notes on Numerical Fluid Mechanics and Multidisciplinary Design, 139, Springer, 2018, pp. 3–41, https://doi.org/10.1007/978-3-319-73411-8_1.
[44] J. Reed, Energy-losses due to elastic wave-propagation during an elastic impact, J. Phys. D Appl. Phys. 18 (1985) 2329–2337, https://doi.org/10.1088/0022-3727/18/12/004.
[45] H. Ouyang, W. Nack, Y. Yuan, F. Chen, Numerical analysis of automotive disc brake squeal: a review, Int. J. Veh. Noise Vib. 1 (2005) 207, https://doi.org/10.1504/ijvnv.2005.007524.
[46] P.R. Armitage, The use of low-frequency Rayleigh waves to detect gauge corner cracking in railway lines, Insight 44 (2002) 369–372.
[47] A. Pantano, D. Cerniglia, Simulation of laser generated ultrasound with application to defect detection, Appl. Phys. A 91 (2008) 521–528, https://doi.org/10.1007/s00339-008-4442-1.
[48] M. Molodova, Z. Li, A. Nunez, R. Dollevoet, Automatic detection of squats in railway infrastructure, IEEE Trans. Intell. Transp. Syst. 15 (2014) 1980–1990, https://doi.org/10.1109/tits.2014.2307955.
[49] M. Oregui, S. Li, A. Núñez, Z. Li, R. Carroll, R. Dollevoet, Monitoring bolt tightness of rail joints using axle box acceleration measurements, Struct. Control. Health Monit. 24 (2017), https://doi.org/10.1002/stc.1848.
[50] A. Nunez, A. Jamshidi, H. Wang, Pareto-based maintenance decisions for regional railways with uncertain weld conditions using the Hilbert spectrum of axle box acceleration, IEEE Trans. Ind. Inf. 15 (2019) 1496–1507, https://doi.org/10.1109/tii.2018.2847736.
[51] P.F. Westeon, C.S. Ling, C. Roberts, C.J. Goodman, P. Li, R.M. Goodall, Monitoring vertical track irregularity from in-service railway vehicles, Proc. Inst. Mech. Eng. F J. Rail Rapid Transit 221 (2016) 75–88, https://doi.org/10.1243/0954409jrrt65.
[52] A.M. Remennikov, S. Kaewunruen, A review of loading conditions for railway track structures due to train and track vertical interaction, Struct. Control. Health Monit. 15 (2008) 207–234, https://doi.org/10.1002/stc.227.
[53] Z. Wei, A. Núñez, Z. Li, R. Dollevoet, Evaluating degradation at railway crossings using axle box acceleration measurements, Sensors 17 (2017), https://doi.org/10.3390/s17102236.

[54] P. Gullers, L. Andersson, R. Lundén, High-frequency vertical wheel–rail contact forces—field measurements and influence of track irregularities, Wear 265 (2008) 1472–1478, https://doi.org/10.1016/j.wear.2008.02.035.
[55] J.R. Edwards, A. Cook, M.S. Dersch, Y. Qian, Quantification of rail transit wheel loads and development of improved dynamic and impact loading factors for design, Proc. Inst. Mech. Eng. F J. Rail Rapid Transit 232 (2018) 2406–2417, https://doi.org/10.1177/0954409718770924.
[56] B. Stratman, Y. Liu, S. Mahadevan, Structural health monitoring of railroad wheels using wheel impact load detectors, J. Fail. Anal. Prev. 7 (2007) 218–225, https://doi.org/10.1007/s11668-007-9043-3.
[57] A. Alemi, Railway Wheel Defect Identification (Ph.D. thesis), Delft University of Technology, 2019.
[58] Z. Li, M. Molodova, A. Nunez, R. Dollevoet, Improvements in axle box acceleration measurements for the detection of light squats in railway infrastructure, IEEE Trans. Ind. Electron. 62 (2015) 4385–4397, https://doi.org/10.1109/tie.2015.2389761.
[59] M. Oregui, Z. Li, R. Dollevoet, Identification of characteristic frequencies of damaged railway tracks using field hammer test measurements, Mech. Syst. Signal Process. 54–55 (2015) 224–242, https://doi.org/10.1016/j.ymssp.2014.08.024.
[60] S. Kaewunruen, A.M. Remennikov, Field trials for dynamic characteristics of railway track and its components using impact excitation technique, NDT&E Int. 40 (2007) 510–519, https://doi.org/10.1016/j.ndteint.2007.03.004.
[61] M.P.N. Burrow, A.H.C. Chan, A. Shein, Deflectometer-based analysis of ballasted railway tracks, Proc. Inst. Civ. Eng. Geotech. Eng. 160 (2007) 169–177, https://doi.org/10.1680/geng.2007.160.3.169.
[62] S. Kaewunruen, A.M. Remennikov, Progressive failure of prestressed concrete sleepers under multiple high-intensity impact loads, Eng. Struct. 31 (2009) 2460–2473, https://doi.org/10.1016/j.engstruct.2009.06.002.
[63] E.G. Berggren, A. Nissen, B.S. Paulsson, Track deflection and stiffness measurements from a track recording car, Proc. Inst. Mech. Eng. F J. Rail Rapid Transit 228 (2014) 570–580, https://doi.org/10.1177/0954409714529267.
[64] K. Knothe, Y. Wu, Receptance behaviour of railway track and subgrade, Arch. Appl. Mech. 68 (1998) 457–470, https://doi.org/10.1007/s004190050179.
[65] Z. Wei, A. Boogaard, A. Nunez, Z. Li, R. Dollevoet, An integrated approach for characterizing the dynamic behavior of the wheel–rail interaction at crossings, IEEE Trans. Instrum. Meas. 67 (2018) 2332–2344, https://doi.org/10.1109/tim.2018.2816800.
[66] N. Vincent, D.J. Thompson, Track dynamic behaviour at high frequencies. Part 2: Experimental results and comparisons with theory, Veh. Syst. Dyn. 24 (1995) 100–114, https://doi.org/10.1080/00423119508969618.
[67] N. Zong, M. Dhanasekar, Sleeper embedded insulated rail joints for minimising the number of modes of failure, Eng. Fail. Anal. 76 (2017) 27–43, https://doi.org/10.1016/j.engfailanal.2017.02.001.
[68] I. Papaioannou, Experimental and Numerical Study on the Optimisation of Insulated Rail Joint Dynamic Behaviour (M.Sc. thesis), Delft University of Technology, 2018.
[69] R.H. Plaut, H. Lohse-Busch, A. Eckstein, S. Lambrecht, D.A. Dillard, Analysis of tapered, adhesively bonded, insulated rail joints, Proc. Inst. Mech. Eng. F J. Rail Rapid Transit 221 (2007) 195–204, https://doi.org/10.1243/0954409jrrt107.
[70] C. Wan, V. Markine, R. Dollevoet, Robust optimisation of railway crossing geometry, Veh. Syst. Dyn. 54 (2016) 617–637, https://doi.org/10.1080/00423114.2016.1150495.

[71] C. Wan, V.L. Markine, I.Y. Shevtsov, Improvement of vehicle–turnout interaction by optimising the shape of crossing nose, Veh. Syst. Dyn. 52 (2014) 1517–1540, https://doi.org/10.1080/00423114.2014.944870.

[72] X. Lu, T.W. Makowsky, D.T. Eadie, K. Oldknow, J. Xue, J. Jia, G. Li, X. Meng, Y. Xu, Y. Zhou, Friction management on a Chinese heavy haul coal line, Proc. Inst. Mech. Eng. F J. Rail Rapid Transit 226 (2012) 630–640, https://doi.org/10.1177/0954409712447170.

[73] R. Stock, D.T. Eadie, D. Elvidge, K. Oldknow, Influencing rolling contact fatigue through top of rail friction modifier application – a full scale wheel–rail test rig study, Wear 271 (2011) 134–142, https://doi.org/10.1016/j.wear.2010.10.006.

[74] D.T. Eadie, M. Santoro, J. Kalousek, Railway noise and the effect of top of rail liquid friction modifiers: changes in sound and vibration spectral distributions in curves, Wear 258 (2005) 1148–1155, https://doi.org/10.1016/j.wear.2004.03.061.

[75] G. Idárraga Alarcón, N. Burgelman, J.M. Meza, A. Toro, Z. Li, The influence of rail lubrication on energy dissipation in the wheel/rail contact: a comparison of simulation results with field measurements, Wear 330–331 (2015) 533–539, https://doi.org/10.1016/j.wear.2015.01.008.

[76] O. Arias-Cuevas, Z. Li, Field investigations into the adhesion recovery in leaf-contaminated wheel–rail contacts with locomotive sanders, Proc. Inst. Mech. Eng. F J. Rail Rapid Transit 225 (2011) 443–456, https://doi.org/10.1177/2041301710394921.

[77] R.P.B.J. Dollevoet, Design of an Anti Head Check Profile Based on Stress Relief (Ph.D. thesis), University of Twente, 2010.

Wheel-rail interface under extreme conditions

Milan Omasta[a] and Hua Chen[b]
[a]Brno University of Technology, Brno, Czech Republic, [b]Railway Technical Research Institute, Tokyo, Japan

8.1 Introduction into the wheel-rail interface

Wheel-rail interface is essential for the safety and operational performance of rail transport. The contact between steel wheel and steel rail occurs on a small area of approximately $1-2\,cm^2$ and this contact zone is responsible for the transfer of nearly all forces between the train and the track. This is especially important when traction and braking are carried out as this contact determines how much the mechanical energy can be transferred from the train to the track.

The issues related to the wheel-rail interface are outlined in Fig. 8.1. This interface represents a complex system whose behavior is influenced by many factors. These are mainly operating conditions such as speed and vertical wheel load, dynamics of rail vehicle and track, maintenance parameters including wheel and rail head profile, surface roughness and materials. Tribological phenomena taking place in the interface determine its behavior. The most important aspect is the ability to transfer traction and braking forces in the longitudinal direction and lateral forces in the perpendicular direction. High contact pressure and shear stress is a source of various types of wear and damage. Rail head and wheel thread are vulnerable to the rolling contact fatigue, wheel flange is exposed to severe sliding wear and the wheel and rail materials undergo cyclic plastic deformation. This interface is also the source of most of the negative effects of rail transport such as rolling and squealing noise and ground-borne vibration.

Wheel-rail interface is a typical open tribological system. This means that the conditions are not stable and well-defined as in the case of the closed system, such as, e.g., a gearbox. The wheel-rail contact is subject to many kinds of environmental influences, especially weather conditions like temperature, humidity and precipitation. These conditions are constantly varying as night and day and different seasons alternate and with a change in the environment in which the train moves. On the other hand, the openness of this system also means that the substances formed in the interface or applied to the contact are dispersed into the surrounding environment. This applies in particular to products for friction management such as wheel flange lubricants, top-of-rail friction modifiers and traction enhancers. There is a growing emphasis on minimizing the negative impact of rail transport through noise and vibration mitigations, soil pollution mitigation and so on.

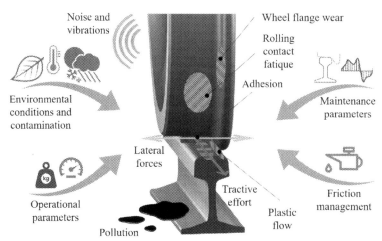

Fig. 8.1 Issues related to the wheel-rail interface.
Credit: Original artwork.

8.2 Basics of the wheel-rail contact

One of the most important aspects of the wheel-rail interface is adhesion. In the rail transport community "adhesion" is defined as the ability of the contact to transfer tangential forces. In the longitudinal direction, these forces allow traction and braking. Schematic representation of a rolling-sliding contact between the rail and the wheel under acceleration is shown in Fig. 8.2. This contact is subject to a normal force F_N representing the force of gravity acting on the wheel. The wheel moves along the rail at velocity v. The wheel rotates with angular velocity ω thanks to torque T. The result is a reaction tangential force F_T acting between the wheel and the rail. Their ratio corresponds to the coefficient of adhesion according to Eq. (8.1). This definition is similar to the coefficient of friction for pure-sliding contact, but the physical meaning is different. Under pure-rolling conditions, circumferential velocity $\omega \times r$ corresponds to the forward velocity v. However, during acceleration, circumferential velocity is higher resulting in sliding velocity in the contact v_s defined using Eq. (8.2). The ratio between the velocity difference and forward velocity is defined as creep or creepage according to Eq. (8.3). When multiplied by 100 we get it in %.

$$\mu = \frac{F_T}{F_N} \tag{8.1}$$

$$v_s = \omega \times r - v \tag{8.2}$$

$$\xi = \frac{v_s}{v} = \frac{\omega \times r - v}{\omega \times r} \tag{8.3}$$

Wheel-rail interface under extreme conditions 139

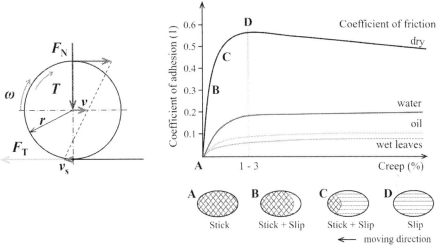

Fig. 8.2 Schematic view of rolling-sliding contact under acceleration and the relationship between the coefficient of adhesion and creep.
Credit: Original artwork.

The relationship between the coefficient of adhesion and the creep is represented by the creep curve, shown in Fig. 8.2. Point A corresponds to pure rolling, where the surfaces are sticking in the whole elastic contact area and no tangential forces are transferred, so the coefficient of adhesion is zero. When the tangential force appears (point B), the contact is divided into the leading part with a stick and the trailing part where slipping occurs. As the tangential force increases the slipping area extends at the expense of the sticking area (point C). When the coefficient of adhesion reaches saturation point, the stick area disappears, and the contact is fully sliding. At this point, the coefficient of adhesion corresponds to the coefficient of friction. The creep level where the coefficient of adhesion reaches its maximum usually ranges between 1% and 3%. If the applied power exceeded the adhesion limit, wheel slip would occur during acceleration. Similarly, wheel-slide may occur during braking. In both cases, undesirable deterioration of the wheels or rails occurs.

As indicated qualitatively in Fig. 8.2, the shape of the creep curve and maximum value of the coefficient of adhesion strongly depends on the conditions of the contact between the wheel and the rail. The typical value of adhesion coefficient measured using an instrumented train ranges between 0.3 and 0.4 under dry conditions. Water decreases the value to 0.25 and oil to 0.2 [1]. A wide range of field and laboratory devices for friction and adhesion coefficients measurement exists. These tribometers and test rigs are inevitable in the wheel-rail interface research [2], however, their limitations must be considered when interpreting the coefficient of adhesion.

The lowest value of the coefficient of adhesion required for braking is usually given in the range of 0.07–0.15 depending on the specific railroad. For traction, the value is around 0.2. Insufficient adhesion is a typical cause of delays, platform overruns and

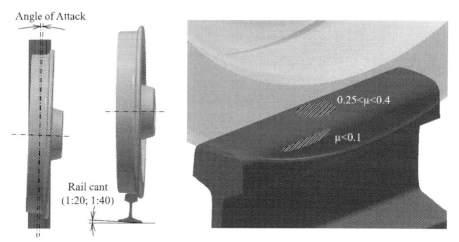

Fig. 8.3 Contact positions in the wheel-rail contact.
Credit: Original artwork.

incidents of Signals Passes at Danger (SPADs). In an extreme case, it leads to wheel burns during acceleration and the formation of wheel flats during sudden braking. On the other hand, too high adhesion/friction results in a higher wear rate and rolling contact fatigue. Higher lateral forces have a negative impact on driving comfort, noise and vibration, risk of flange climb derailment, rail corrugation, and track deterioration. The wheel-rail contact typically occurs in two distinct areas, as shown in Fig. 8.3. At the straight track, the contact appears solely between the wheel tread and the top of the rail. The optimal coefficient of adhesion in this zone is 0.25–0.4. On sharp curves, the contact extends toward the wheel flange and rail gauge. We are talking about two-point contact. A high sliding velocity acts in this contact leading to severe to catastrophic wear [3,4]. So, the lowest possible coefficient of friction is desirable. Maintaining an appropriate adhesion between the wheel and rail is very important concerning safety and efficiency.

Wheel-rail contamination and weather conditions are the most important factors influencing the behavior of the open tribological system. Various substances and materials that are found in the wheel-rail interface are categorized in Fig. 8.4. The first group is substances that naturally originate from tribological processes on the rail or in the interface, namely wear particles from the wheel and the rail and iron oxides and hydroxides that are products of chemical reaction with water and oxygen. Another group can be considered as external contaminants. The most important part covers natural contaminants from the environment like water, snow and fallen leaves. Water can be introduced to the contact through rainfall, moisture condensation, snow melting and flooding. Fallen leaves come from the vegetation along the track and are the most significant cause of low adhesion problems in autumn [5]. The contact can be further contaminated from the environment by the soil and sand, usually carried by the wind in arid and desert areas, and by rock salt or grit used to treat the roads and platforms

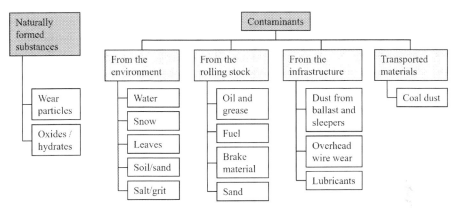

Fig. 8.4 Classification of various contact contaminants.
Credit: Original artwork.

against the winter snow and ice [6]. The source of artificial contaminants is mainly the rail vehicle itself. Operating fluids such as grease, fuel oils and transformers oils pose the greatest risk in terms of the influence on traction and braking. There is a wide range of contamination particles that originates from the other components of the train, such as wear particles from brake pads and brake discs [7]. The sand listed in this category refers to the material used by the rolling stock to restore wheel-rail adhesion. Rail infrastructure can also be a source of contamination. The rail head is exposed to dust from ballast and concrete sleepers, wear particles from overhead wires and a lubricant from way-side lubricators. An example of contamination by the transported medium could be, e.g., coal dust from the trains that carry coal in uncovered rail cars [8]. Understanding the impact and hazards resulting from the contamination is one of the main challenges in the wheel-rail interface research.

8.3 The wheel-rail interface under extreme conditions

8.3.1 Extreme events and their consequences to the wheel-rail interface

There is a number of studies reporting the impact of climate changes or extreme weather conditions on railway transport [9–13]. However, these studies remain on global analysis related to the whole infrastructure, rolling stock and asset management. This chapter discusses, how the predicted changes and events affect the behavior of the wheel-rail interface. Table 8.1 gives an overview of the potential impacts of extreme climate events on wheel-rail interface and consequences to the railway service and operation. The consequences are generally determined by the effect of specific contamination or climate condition on the behavior of the wheel-rail interface. So, the following sub-chapters deal in detail with these effects, especially on the adhesion in the wheel-rail interface.

Table 8.1 Overview of the potential impacts of extreme climate events on the wheel-rail interface and consequences to the railway service and operation.

Climate event	Impact	Consequence to the wheel-rail interface	Consequence to service and operation
Heavy rain	Flooding	Water contamination	Risk of the low adhesion under high speed
		Increased formation of hydroxides	Risk of low the adhesion phenomenon
Coastal storm surges and tsunami waves	Coastal flooding	Contamination with saltwater	Risk of the low adhesion under high speed
		Increased formation of hydroxides	Risk of low the adhesion phenomenon
Cold climate and cold waves	Extremely low temperature	Sub-zero temperature	Increased wear a rolling contact fatigue
			Slower adhesion recovery process
	Snow	Water contamination under sub-zero temperature	Increased wear a rolling contact fatigue
Heatwaves	High ambient and surface temperature	Thermal loading of the wheel and rail	Increased rolling contact fatigue and deterioration
Dessert storms and high winds in arid areas	Contamination by sand and soil	Formation of the third-body layer	Risk of low adhesion due to the "lubrication" action of sand layer
			Risk of the contact insulation and loss of track circuit function
			Increased abrasive wear
Extreme winds	Leaf fall contamination	Formation of the third-body layer	Risk of the low adhesion phenomenon
			Risk of the contact insulation and loss of track circuit function

8.3.2 Water at the wheel-rail contact

Water is one of the most common natural contaminants of the wheel-rail interface. Under normal conditions, water contamination results from rainfall or morning dew. In extreme cases, railway lines can be flooded due to heavy rain and storms, especially in low-lying land with malfunctioning drainage. The threat of flooding is also

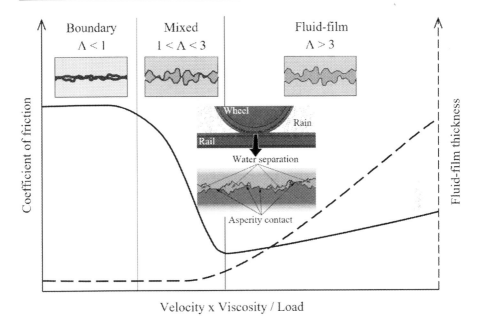

Fig. 8.5 Stribeck curve describing various lubricating regimes.
Credit: Original artwork.

relevant in the underground railway system [14]. In any case, the presence of water has a great influence on adhesion between the rails and running wheels [15].

The effect of water on adhesion in the contact is connected with different mechanisms based on the lubrication regime where the contact operates. Different lubrication regimes are described using a Stribeck's curve in Fig. 8.5. This curve shows qualitatively the coefficient of friction and lubricant-film thickness as a function of dimensionless parameters including velocity, lubricant viscosity, and load. These regimes vary according to the level of surface separation. In the boundary regime, the contact bodies are separated by a boundary friction layer that naturally occurs on the rail in an open environment. As the velocity increases and the sufficient amount of water or other fluid lubricant appears on the rail, the surfaces start to be separated because of the hydrodynamic action of the liquid. In the mixed regime, this separation is only a local while the rest of the contact carried by the surface asperities remains in the boundary regime. The hydrodynamic action increases with velocity, resulting in full separation of the bodies with a full fluid lubricating film and we are talking about fluid-film regime. The action of water in the contact depends on the regime.

8.3.2.1 Boundary lubrication

In the boundary regime, water is unlikely to act in the contact on its own but forms a mixture with other substances and contaminants. Naturally, this is mainly a combination of wear particles. The early experiments by Beagley and Pritchard [16] have demonstrated that a small amount of water mixed with wear debris forms a lubricating paste that results in a very low friction coefficient (0.05). The decrease in adhesion is more pronounced the less water is applied. This action of water is called "Wet Rail Phenomenon" and is associated with dew on the head of the rail, misty conditions and very light rain [17]. The practical consequence of this phenomenon is that a small amount of water leads to lower adhesion than large amount that washes the contaminants away. This phenomenon usually occurs in the mornings and evenings due to dew formation.

Water is also involved in the formation of a very thin friction layer, hardly visible to the naked eye, often called "third-body layer." The third-body concept [18] is often used in wheel-rail tribology to describe the behavior of the layer based on its rheology [19,20]. This layer is a result of mechanical, chemical and thermal interactions of the surfaces with the contaminants in the high-pressure rolling-sliding contact. In addition to water and wear particles, other natural contaminants like leaves contribute to its formation. In this case, water often acts as a trigger for the low adhesion problem in an already formed third-body layer [21–26].

8.3.2.2 Fluid-film lubrication

In fluid-film regime, liquid water acts as a lubricant between the contacting bodies. The wheel and the rail form a converging–diverging geometry, which together with a sufficient relative velocity creates the conditions for the development of hydrodynamic pressure. Because of the high contact pressure, elastic deformations of the contacting surfaces are of the same order of magnitude as the film thickness separating them. So, we are talking about elastohydrodynamic lubrication (EHL). Someone could look at the analogy to aquaplaning by the tires of a road vehicle when a layer of water builds between the wheels of the vehicle and the road surface, leading to the loss of control. An elastic deformation of compliant rubber tire and lower car weight leads to a low contact pressure in the order of 0.1–1 MPa. This lubrication regime is often referred to as "soft" EHL and the velocity required to develop the hydrodynamic action of water film is relatively low. However, the contact pressure between the steel wheel and rail is much higher, in order of 1 GPa. Under the "hard" EHL regime, the hydrodynamic action of water is usually too low to fully separate the wheel and the rail. The partial separation, however, becomes extremely important, especially for light-weight vehicles operated at high speed.

Water film thickness can be estimated using EHL theory. A wide range of film thickness equations and correction factors was established in last decades for various regimes and conditions [27]. These equations usually represent last-square fit of the results determined using a numerical solution for given parameters. Originally, they were intended and validated for the contacts lubricated with oil whose viscosity is highly dependent on the contact pressure. Piezo-viscosity of water is not as strong, so one option is to consider the isoviscous elastic lubrication regime. This regime is characterized as the regime where the elastic deformation of the contact bodies has a significant contribution to the fluid film, but the pressure in the contact is insufficient to cause a substantial increase in fluid viscosity. Nevertheless, it is also often interpreted as a regime where the pressure is sufficient but the pressure-viscosity

Table 8.2 Empirical equations for the lubricant-film thickness prediction.

Author	Regime and contact	Lubricant-film thickness equation
Esfahanian and Hamrock [28]	Iso-viscous elastic, elliptical	$H_{min} = 8.70 U^{0.66} W^{-0.21}(1 - 0.85 e^{-31k})$
Hamrock and Dowson [29]	Piezo-viscous elastic, elliptical	$H_{min} = 3.63 U^{0.67} W^{-0.073} G^{0.49}(1 - e^{-68k})$
Chen et al. [30]	Piezo-viscous elastic, line	$H_{min} = 2.578 U^{0.59} W^{-0.211} G^{0.002}$
Wu et al. [31]	Piezo-viscous elastic, elliptical	$H_{min} = 9.36 U^{0.72} W^{-0.29} G^{0.007}(1 - e^{-68k})$

dependence of the lubricant is low. For the elliptical contact, frequently used equation by Esfahanian can be utilized (see Table 8.2), where non-dimensional parameters are given by the following: $H_{min} = \frac{h_{min}}{R_x}$; $U = \frac{\eta_0 v}{E R_x}$; $W = \frac{F}{E R_x^2}$ for point contact and $W = \frac{F_l}{E R_x}$ for elliptical contact; $G = \alpha E$; $k = 1.03 \left(\frac{R_y}{R_x}\right)^{0.64}$.

Another option is to consider water as a piezo-viscous fluid. The dependence of water viscosity on pressure and temperature in the range relevant for the wheel-rail contact was published, e.g., by Bett and Cappi [32] (see Fig. 8.6). Piesoviscosity of fluid is accounted for in film thickness prediction formulas using pressure-viscosity coefficient α. The simple value is not suitable for water because of its anomalous behavior. So, the equivalent values of pressure-viscosity coefficient depending on contact pressure and temperature were calculated [30]. Then, an equation for piezo-viscous elastic regime and elliptical contact, e.g., by Hamrock and Dowson can be used. Specific empirical equations were also developed based on the numerical simulation of the water-lubricated wheel-rail contact. Chen et al. [30] developed a

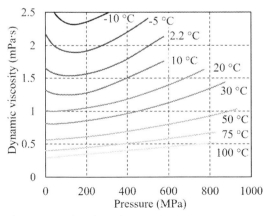

Fig. 8.6 Viscosity of water as a function of temperature and pressure, according to Bett and Cappi [32].
Credit: Adapted from Springer Nature, K. Bett, J. Cappi, Effect of pressure on the viscosity of water, Nature 207 (4997) (1965) 620–621, https://doi.org/10.1038/207620a0, copyright (1965).

formula for the line contact using temperature–pressure-viscosity parameters of water according to Bett and Cappi. Recently, similar solution has been extended to 3D simulation [31]. It should be noted that all the predictions are only valid for the smooth surface case and isothermal conditions. The calculated value expresses the film thickness that can be formed under the given conditions if the contact is sufficiently supplied with liquid. If there is not enough water on the rail, the contact is starving and a full liquid film does not develop.

8.3.2.3 Effect of speed

It is well known, that coefficient of adhesion is strongly affected by the train velocity even for dry contact. For the purpose of the effective application of tractive effort, equations predicting the maximum adhesion, i.e., coefficient of friction, as a function of train velocity were developed based on experimentation. These equations describe a curve that is actually the envelope of peaks of adhesion characteristics for different velocities. Well, known equation by Curtius and Kniffler adopted by Deutsche Bahn AG, equation used by the French National Railways (SNCF) and equation developed for high-speed Shinkansen vehicles are summarized in Table 8.3. The equations are available for dry and wet conditions, which usually means the best and the worst case, so the upper and lower limit of peak adhesion.

The comparison of these empirical equations is shown in graph in Fig. 8.7. The curves for wet conditions predict lower adhesion, nevertheless, the differences between the individual predictions are large. It should be noted that the classic Curtius and Kniffler equation is based on adhesion measurement using a measuring train, Ohyama's equation was determined using a large-sized high-speed rolling contact testing machine with targeted application of water to the contact. It is evident, that the slope of the curves is the same for dry and wet contact, so the effect of velocity is not necessarily associated with a hydrodynamic action of water.

8.3.2.4 Effect of surface roughness

Lubricant-film thickness is often interpreted in relation to surface roughness using a lubrication parameter Λ [33]. This parameter is defined by Eq. (8.4) as the ratio of the corresponding minimum film thickness $h_{\min}$ to the composite surface roughness Rq of the rail and wheel. Although the parameter does not take into account, e.g., deformation of surface asperities in the contact, it can be used for an estimation of the

Table 8.3 Empirical equations for the prediction of the coefficient of adhesion as a function of train velocity.

	Dry	Wet
Deutsche Bahn AG (Curtius–Kniffler)	$\mu = \dfrac{7.5}{v+44} + 0.161$	$\mu = \dfrac{7.5}{v+44} + 0.13$
French National Railways (SNCF)	$\mu = 0.33 \cdot \dfrac{8+0.1v}{8+0.2v}$	$\mu = 0.24 \cdot \dfrac{8+0.1v}{8+0.2v}$
Japanese Railway (Ohyama)	—	$\mu = \dfrac{13.6}{v+85}$

Wheel-rail interface under extreme conditions

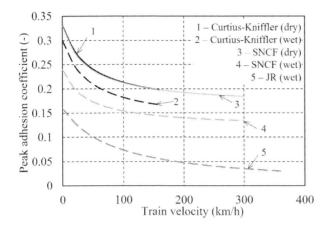

Fig. 8.7 Comparison of empirical equations for the prediction of the coefficient of adhesion as a function of train velocity. Credit: Original artwork.

lubricating regime according to Fig. 8.5. The figure also shows that for rough surfaces in the mixed regime a part of the load is carried by the contact between surface asperities where boundary lubrication occurs, and a part by hydrodynamic action of fluid film. This is one of the most common assumptions in the theoretical solution of the problem [34].

$$\Lambda = \frac{h_{min}}{\sqrt{Rq_w^2 + Rq_r^2}} \qquad (8.4)$$

It is well known that increasing surface roughness moves the lubrication regime toward a boundary lubrication. The effect of surface roughness on wheel-rail adhesion has been extensively studied at RTRI, Japan [35–39]. Experimental results confirm that in the water-lubricated contact the maximum traction coefficient increases with surfaces roughness, as described in Fig. 8.8 [35]. Fig. 8.9 shows results of the theoretical analysis with EHL model and stochastic distribution of the surface asperity heights using Greenwood–Williamson's model [38]. It is evident and not surprising that the adhesion coefficient increases with an increase of the contact pressure, but the effect of the contact pressure (i.e., the axle load) depends on the surface roughness. For the surfaces with low roughness, the adhesion coefficient slightly increases with the pressure, but for rough surfaces, it decreases significantly. Water film thickness increases with surface roughness, but the effect nearly disappears at high contact pressure. It can be concluded that the effect of contact pressure on the adhesion of the wheel and rail depends on the surface roughness and the running speed of a vehicle [38].

8.3.2.5 Effect of water temperature

Water temperature has been identified as a parameter strongly influencing the adhesion coefficient under water-lubricated conditions, as indicated by the experimental results in Fig. 8.10. A rise in the water temperature causes an increase in the

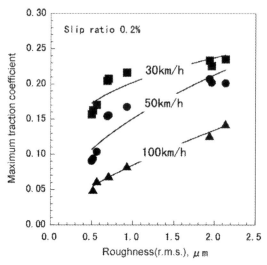

Fig. 8.8 Relationship between the maximum traction coefficient and surface roughness, from Chen et al. [35].
Credit: Reprinted from H. Chen, T. Ban, M. Ishida, T. Nakahara, Experimental investigation of influential factors on adhesion between wheel and rail under wet conditions, Wear 265 (9–10) (2008) 1504–1511, https://doi.org/10.1016/j.wear.2008.02.034, copyright (2008), with permission from Elsevier.

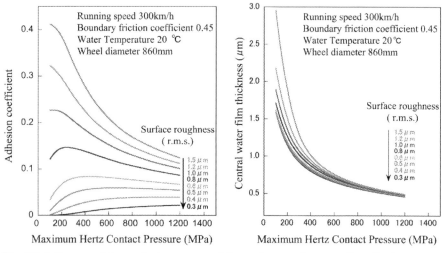

Fig. 8.9 Influences of the contact pressure and the surface roughness on the adhesion coefficient and water film thickness, from Chen et al. [38].
Credit: Reprinted from H. Chen, A. Namura, M. Ishida, T. Nakahara, Influence of axle load on wheel/rail adhesion under wet conditions in consideration of running speed and surface roughness, Wear 366–367 (2016) 303–309, https://doi.org/10.1016/j.wear.2016.05.012, copyright (2016), with permission from Elsevier.

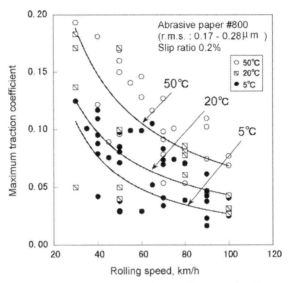

Fig. 8.10 Relationship between maximum traction coefficient and rolling speed at different water temperatures, from Chen et al. [35].
Credit: Reprinted from Wear, H. Chen, T. Ban, M. Ishida, T. Nakahara, Experimental investigation of influential factors on adhesion between wheel and rail under wet conditions, Wear 265 (9–10) (2008) 1504–1511, https://doi.org/10.1016/j.wear.2008.02.034, copyright (2008), with permission from Elsevier.

coefficient of adhesion [35]. Such experimental results are consistent with the theoretical analyses, where the change in water viscosity is considered as the main cause of the effect. On the other hand, recent analysis indicates that the effect is strong also in the boundary regime, where the water viscosity is of less importance. In this regime, the effect of temperature cannot simply be related to the lubrication parameter Λ [39]. The above-mentioned research suggests that under the wet conditions the coefficient of adhesion is maintainable at a relatively high level by raising water temperature or increasing the surface roughness.

8.3.3 Humidity and oxidation

The action of humid air, its temperature and surface oxidation are connected and it is not possible to clearly separate the influence of individual factors. Iron oxides are the most important contaminants occurring naturally in the wheel-rail interface [40,41]. The general designation "iron oxides" includes chemical compounds with different content in Fe cations, oxygen, hydroxyl and water. In situ measurements of rail head oxides have shown that five types of oxides are typically present on a rail head: magnetite (Fe_3O_4), hematite (Fe_2O_3), geothite (a-FeOOH), lepidocrocite (g-FeOOH) and akaganeite (b-FeOOHCl) [42]. Each oxide has different effect on the wheel-rail adhesion; however, usually, several oxides act together. Magnetite, known as "black oxide," has a tendency to decrease the friction while hematite increases the friction.

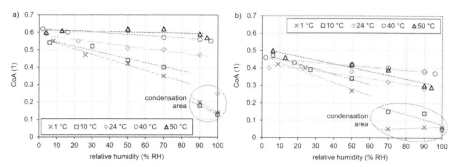

Fig. 8.11 Effect of relative humidity and temperature on coefficient of adhesion for clean disc (A) and contaminated disc (B), from Galas et al. [45].
Credit: Reprinted from Tribology International, R. Galas, M. Omasta, L. Shi, H. Ding, W. Wang, I. Krupka, M. Hartl, The low adhesion problem: the effect of environmental conditions on adhesion in rolling-sliding contact, Tribol. Int. 151 (2020), 106521, https://doi.org/10.1016/j.triboint.2020.106521, copyright (2020), with permission from Elsevier.

It is believed that under high humidity the normal atmospheric oxidation is inhibited because of the effect of water molecules in the air [43]. If the surface is covered with a thick layer of hydrates (rust), the friction becomes independent of the presence of liquid water [25,26].

Daytime evolution of relative humidity and temperature have a substantial impact on the adhesion coefficient. It has been shown that the coefficient of adhesion reduces from 0.55 to 0.22 with increasing relative humidity, while the effect of temperature is rather negligible [16]. Similar findings were reported from pin-on-disc experiments [24,25,40,44]. The effect of temperature and relative humidity on the coefficient of adhesion in the rolling-sliding contact is shown in Fig. 8.11 [45]. The first graph represents "clean" disc conditions, while the second was determined for the disc contaminated with dry leaf-based layer. In both cases, there is a clear trend in the decrease in adhesion with increasing humidity and decreasing temperature. This abrupt drop in coefficient of adhesion occurs as a result of the softening of the leaf layer due to the small amount of condensation water. Based on the results, the coefficient of adhesion can be predicted using Eqs. (8.5) and (8.6) for the clean and contaminated contact, respectively [45].

$$\mu_{clean} = 0.5461 + 0.003029 \cdot T - 0.002844 \cdot RH - 3.346 \cdot 10^{-5} \cdot T^2 + 5.401 \cdot 10^{-5} \cdot T \cdot RH \quad (8.5)$$

$$\mu_{contaminated} = 0.4153 + 0.001587 \cdot T - 0.001662 \cdot RH \quad (8.6)$$

8.3.3.1 Salt environment

Low-lying coastal areas are prone to seawater flooding. The ongoing climate changes are responsible for increasing the exposure of coastal infrastructure to coastal flooding. Railway infrastructure may be significantly flooded as the result of extremely high tides, coastal storm surges and tsunami waves, causing seawater to spill onto land. Even without such extreme weather phenomena, the salty environment affects tribology of the wheel-rail interface. Generally, salt/water solution increases adhesion level above that for wet conditions. On the other hand, the presence of salt on rail head affects the formation of oxides. Experience from Japan Railways shows that a small amount of β-FeOOH can be found specifically in salty-environment tunnels [46]. This ferric oxyhydroxide provides a low coefficient of friction, so the roll-slip phenomenon can be expected. Increased levels of oxidation together with wet rails may cause significant low adhesion problems [6]. Deposition of saltwater is an important factor initiating corrosion and rail track degradation [47].

8.3.4 Extreme temperatures

8.3.4.1 Extremely high temperature

With changes in the global climate, severe weather events such as heatwaves are increasingly common. Temperature affects the tribological processes by influencing the properties of the contacting surfaces. It should be noted that ambient temperature is only one of the temperature-related effects. The temperature of the steel rails in direct sunlight can be more than 20 °C above ambient air temperature. On the other hand, the temperate required to soften wheel material starts at 200–250 °C. Even much important is the flash temperature acting in the contact under the high pressure and shear stress [48]. Extremely high ambient temperatures cause the critical deterioration processes on railway infrastructure like rail buckling. Heat-related deterioration processes also apply to the wheels [49,50]. However, there are no reports on the risk associated with the adhesion or deterioration specifically in the wheel-rail interface.

8.3.4.2 Sub-zero temperatures

Resilience of railway transport to extremely low temperatures is of increasing interest and importance in various parts of the world. The railway industry faces new challenges as the rail network is extended to cold alpine regions. For example, the high-speed Harbin–Dalian passenger dedicated line in Northeast China is operated at the mean air temperature in the coldest month of about −13.5 to −17.5 °C and an extreme as low as −36.5 °C [51]. The operating temperature at the high-elevation Qinghai-Tibet railway even falls to −45 °C [52]. Other challenges are related to cold waves that occur more frequently and with greater intensity. An example is the February 2021 North American cold wave that brought record-low temperatures to a significant portion of Canada and the United States.

Extremely low temperature influences the behavior of a wide range of track and train equipment depending on the type of asset. From a wheel-rail interface perspective, the effect of sub-zero temperatures on adhesion and wear is important. There are only a few studies dealing with the effects. The effect of sub-zero temperature and various contaminants on adhesion coefficient and wear was recently studied using

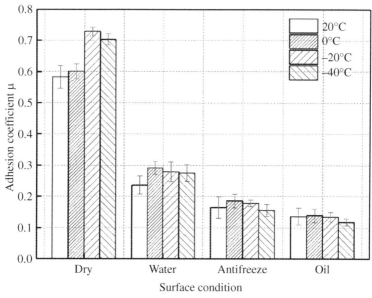

Fig. 8.12 Comparison of adhesion coefficient under different surface conditions, from Shi et al. [55].
Credit: Reprinted from L. Shi, L. Ma, J. Guo, Q. Liu, Z. Zhou, W. Wang, Influence of low temperature environment on the adhesion characteristics of wheel-rail contact, Tribol. Int. 127 (2018) 59–68, https://doi.org/10.1016/j.triboint.2018.05.037, copyright (2018), with permission from Elsevier.

a small-scale twin-disc test rig with environmental chamber and simulations at Southwest Jiaotong University, China [53–56]. The adhesion measurement shows that the adhesion coefficient at extremely low temperature is higher than that at room temperature, as shown in Fig. 8.12. The effect of contaminants is roughly the same as at room temperature, however, the adhesion recovery process is significantly slower [55]. The low temperature has a significantly negative effect on wear and rolling contact fatigue [54,56].

8.3.5 Solid particles at the wheel-rail contact

Solid particles contaminating the wheel-rail interface naturally originate from the railway superstructure such as ballast, concrete sleepers, etc. The internal source of solid particles is also sanding used by traction rail vehicles to restore adhesion during traction and braking under degraded adhesion conditions. These sources of contamination are common and usually not problematic. Extreme conditions can be caused by particles windblown from the surrounding environment. This is in the context of the spread of arid areas due to the climate change and the expansion of rail networks in the deserts of Middle and Far East and North Africa regions [57].

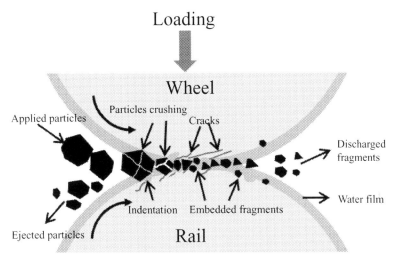

Fig. 8.13 Schematic of the behavior of solid particle in the twin-disc contact, from Wang [58]. Credit: Reprinted from C. Wang, L. Shi, H. Ding, W. Wang, R. Galas, J. Guo, Q. Liu, Z. Zhou, M. Omasta, Adhesion and damage characteristics of wheel/rail using different mineral particles as adhesion enhancers, Wear (2021) 203796, https://doi.org/10.1016/j.wear.2021.203796, copyright (2021), with permission from Elsevier.

When the sand or other mineral particles laying on the rail come into the contact, several phases occur, as illustrated for twin-disc configuration in Fig. 8.13 [58]. At first, some of the particles are ejected from the contact and the others are drawn into the converging gap between the contacting bodies. During this phase, the particles are crushed into small fragments by the normal force. An indentation of the wheel and rail appears, and some fragments can be embedded into the surfaces. As the particles enter the contact region, they are further crushed and subjected to shear stress due to the tangential force. Some particles undergo shear failure and some cause plowing of contacting bodies. If the particles are embedded in the surface, we are talking about two-body abrasion and if they are free to roll or slide, we refer to this as a three-body abrasive wear. The consequence of the particles action is increased wear. The results of the twin-disc tests simulating the effect of a desert environment show 1.4–2.2 times greater wear rate under sand conditions compared to "clean" conditions. The presence of sand also causes strong ratcheting and crack formations, whereby the particles can be pushed into existing cracks and assist their propagation [59]. Since the hard particles are usually embedded in the softer wheel material, the two-body abrasion leads to 2.5 times more wear on the harder rail material [60]. The authors of this study suggest that the initial crushing of large particles causes surface damage that is insignificant compared to the roughness of the rail surface and in-contact wear mechanism [60]. On the other hand, recent research has indicated that the sand fragments used instead of regular sand cause much milder negative effects such as wear, surface damage and plastic deformation. This is thanks to the absence of the crushing process [61].

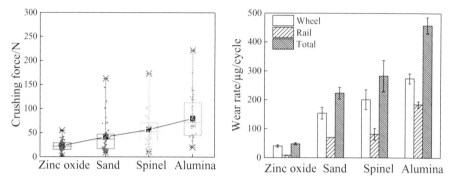

Fig. 8.14 Crushing stress and wear rate for four types of mineral particles, from Wang [58]. Credit: Reprinted from C. Wang, L. Shi, H. Ding, W. Wang, R. Galas, J. Guo, Q. Liu, Z. Zhou, M. Omasta, Adhesion and damage characteristics of wheel/rail using different mineral particles as adhesion enhancers, Wear (2021) 203796, https://doi.org/10.1016/j.wear.2021.203796, copyright (2021), with permission from Elsevier.

It is generally accepted that the wear rate caused by the contamination particles is related to their hardness. The graph in Fig. 8.14 compares crushing stress and wear rate measured at twin-disc set-up with wheel and rail steel for four types of mineral particles showing that there is an obvious correlation [58].

Although silica sand is commonly applied to the contact to restore adhesion under degraded adhesion conditions, under dry and light-wet conditions solid particles may act as a solid lubricant within the meaning of a reduction of adhesion [60,62–64]. This is clear from Fig. 8.15, where the coefficient of friction is compared for different wheel steels with and without sand application during the long-term test [59]. The effect of single application is indicated in Fig. 8.16 under wet conditions. There is a drop in the coefficient of adhesion immediately after the sand application, and this drop is more pronounced for larger applied quantity [64]. This is probably because the individual sand fragments slide over each other in the contact. A thick layer of dry sand has a relatively low shear strength.

A significant negative effect of the contact contamination is a risk of loss of the function of the track circuit used to detect the position of a train on the track. As studied by many researchers [62,63,65], too large a quantity of sand between the wheels and the rails leads to an electrical insulation of the contact. Several models have been developed to predict a critical sand density above which isolation would occur [63,65,66]. The most conservative value found is 7.5 g/m [67]. The level of insulation also depends on the size of sand particles, whereby larger particles seem to cause less electrical insulation [62]. This may be because smaller particles have a stronger tendency to stick to the surface and to build up an insulating layer that resists being pushed out of the contact. For a similar reason, a critical sand density was found to be lower when water or leaves are added into the interface [62,63,65].

Wheel-rail interface under extreme conditions

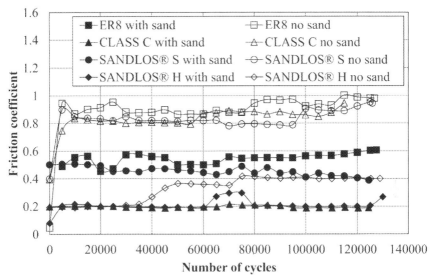

Fig. 8.15 Friction coefficients of the tests with and without sand., from Faccoli et al. [59].
Credit: Reprinted from M. Faccoli, C. Petrogalli, M. Lancini, A. Ghidini, A. Mazzù, Effect of desert sand on wear and rolling contact fatigue behaviour of various railway wheel steels, Wear 396-397 (2018) 146–161, https://doi.org/10.1016/j.wear.2017.05.012, copyright (2018), with permission from Elsevier.

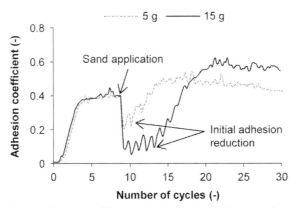

Fig. 8.16 The effect of adhesion coefficient after the single application of sand particles under light-wet conditions, from Omasta [64].
Credit: Reprinted from M. Omasta, M. Machatka, D. Smejkal, M. Hartl, I. Křupka, Influence of sanding parameters on adhesion recovery in contaminated wheel–rail
contact, Wear 322-323 (2015) 218–225, https://doi.org/10.1016/j.wear.2014.11.017, copyright (2015), with permission from Elsevier.

References

[1] C. Logston, G. Itami, Locomotive friction-creep studies, J. Eng. Ind. 102 (3) (1980) 275–281, https://doi.org/10.1115/1.3183865.

[2] M. Harmon, J. Santa, J. Jaramillo, A. Toro, A. Beagles, R. Lewis, Evaluation of the coefficient of friction of rail in the field and laboratory using several devices, Tribol. Mater. Surf. Interfaces 14 (2) (2020) 119–129, https://doi.org/10.1080/17515831.2020.1712111.

[3] R. Lewis, U. Olofsson, Mapping rail wear regimes and transitions, Wear 257 (7–8) (2004) 721–729, https://doi.org/10.1016/j.wear.2004.03.019.

[4] R. Lewis, R. Dwyer-Joyce, U. Olofsson, J. Pombo, J. Ambrósio, M. Pereira, C. Ariaudo, N. Kuka, Mapping railway wheel material wear mechanisms and transitions, Proc. Inst. Mech. Eng. F J. Rail Rapid Transit 224 (3) (2010) 125–137, https://doi.org/10.1243/09544097JRRT328.

[5] A. Meierhofer, G. Trummer, C. Bernsteiner, K. Six, Vehicle tests showing how the weather in autumn influences the wheel–rail traction characteristics, Proc. Inst. Mech. Eng. F J. Rail Rapid Transit 234 (4) (2020) 426–435, https://doi.org/10.1177/0954409719863643.

[6] C. Hardwick, R. Lewis, U. Olofsson, Low adhesion due to oxide formation in the presence of salt, Proc. Inst. Mech. Eng. F J. Rail Rapid Transit 228 (8) (2013) 887–897, https://doi.org/10.1177/0954409713495666.

[7] S. Abbasi, L. Olander, C. Larsson, U. Olofsson, A. Jansson, U. Sellgren, A field test study of airborne wear particles from a running regional train, Proc. Inst. Mech. Eng. F J. Rail Rapid Transit 226 (1) (2011) 95–109, https://doi.org/10.1177/0954409711408774.

[8] D. Jaffe, J. Putz, G. Hof, G. Hof, J. Hee, D. Lommers-Johnson, F. Gabela, J. Fry, B. Ayres, M. Kelp, M. Minsk, Diesel particulate matter and coal dust from trains in the Columbia River Gorge, Washington State, USA, Atmos. Pollut. Res. 6 (6) (2015) 946–952, https://doi.org/10.1016/j.apr.2015.04.004.

[9] P. Chinowsky, J. Helman, S. Gulati, J. Neumann, J. Martinich, Impacts of climate change on operation of the US rail network, Transp. Policy 75 (2019) 183–191, https://doi.org/10.1016/j.tranpol.2017.05.007.

[10] M. Koetse, P. Rietveld, The impact of climate change and weather on transport: an overview of empirical findings, Transp. Res. Part D: Transp. Environ. 14 (3) (2009) 205–221, https://doi.org/10.1016/j.trd.2008.12.004.

[11] P. Leviäkangas, A. Tuominen, R. Molarius, J. Ludvigsen, et al., Extreme weather impacts on transport systems, in: EU_EWENT—Extreme Weather impacts on European Networks of Transport, 2014.

[12] I. Stipanovic, H. ter Maat, A. Hartmann, G. Dewulf, Risk assessment of climate change impacts on railway, in: Engineering Project Organization Conference, 2013.

[13] J. Thornes, B. Davis, Mitigating the impact of weather and climate on railway operations in the UK, in: ASME/IEEE Joint Railroad Conference, ASME, 2002, pp. 29–38, https://doi.org/10.1109/RRCON.2002.1000089.

[14] E. Forero-Ortiz, E. Martínez-Gomariz, M. Cañas Porcuna, L. Locatelli, B. Russo, Flood risk assessment in an underground railway system under the impact of climate change—a case study of the Barcelona metro, Sustainability 12 (13) (2020), https://doi.org/10.3390/su12135291.

[15] K. Nagase, A study of adhesion between the rails and running wheels on main lines: results of investigations by slipping adhesion test bogie, Proc. Inst. Mech. Eng. F J. Rail Rapid Transit 203 (1) (1989) 33–43, https://doi.org/10.1243/PIME_PROC_1989_203_206_02.

[16] T. Beagley, C. Pritchard, Wheel/rail adhesion—the overriding influence of water, Wear 35 (2) (1975) 299–313, https://doi.org/10.1016/0043-1648(75)90078-2.
[17] Adhesion Research Group, Investigation Into the Effect of Moisture on Rail Adhesion (T1042), RSSB, 2014. https://www.sparkrail.org/Lists/Records/DispForm.aspx?ID=11364.
[18] M. Godet, The third-body approach: a mechanical view of wear, Wear 100 (1–3) (1984) 437–452, https://doi.org/10.1016/0043-1648(84)90025-5.
[19] S. Descartes, C. Desrayaud, E. Niccolini, Y. Berthier, Presence and role of the third body in a wheel–rail contact, Wear 258 (7–8) (2005) 1081–1090, https://doi.org/10.1016/j.wear.2004.03.068.
[20] G. Trummer, L. Buckley-Johnstone, P. Voltr, A. Meierhofer, R. Lewis, K. Six, Wheel-rail creep force model for predicting water induced low adhesion phenomena, Tribol. Int. 109 (2017) 409–415, https://doi.org/10.1016/j.triboint.2016.12.056.
[21] P. Cann, The "leaves on the line" problem—a study of leaf residue film formation and lubricity under laboratory test conditions, Tribol. Lett. 24 (2) (2006) 151–158, https://doi.org/10.1007/s11249-006-9152-2.
[22] H. Chen, T. Furuya, S. Fukagai, S. Saga, J. Ikoma, K. Kimura, J. Suzumura, Wheel slip/slide and low adhesion caused by fallen leaves, Wear 446–447 (2020), https://doi.org/10.1016/j.wear.2020.203187.
[23] K. Ishizaka, S. Lewis, D. Hammond, R. Lewis, Chemistry of black leaf films synthesised using rail steels and their influence on the low friction mechanism, RSC Adv. 8 (57) (2018) 32506–32521, https://doi.org/10.1039/C8RA06080K.
[24] U. Olofsson, K. Sundvall, Influence of leaf, humidity and applied lubrication on friction in the wheel-rail contact: pin-on-disc experiments, Proc. Inst. Mech. Eng. F J. Rail Rapid Transit 218 (3) (2004) 235–242, https://doi.org/10.1243/0954409042389364.
[25] Y. Zhu, U. Olofsson, H. Chen, Friction between wheel and rail: a pin-on-disc study of environmental conditions and iron oxides, Tribol. Lett. 52 (2) (2013) 327–339, https://doi.org/10.1007/s11249-013-0220-0.
[26] Y. Zhu, U. Olofsson, R. Nilsson, A field test study of leaf contamination on railhead surfaces, Proc. Inst. Mech. Eng. F J. Rail Rapid Transit 228 (1) (2013) 71–84, https://doi.org/10.1177/0954409712464860.
[27] M. Marian, M. Bartz, S. Wartzack, A. Rosenkranz, Non-dimensional groups, film thickness equations and correction factors for elastohydrodynamic lubrication: a review, Lubricants 8 (10) (2020), https://doi.org/10.3390/lubricants8100095.
[28] M. Esfahanian, B. Hamrock, Fluid-film lubrication regimes revisited, Tribol. Trans. 34 (4) (1991) 628–632, https://doi.org/10.1080/10402009108982081.
[29] B. Hamrock, D. Dowson, Isothermal elastohydrodynamic lubrication of point contacts: part III—fully flooded results, J. Lubr. Technol. 99 (2) (1977) 264–275, https://doi.org/10.1115/1.3453074.
[30] H. Chen, A. Yoshimura, T. Ohyama, Numerical analysis for the influence of water film on adhesion between rail and wheel, Proc. Inst. Mech. Eng. J. J. Eng. Tribol. 212 (5) (1998) 359–368, https://doi.org/10.1243/1350650981542173.
[31] B. Wu, M. Chen, T. Wu, Z. Wen, X. Jin, A simple 3-dimensional model to analyse wheel-rail adhesion under wet condition, Lubr. Sci. 30 (2) (2018) 45–55, https://doi.org/10.1002/ls.1402.
[32] K. Bett, J. Cappi, Effect of pressure on the viscosity of water, Nature 207 (4997) (1965) 620–621, https://doi.org/10.1038/207620a0.
[33] T. Tallian, On competing failure modes in rolling contact, ASLE Trans. 10 (4) (1967) 418–439, https://doi.org/10.1080/05698196708972201.

[34] D. Kvarda, R. Galas, M. Omasta, L. Shi, H. Ding, W. Wang, I. Krupka, M. Hartl, Asperity-based model for prediction of traction in water-contaminated wheel-rail contact, Tribol. Int. 157 (2021), https://doi.org/10.1016/j.triboint.2021.106900.

[35] H. Chen, T. Ban, M. Ishida, T. Nakahara, Experimental investigation of influential factors on adhesion between wheel and rail under wet conditions, Wear 265 (9–10) (2008) 1504–1511, https://doi.org/10.1016/j.wear.2008.02.034.

[36] H. Chen, M. Ishida, T. Nakahara, Analysis of adhesion under wet conditions for three-dimensional contact considering surface roughness, Wear 258 (7–8) (2005) 1209–1216, https://doi.org/10.1016/j.wear.2004.03.031.

[37] H. Chen, T. Ban, M. Ishida, T. Nakahara, Effect of water temperature on the adhesion between rail and wheel, Proc. Inst. Mech. Eng. J. J. Eng. Tribol. 220 (7) (2006) 571–579, https://doi.org/10.1243/13506501JET75.

[38] H. Chen, A. Namura, M. Ishida, T. Nakahara, Influence of axle load on wheel/rail adhesion under wet conditions in consideration of running speed and surface roughness, Wear 366–367 (2016) 303–309, https://doi.org/10.1016/j.wear.2016.05.012.

[39] H. Chen, H. Tanimoto, Experimental observation of temperature and surface roughness effects on wheel/rail adhesion in wet conditions, Int. J. Rail Transp. 6 (2) (2017) 101–112, https://doi.org/10.1080/23248378.2017.1415772.

[40] Y. Zhu, X. Chen, W. Wang, H. Yang, A study on iron oxides and surface roughness in dry and wet wheel–rail contacts, Wear 328–329 (2015) 241–248, https://doi.org/10.1016/j.wear.2015.02.025.

[41] Y. Zhu, The influence of iron oxides on wheel–rail contact: a literature review, Proc. Inst. Mech. Eng. F J. Rail Rapid Transit 232 (3) (2017) 734–743, https://doi.org/10.1177/0954409716689187.

[42] J. Suzumura, Y. Sone, A. Ishizaki, D. Yamashita, Y. Nakajima, M. Ishida, In situ X-ray analytical study on the alteration process of iron oxide layers at the railhead surface while under railway traffic, Wear 271 (1–2) (2011) 47–53, https://doi.org/10.1016/j.wear.2010.10.054.

[43] E. Leheup, R. Pendlebury, Unlubricated reciprocating wear of stainless steel with an interfacial air flow, Wear 142 (2) (1991) 351–372, https://doi.org/10.1016/0043-1648(91)90174-S.

[44] Y. Zhu, Y. Lyu, U. Olofsson, Mapping the friction between railway wheels and rails focusing on environmental conditions, Wear 324-325 (2015) 122–128, https://doi.org/10.1016/j.wear.2014.12.028.

[45] R. Galas, M. Omasta, L. Shi, H. Ding, W. Wang, I. Krupka, M. Hartl, The low adhesion problem: the effect of environmental conditions on adhesion in rolling-sliding contact, Tribol. Int. 151 (2020), https://doi.org/10.1016/j.triboint.2020.106521.

[46] M. Ishida, Managing the wheel–rail interface: the Japanese experience, in: Wheel–Rail Interface Handbook, Elsevier, 2009, pp. 701–758, https://doi.org/10.1533/9781845696788.2.701.

[47] W. Xu, B. Zhang, Y. Deng, Z. Wang, Q. Jiang, L. Yang, J. Zhang, Corrosion of rail tracks and their protection, Corros. Rev. 39 (1) (2021) 1–13, https://doi.org/10.1515/corrrev-2020-0069.

[48] C. Tomberger, P. Dietmaier, W. Sextro, K. Six, Friction in wheel–rail contact: a model comprising interfacial fluids, surface roughness and temperature, Wear 271 (1–2) (2011) 2–12, https://doi.org/10.1016/j.wear.2010.10.025.

[49] S. Dedmon, Effect of temperature on the performance of railroad wheels, Proc. Inst. Mech. Eng. F J. Rail Rapid Transit 231 (7) (2017) 786–793, https://doi.org/10.1177/0954409717712072.

[50] T. Kato, H. Kato, T. Makino, Effect of elevated temperature on shelling property of railway wheel steel, Wear 366–367 (2016) 359–367, https://doi.org/10.1016/j.wear.2016.04.015.

[51] H. Liu, F. Niu, Y. Niu, J. Xu, T. Wang, Effect of structures and sunny–shady slopes on thermal characteristics of subgrade along the Harbin–Dalian passenger dedicated line in Northeast China, Cold Reg. Sci. Technol. 123 (2016) 14–21, https://doi.org/10.1016/j.coldregions.2015.11.007.

[52] Y. Wang, H. Zhou, Y. Shi, B. Feng, Mechanical properties and fracture toughness of rail steels and thermite welds at low temperature, Int. J. Miner. Metall. Mater. 19 (5) (2012) 409–420, https://doi.org/10.1007/s12613-012-0572-8.

[53] R. Luo, W. Teng, X. Wu, H. Shi, J. Zeng, Dynamics simulation of a high-speed railway car operating in low-temperature environments with stochastic parameters, Veh. Syst. Dyn. 58 (12) (2020) 1914–1934, https://doi.org/10.1080/00423114.2019.1662922.

[54] L. Ma, L. Shi, J. Guo, Q. Liu, W. Wang, On the wear and damage characteristics of rail material under low temperature environment condition, Wear 394–395 (2018) 149–158, https://doi.org/10.1016/j.wear.2017.10.011.

[55] L. Shi, L. Ma, J. Guo, Q. Liu, Z. Zhou, W. Wang, Influence of low temperature environment on the adhesion characteristics of wheel-rail contact, Tribol. Int. 127 (2018) 59–68, https://doi.org/10.1016/j.triboint.2018.05.037.

[56] L. Zhou, Y. Hu, H. Ding, Q. Liu, J. Guo, W. Wang, Experimental study on the wear and damage of wheel-rail steels under alternating temperature conditions, Wear (2021), https://doi.org/10.1016/j.wear.2021.203829.

[57] L. Bruno, M. Horvat, L. Raffaele, Windblown sand along railway infrastructures: a review of challenges and mitigation measures, J. Wind Eng. Ind. Aerodyn. 177 (2018) 340–365, https://doi.org/10.1016/j.jweia.2018.04.021.

[58] C. Wang, L. Shi, H. Ding, W. Wang, R. Galas, J. Guo, Q. Liu, Z. Zhou, M. Omasta, Adhesion and damage characteristics of wheel/rail using different mineral particles as adhesion enhancers, Wear (2021), https://doi.org/10.1016/j.wear.2021.203796.

[59] M. Faccoli, C. Petrogalli, M. Lancini, A. Ghidini, A. Mazzù, Effect of desert sand on wear and rolling contact fatigue behaviour of various railway wheel steels, Wear 396-397 (2018) 146–161, https://doi.org/10.1016/j.wear.2017.05.012.

[60] D. Grieve, R. Dwyer-Joyce, J. Beynon, Abrasive wear of railway track by solid contaminants, Proc. Inst. Mech. Eng. F J. Rail Rapid Transit 215 (3) (2005) 193–205, https://doi.org/10.1243/0954409011531512.

[61] L. Shi, C. Wang, D. Ding, D. Kvarda, R. Galas, M. Omasta, W. Wang, Q. Liu, M. Hartl, Laboratory investigation on the particle-size effects in railway sanding: comparisons between standard sand and its micro fragments, Tribol. Int. 146 (2020) 106259, https://doi.org/10.1016/j.triboint.2020.106259.

[62] O. Arias-Cuevas, Z. Li, R. Lewis, Investigating the lubricity and electrical insulation caused by sanding in dry wheel–rail contacts, Tribol. Lett. 37 (3) (2010) 623–635, https://doi.org/10.1007/s11249-009-9560-1.

[63] R. Lewis, R. Dwyer-Joyce, J. Lewis, Disc machine study of contact isolation during railway track sanding, Proc. Inst. Mech. Eng. F J. Rail Rapid Transit 217 (1) (2003) 11–24, https://doi.org/10.1243/095440903762727311.

[64] M. Omasta, M. Machatka, D. Smejkal, M. Hartl, I. Křupka, Influence of sanding parameters on adhesion recovery in contaminated wheel–rail contact, Wear 322-323 (2015) 218–225, https://doi.org/10.1016/j.wear.2014.11.017.

[65] R. Lewis, J. Masing, Static wheel/rail contact isolation due to track contamination, Proc. Inst. Mech. Eng. F J. Rail Rapid Transit 220 (1) (2006) 43–53, https://doi.org/10.1243/095440906X77919.

[66] S. Descartes, M. Renouf, N. Fillot, B. Gautier, A. Descamps, Y. Berthier, P. Demanche, A new mechanical–electrical approach to the wheel-rail contact, Wear 265 (9–10) (2008) 1408–1416, https://doi.org/10.1016/j.wear.2008.02.040.

[67] W. Skipper, A. Chalisey, R. Lewis, A review of railway sanding system research: wheel/rail isolation, damage, and particle application, Proc. Inst. Mech. Eng. F J. Rail Rapid Transit 234 (6) (2020) 567–583, https://doi.org/10.1177/0954409719851634.

Train and track interactions

Wanming Zhai[a], Shengyang Zhu[a], and Stefano Bruni[b]
[a]State Key Laboratory of Traction Power, Southwest Jiaotong University, Chengdu, People's Republic of China, [b]Department of Mechanical Engineering, Politecnico di Milano, Milano, Italy

9.1 Introduction: An overview of train and track interactions

Railways are major transportation arteries in many countries and play a very important role in social and economic development. Rolling stocks (including locomotives, passenger cars, and freight wagons) and tracks are essential components of the railway system [1]. For a long time, studies on railway vehicle dynamics and track structure vibration were carried out separately. This resulted in two relatively independent disciplines, i.e., vehicle dynamics [2,3] and track dynamics [4,5]. Thanks to the long-term studies and practices by railway scientists all over the world, the theories of vehicle dynamics and track dynamics are becoming more and more complete. Significant achievements of these systematic studies have been reported from many fields, including vehicle dynamics modeling, wheel-rail contact geometry, model of creep forces, vehicle hunting stability, track dynamics modeling, vibration characteristics of track structure, track loading, and deformation characteristics, etc. These research outputs have laid the theoretical foundation in revealing and understanding vehicle dynamics performances and track dynamics characteristics. These outputs have also played a pivotal role in the development of modern railway transportation systems.

The rapid development of modern railway transportation, especially the dramatic increases of operating speed, hauling mass, and transportation density, makes the dynamics problems of the railway vehicle and track systems more prominent and complicated. On the one hand, the dynamic effect of trains with a higher speed and heavier wheel-axle load on track structures was intensified, which directly affected the fatigue life of the infrastructure and thus increased the cost for maintenance and repairs. On the other hand, the track geometry deformation and the subgrade settlement of the railway lines were increased, which led to increasing detrimental effects on the dynamic behavior of running trains. In particular, the vibrations and impacts that resulted from the damaged and worn wheel-rail interface have become even more severe, which can lead to severe safety problems of the wheel-rail system. To sum up, train and track interactions have been a quite important research topic in railway community under the background of higher train running speed and greater wheel-axle load. Therefore, it is necessary to conduct dedicated and in-depth research on the dynamic interaction between train and track systems.

Given this situation, Zhai proposed the new concept of "vehicle–track coupled dynamics" from the perspective of an overall integrated vehicle and track system in the late 1980s, and the theory was put into practice in the early 1990s [1,6–8]. Many nonlinear factors and time-varying parameters are involved in the established large-scale dynamics system, which can be efficiently solved by a fast integration algorithm, namely Zhai method [9]. Subsequently, railway researchers had carried out a large number of theoretical and practical research projects in the field of vehicle-track coupled dynamics over the past 30 years. The theoretical and analysis framework of the vehicle-track coupled dynamics has matured in the early 21st century after a long time of developments [8], and the theory has gradually become a fundamental method for the dynamic analysis of vehicle-track interactions. Recently, there were many improvements and extensions developed in the model, i.e.,

- Vehicle-track interactions with various infrastructures, such as vehicle-track-bridge coupled dynamics [10–12]; vehicle-track-tunnel coupled dynamics [13]; vehicle-turnout interactions [14–16]; vehicle-track-subgrade coupled dynamics [17].
- Refined modeling of vehicle and track subsystems according to the analysis emphasis and structural features, such as the modeling of heavy-haul coupler and draft gear system and its application to train-track longitudinal dynamics [18]; establishment of a locomotive-track coupled dynamics model with gear transmissions [19,20]; consideration of shear effect and in-plane vibrations of the slab track system [21,22]; implementation of a nonlinear and fractional derivative viscoelastic model for rail pads [23]; integration of car-body [24] and wheelset [25] flexibility into the vehicle-track system.
- Environmental vibration induced by vehicle operation [26–28], vibration attenuation [29,30], and acoustic radiation [31] for urban rail transit.
- Stochastic analysis of vehicle-track coupled system in the time-space domain based on the probability density evolution method [32].
- Wheel-rail wear and rolling contact fatigue [33–36], wheel-rail noise [37,38], and wheel-rail contact algorithm [39].
- Vehicle-track long-term performance degradation, such as the influence of infrastructural defects on dynamic responses of vehicle-track system and running safety [40–43]; failure mechanisms of structural damage and its effect on dynamic performance of ballastless tracks [44–48]; and the tooth root crack propagation of gear transmission system [49].
- Vehicle-track interactions under various environment conditions, such as the effect of extreme high temperature [50–53]; extreme low temperature, ice and snow [54,55]; and so on [56–67].

In this chapter, modeling strategies of train and track interactions will be introduced in Section 9.2, based on which two specific examples regarding long-term performance degradation in high-speed railways will be illustrated in Sections 9.3 and 9.4, respectively. Finally, Section 9.5 presents a review of the effects of weather conditions on train-track interaction.

9.2 Models of train and track interactions

9.2.1 Vehicle-track coupled dynamics model

This modeling details of the vehicle-track spatially coupled dynamics model for ballasted track and nonballasted track are presented in this section, which are established based on the vehicle-track coupled dynamics theory [1].

9.2.1.1 Vehicle-ballasted track coupled dynamics model

Figs. 9.1–9.3 show a vehicle-track spatially coupled dynamics model for a typical passenger vehicle and a ballasted track. In the vehicle model, the car body is supported on two double-axle bogies through the secondary suspensions, and the bogie frames are linked with the wheelsets through the primary suspensions. The primary and the secondary suspensions are represented by three-dimensional spring-damper elements, and yaw dampers and antiroll springs are considered in the secondary suspensions. Additionally, the lateral clearances between the car body and the stop-blocks on

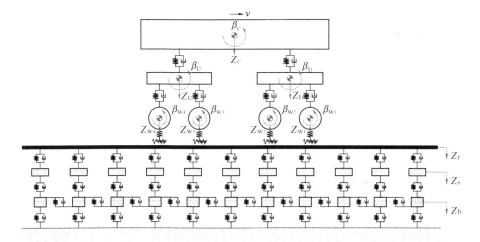

Fig. 9.1 Passenger vehicle-track spatially coupled model (side view).
Modified after W. Zhai, K. Wang, C. Cai, Fundamentals of vehicle–track coupled dynamics, Veh. Syst. Dyn. 47 (11) (2009) 1349–1376.

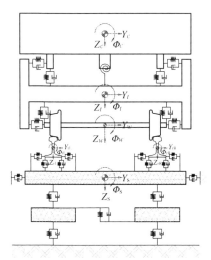

Fig. 9.2 Passenger vehicle-track spatially coupled model (end view) [8].

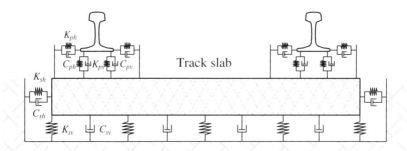

Fig. 9.3 Vertical and lateral dynamics model for the typical slab track [8].

the bogie frames are also modeled in the secondary suspensions. The vehicle is assumed to move along the track with a constant traveling speed. Totally 35 degree of freedoms (DOFs) are taken into account in the passenger vehicle model including the vertical displacement Z, the lateral displacement Y, the roll angle Φ, the yaw angle ψ, and the pitch angle β.

A typical ballasted track model is mainly composed of rails, rail pads, sleepers, ballast, and subgrade. The continuous Euler or Timoshenko beams are usually adopted to model the rails that are discretely supported at rail-sleeper junctions. The vertical, lateral, and torsional vibrations are all considered in the rail dynamics model, and lateral springs and dampers are employed to represent the shear elasticity and damping properties of the fastening system. The sleeper is regarded as a rigid body with considering the vertical, lateral, and roll angle motions. Similarly, the elasticity and damping property between the sleeper and the ballast in lateral direction are represented by lateral springs and dampers. The vertical elasticity and damping of rail fastening system, ballast, and subgrade are simulated by three layers of springs and dampers. A five-parameter model of ballast under each rail supporting point is employed [68] with only accounting for the vertical motion of the ballast mass. To reflect the continuity and the coupling effects of interlocking ballast granules, a couple of shear stiffness and damping are introduced between adjacent ballast masses in the ballast model.

9.2.1.2 Vehicle-ballastless track coupled dynamics model

If the analyzed subject is a nonballasted track, the above-ballasted track dynamics model can be replaced correspondingly. Generally, there are mainly four typical nonballasted track structures, including the long-sleeper embedded track (with the sleepers firmly poured into an in-situ concrete track slab), the elastic supporting-block track (with the elastically encased supporting-blocks between rails and slabs), the slab track (with the slabs supported by elastic cement asphalt mortar layer), and the floating slab track (with the floating slab supported by steel/rubber springs). Here, we only introduce the dynamics model of the slab track for example, more details of the other nonballasted track models can be referred to the monograph [1].

The slab track widely used in high-speed railways mainly consists of rails, rail pads, concrete slabs, cement asphalt mortar (CAM) layer and concrete base. In the slab track model, the rails could be modeled as the Euler or Timoshenko beams with taking the vertical, lateral, and torsional motions into account, while the track slabs could be treated as elastic rectangle plates supported on viscoelastic foundation, as shown in Fig. 9.3, where K_{sv} and C_{sv} denote the vertical stiffness and damping of the CAM layer, respectively, and K_{sh} and C_{sh} are the lateral stiffness and damping of the CAM layer, respectively. Due to the bending stiffness of the slab is very large in its lateral direction, it is sufficient to simply consider the rigid mode for the lateral vibration of the slab.

9.2.2 Train-track coupled dynamics model

Based on the established vehicle-track coupled dynamics model, a three-dimensional dynamics model for train-track coupled system will be introduced in this section. Here, the internal dynamic interactions between adjacent vehicles in a train are considered, which is especially important for heavy-haul trains with long formation under the traction/braking conditions where the longitudinal shock and impact are much more intensified, and could even threaten the train operation safety.

9.2.2.1 Physical model

Fig. 9.4 presents the main components of the train and the track system as well as their interrelationship [18]. The adjacent vehicles of a train would most likely exhibit different running statuses when the traction/braking control is implemented, or when the

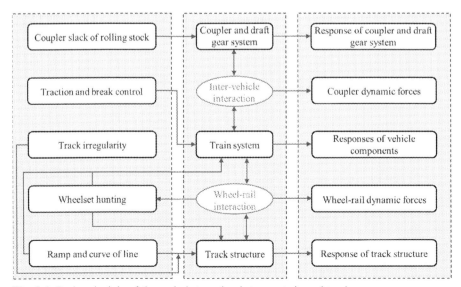

Fig. 9.4 Basic principle of dynamic interaction between train and track.

train passes through curves and ramps. In these cases, various postures of the coupler and draft gear systems, such as a large tilt angle of the coupler and/or a mismatch of coupling heights, would be inevitably emerged, leading to unfavorable dynamic inter-vehicle interactions. The induced in-train forces could further transmit to the wheelsets through the bogie suspension systems, which will probably affect the wheel-rail contact behaviors and the track structure vibrations. In turn, the track vibrations will have a certain effect on the vehicle dynamic responses, and finally influence the working conditions of the coupler and the draft gear packages. Therefore, the coupler and draft gear subsystem, the train subsystem and the track subsystem are closely interconnected through their coupling effects.

For modeling of the train-track spatially coupled dynamics system [18], a modular modeling method has been proposed which is illustrated in Fig. 9.5. As can be seen, the simulation system mainly includes the train control submodule, the train submodule, the wheel-rail contact submodule, and the railway line submodule. The train and the track subsystems are coupled through the wheel-rail nonlinear contact. The train control module provides the control strategies (e.g., determination of driving torque, running resistance, traction, and braking characteristics) for the train operation under different railway line conditions.

The train-track coupled dynamics model could be established based on the basic principle of the train-track interactions and the modular modeling method. Firstly, the structure characteristics of train and track need to be understood well so as to simulate their dynamic behavior. The basic components of a certain type of railway vehicles are usually based on well-established schemes. For example, a locomotive is generally composed of the carbody, bogie frames, traction motors, wheelsets, and suspension systems, while a freight wagon usually consists of the carbody, side frames, bolsters, wheelsets, and suspension systems. The ballasted track structure that widely used in the heavy-haul railways also has a standard structure. Fig. 9.6 shows a

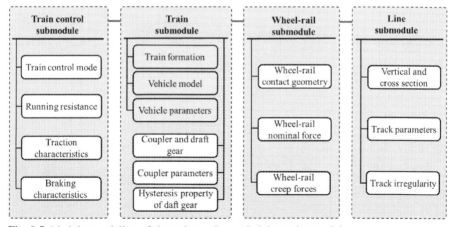

Fig. 9.5 Modular modeling of the train-track coupled dynamics model.

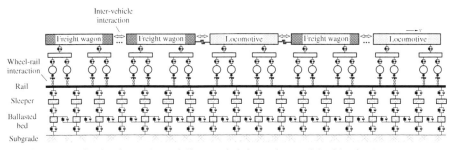

Fig. 9.6 Heavy-haul train-track spatially coupled dynamics model (side view).

schematic diagram of a typical heavy-haul train-track coupled dynamics model, where the locomotives are distributed in different positions of the train (with the distributed power mode), and the track is the commonly used ballasted track. Specially, the forces such as the traction force, the braking force, the coupler force and the running resistance, and the possible large creepage between the wheel and the rail are all considered in the vehicle model.

9.2.2.2 Interactions of intervehicles

The nonrigid automatic couplers with a self-centering ability are commonly adopted in heavy-haul locomotives and freight wagons in China. There is usually a free clearance between two couplers connected with each other, which permits them to move relatively in the vertical direction. The coupler could also sway within a small angle range relative to its draft key in the horizontal direction.

In the longitudinal train dynamics, a pair of connected couplers is usually considered as massless entity. Fig. 9.7 shows a typical simplified mathematic model [2], where K_{buf} and C_{buf} represent the stiffness and the damping of the draft gear, respectively; K_s denotes the structural stiffness of the carbody; K_{buf} and C_{buf} have the nonlinear characteristics, rather than two constants. For convenience in simulation, the stiffness and the damping of the draft gear are usually described by the hysteresis curves. Note that the intervehicle interactions in the 3-D train model will be affected

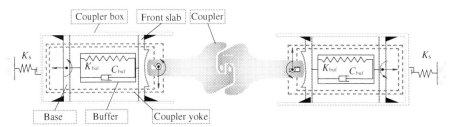

Fig. 9.7 Dynamic model of heavy-haul coupler and draft gear system.

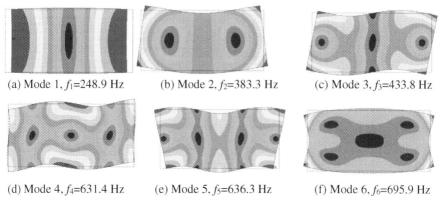

Fig. 9.8 In-plane natural frequencies and mode shapes of a track slab.

comprehensively by the coupler forces in the longitudinal, lateral, and vertical directions, which are detailed in the monograph [1].

9.2.2.3 In-plane vibration of track slab

In the majority of previously proposed train-slab track dynamics models, flexural vibration of the track slab is of main concern, while the longitudinal and lateral (in-plane) vibrations are often neglected or assumed as rigid behavior. Actually, apart from the wheel-rail vertical interaction forces, the longitudinal and lateral counterparts can also be transmitted to the track structures through the fasteners, which will inevitably induce in-plane vibration of the substructures [21]. The first six in-plane natural frequencies and modes of a track slab are illustrated in Fig. 9.8.

9.3 Vehicle-track interaction due to differential subgrade settlement

The use of ballastless tracks on soil subgrade is a major concern in view of the post-construction settlement, especially the differential subgrade settlement. Since soft soil ground with large compressibility and low permeability is widely distributed in China [69], the differential settlement on soft soil subgrade along the high-speed railway lines is a common occurrence due to the nonuniform soil properties, variation of groundwater-level and other defects of subgrade [70,71]. Relevant monitoring has shown that in some regions, subgrade settlements can be quite serious, sharply changing the original gradient of the railway line so that speed restrictions has to be implemented. Hence, the control of differential settlement becomes a key subject in the high-speed railways.

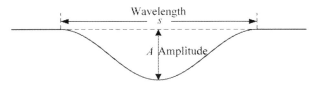

Fig. 9.9 Sketch of the differential subgrade settlement.

9.3.1 Modeling of differential subgrade settlement

The differential subgrade settlement is introduced into the vehicle-slab track dynamics model as a given boundary condition [40]. The Heaviside function is employed here to describe the compression-only contact state between the track and the subgrade as follows:

$$H_{di}(t) = \begin{cases} 1, & \text{for } Z_b(x_i, t) > Z_0(x_i) \text{ (supported)} \\ 0, & \text{for } Z_b(x_i, t) \leq Z_0(x_i) \text{ (unsupported)} \end{cases} \quad (9.1)$$

where $Z_b(x, t)$ is the vertical displacement of the concrete base; and $Z_0(x)$ is the differential subgrade settlement. The force of the subgrade spring can be written as

$$F_{bi}(t) = H_{di}\{k_b[Z_b(x_i, t) - Z_0(x_i)] + c_{bi}\dot{Z}_b(x_i, t)\} \quad (9.2)$$

Commonly, the cosine curve is adopted as the typical settlement pattern, as shown in Fig. 9.9. The expression for the subgrade displacement $Z_0(x_i)$ is defined as

$$Z_0(x_i) = \begin{cases} \dfrac{A}{2}\left[1 + \cos\left(\dfrac{2\pi}{s} \cdot \left(x_i - \dfrac{l}{2}\right)\right)\right] & \text{for } \dfrac{l}{2} - \dfrac{s}{2} < x_i < \dfrac{l}{2} + \dfrac{s}{2} \\ 0, & \text{others} \end{cases} \quad (9.3)$$

9.3.2 Effect of subgrade settlement on dynamic response

Different combinations of settlement wavelength and amplitude are discussed with a train speed of 300 km/h as an example. To highlight the influence of the differential settlement, the track irregularities is omitted in the simulation.

9.3.2.1 Influence of settlement wavelength

Fig. 9.10 shows the calculated effect of the settlement wavelength on the dynamic responses of the coupled system. The settlement wavelength has been increased from 10 to 30 m, and the settlement amplitude is set as 10 mm. It can be seen in Fig. 9.10A that the settlement wavelength of 10 m exacerbates the vehicle vibration the most, and the carbody acceleration gradually decreases with the increase in settlement wavelength. Fig. 9.10B shows that the initial track-subgrade contact force for the settlement

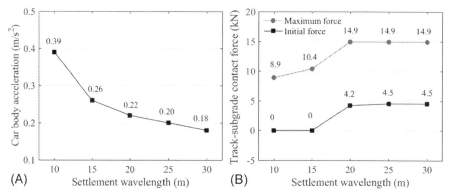

Fig. 9.10 Effect of settlement wavelength on system dynamic responses: (A) carbody acceleration; (B) track-subgrade contact force [1].

wavelength of 10 m is 0, which indicates the presence of the void. As the wavelength increases, the unsupported areas are gradually eliminated, and the difference between the vehicle-induced contact force and the settlement-induced one stabilizes around 10 kN.

9.3.2.2 Influence of settlement amplitude

The influence of the settlement amplitude is illustrated in Fig. 9.11 [1], with a constant settlement wavelength of 15 m. Fig. 9.11A shows almost linear growth of the carbody acceleration as the settlement amplitude increases from 10 to 30 mm, which implies that the ride comfort will be likely to exceed the critical limit if the settlement amplitude continues to worsen. It can be seen from Fig. 9.11B that the initial track-subgrade contact force is always nil, it indicates that there is always a separation between the track and the subgrade. Due to the high rigidity of track structure, the gap between the track and the subgrade widens with the increasing settlement amplitude, which leads to significant attenuation of contact forces induced by the passing train, as depicted by the red dotted line.

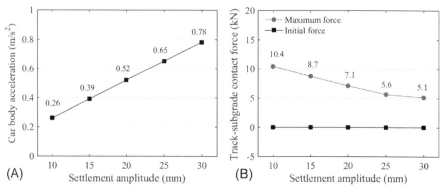

Fig. 9.11 Effect of settlement amplitude on system dynamic responses: (A) carbody acceleration; (B) track-subgrade contact force.

9.4 Vehicle-track interaction due to polygonal wheel under traction condition

Traction or braking operations are usually applied to the powered cars or locomotives for acceleration and deceleration, especially in mountain high-speed railways that have complex and large gradient sections. Under such a circumstance, the wheel-rail longitudinal interactions may become more intense, which could induce longitudinal impact on the track structure. By introducing system longitudinal vibrations and implementing the interface interactions into the classical vehicle-track system, a novel vehicle-slab track vertical-longitudinal coupled dynamics model is established in this section, and investigated under the combined excitation of polygonal wheel wear and traction conditions.

9.4.1 Modeling of polygonal wheel wear

Wheel polygon refers to periodic radial deviation formed by wheel nonuniform wear. Fig. 9.12 shows the field measured wheel radial profile of a Chinese high-speed train. In fact, most wheel polygons present the main harmonic components within the 40th order; components with higher-order usually have a small amplitude. Considering the first N-order dominant components and ignoring the component A_0, the wheel polygon can be represented by the following equation using Fourier series:

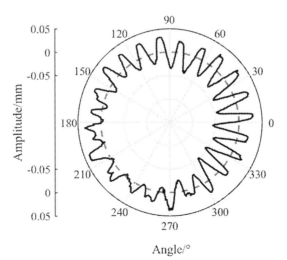

Fig. 9.12 Field measured wheel polygon wear.
Reprinted from J. Luo, S. Zhu, W. Zhai, Theoretical modelling of a vehicle-slab track coupled dynamics system considering longitudinal vibrations and interface interactions, Veh. Syst. Dyn. (2020), https://doi.org/10.1080/00423114.2020.1751860. Copyright 2022, with permission from Taylor & Francis.

$$Z_0(t) = \sum_{i=1}^{N} A_i \sin\left[i\left(\frac{v}{R}\right)t + \varphi_i\right] \tag{9.4}$$

where i is the order of the wheel polygon, A_i is the amplitude of the ith harmonic wave, φ_i is the corresponding phase, A_i and φ_i can be obtained by performing discrete Fourier transform of the measured radial deviation of the wheel along its circumference.

9.4.2 Modeling of longitudinal nonlinear interactions of slab track

9.4.2.1 Longitudinal resistance model of the fastener

Essentially, the longitudinal resistance of the fasteners is a kind of friction force generated from the interface between the rail and rubber pad, which can be well characterized by the Dahl friction model [72], based on which the following explicit expressions for the longitudinal resistance of fasteners are presented [48].

$$\begin{cases} F_{Lf}(x) = F_{Lfs} & \text{for } x = x_s \\ F_{Lf}(x) = F_{Lm} - (F_{Lm} - F_{Lfs}) \cdot \exp\left(-\dfrac{\sigma_0 \cdot (x - x_s)}{F_{Lm}}\right) & \text{for } x > x_s \text{ or increasing } x \\ F_{Lf}(x) = -F_{Lm} + (F_{Lm} + F_{Lfs}) \cdot \exp\left(\dfrac{\sigma_0 \cdot (x - x_s)}{F_{Lm}}\right) & \text{for } x < x_s \text{ or decreasing } x \end{cases} \tag{9.5}$$

where (x_s, F_{Lfs}) is defined as a reference state and needs to be updated during the motion; F_{Lm} denotes the ultimate resistance force of the fasteners.

9.4.2.2 Discrete cohesive zone model

A series of nonlinear cohesive springs, whose constitutive law follows the bilinear discrete cohesive zone model, is adopted to simulate the interfacial bond-slip behavior between the track slab and CA mortar [48]. An evolution of the damage variable D, proposed by Camanho and Davia [73], is used to describe the linear softening process and track the extent of damage accumulated at the interface. For the jth ($j = 1, 2, \ldots$) time step t_j, it can be expressed as

$$D(t_j) = \frac{\delta_{tf}\left(\delta_{tmax}(t_j) - \delta_{t0}\right)}{\delta_{tmax}(t_j)\left(\delta_{tf} - \delta_{t0}\right)} \tag{9.6}$$

where δ_{tmax} refers to the maximum effective displacement attained during the loading history; δ_{t0} denotes the effective displacement at damage initiation; and δ_{tf} denotes the effective displacement at complete failure. Considering the irreversibility of damage, the nonlinear longitudinal cohesive force can be calculated by

$$F_t(\delta_t(t_j)) = \begin{cases} k_t \delta_t(t_j) & \delta_{tmax}(t_j) < \delta_{t0} \\ k_t(1-D(t_j))\delta_t(t_j) & \delta_{t0} \leq \delta_{tmax}(t_j) < \delta_{tf} \\ 0 & \delta_{tmax}(t_j) \geq \delta_{tf} \end{cases} \quad (9.7)$$

where k_t is the initial stiffness of the cohesive spring without damage, defined by $k_t = F_{tm}/\delta_{t0}$; F_{tm} is the interfacial shear strength.

9.4.3 Characteristics of system dynamic responses

Two representative excitations originated from the wheel and rail are selected for the dynamics system involving the polygonal wheel wear and track random irregularities. The track irregularities characterized by wavelengths between 2 and 200 m, and a measured sample of the polygonal wheel wear are adopted in the simulation. The motor car is accelerated from 0 to 250 km/h considering the following two cases: Case I, the driving torque and track irregularities are introduced into the vehicle-slab track system, denoted by C_1; Case II, the driving torque, track irregularities, and polygonal wheel wear are introduced into the vehicle-slab track system, denoted by C_2. The mechanical parameters of the cohesive zone model and longitudinal resistance model are reported in Ref. [48].

The wheel-rail interaction forces are illustrated in Fig. 9.13. As can be seen in Fig. 9.13A, the wheel-rail vertical force increases gradually during the vehicle speed-up process, and this phenomenon becomes especially obvious for C_2. This is due to the fact that the polygonal wheel wear is essentially a kind of periodic short wave irregularity, which will aggravate the wheel-rail impact to a great extent. When the vehicle speed is lower than 20 km/h, the wheel-rail creep force obtained in C_1 agrees well with that in C_2 and reaches the maximum value of 10.95 kN at 2.5 km/h. As the running speed continues to increase, the fluctuating range of the

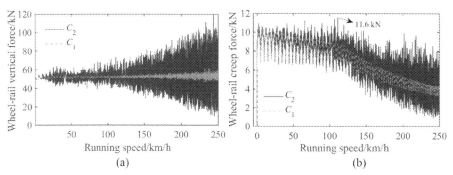

Fig. 9.13 Wheel-rail interaction forces in the time domain: (A) vertical forces; (B) creep forces. Reprinted from J. Luo, S. Zhu, W. Zhai, Theoretical modelling of a vehicle-slab track coupled dynamics system considering longitudinal vibrations and interface interactions, Veh. Syst. Dyn. (2020), https://doi.org/10.1080/00423114.2020.1751860. Copyright 2022, with permission from Taylor & Francis.

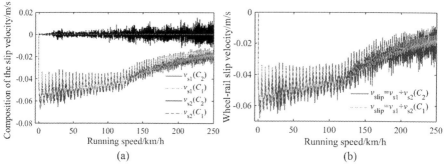

Fig. 9.14 Time histories of wheel-rail slip velocity: (A) composition; (B) superposition.
Reprinted from J. Luo, S. Zhu, W. Zhai, Theoretical modelling of a vehicle-slab track coupled dynamics system considering longitudinal vibrations and interface interactions, Veh. Syst. Dyn. (2020), https://doi.org/10.1080/00423114.2020.1751860. Copyright 2022, with permission from Taylor & Francis.

wheel-rail creep force in C_2 becomes much wider, but the maximum value (11.6 kN at around 110 km/h) is not markedly increased, as shown in Fig. 9.13B.

Time histories of the composition and superposition of the wheel-rail slip velocity are shown in Fig. 9.14, where v_{s1} represents the slip velocity induced by the driving torque and wheel longitudinal vibration, and v_{s2} is the one induced by the rail longitudinal vibration. It can be seen that the main differences of the wheel-rail slip velocity for the two cases lay in the v_{s2}. The participation of polygonal wheel wear results in more significant rail longitudinal vibration for higher running speed.

Fig. 9.15 illustrates the damage variable defined by Eq. (9.6) at the track slab-CA mortar interface for C_2 defined in Eq. (9.6), whereas no damage can be found for C_1.

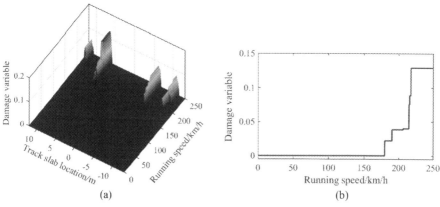

Fig. 9.15 Damage variable at the track slab-CA mortar interface for C_2: (A) time histories and distribution; (B) evolution of the maximum damage variable.
Reprinted from J. Luo, S. Zhu, W. Zhai, Theoretical modelling of a vehicle-slab track coupled dynamics system considering longitudinal vibrations and interface interactions, Veh. Syst. Dyn. (2020), https://doi.org/10.1080/00423114.2020.1751860. Copyright 2022, with permission from Taylor & Francis.

Note that the coordinate origin in Fig. 9.15A is fixed at the track slab beneath the centroid of the car body. It can be observed that the interface damage occurs near the wheel loading point with the maximum damage variable of 0.129, whose evolution process is displayed in Fig. 9.15B. As it can be seen, the interface damage initiates at 180 km/h, undergoes rapid propagation at around 215 km/h, and eventually maintains the maximum value after the running speed exceeds 218 km/h.

9.5 Vehicle-track interaction under extreme weather conditions

Extreme weather conditions strongly affect train-track interaction and may jeopardize the vehicles' running safety, increase the risk of failures in the running gear, accelerate wear and tear of the rolling stock and infrastructure. In this section, a review of the effects of weather conditions on train-track interaction is proposed, considering different types of extreme weather conditions.

9.5.1 Effects of extreme high temperature

Extreme high temperature has two main effects on vehicle dynamics and train-track interaction: on one hand, in a very hot and dry climate the adhesion coefficient between wheels and rails will be particularly high, increasing the risk of flange climb derailment; an examination of the effects of the friction coefficient on derailment can be found in [74,75]. On the other hand, high temperatures induce thermal deformation of the rails and in extreme cases may lead to the thermal buckling of the rails, resulting in increased wheel-rail interaction possibly leading to running safety issues. Methods to assess the effect of thermal loads on rail buckling can be found in [76,77], while the effect of a track buckle on train-track interaction and on the risk of derailment is specifically investigated in [50–53].

9.5.2 Effects of extreme low temperature, snow and ice

Extreme low temperature can also be harmful to train-track interaction. The direct effect of low temperatures on the dynamics of a railway vehicle on the track is not so important, although of course low temperatures will affect the level of adhesion between the wheels and the rails, see, e.g., [54]. The main effect of low temperatures is however to affect the toughness of steel material in the wheels which can then be subject to fragile failures. This is the reason why wheelset manufacturers have developed special steel grades suitable for very low temperatures.

Another effect of low temperatures is when this is accompanied by snow: in this case, snow aggregations can be formed in the vehicle underfloor, possibly impairing the functionality of some components in the running gear and requiring a special design of the running gear for cold climates, see, e.g., [55].

9.5.3 Effects of rainy weather and rail contamination from leaves

Rainy weather and a humid environment again have an influence on the level of adhesion between wheels and rails, in this case causing low adhesion with the risk of wheel slip or skid. Sanding represents the obvious remedy to this issue but is not free from problems. In particular, sand in the wheel/rail contact increases the electrical resistance of the contact area which is critical to the correct functioning of some widely adopted signaling systems [56]. A problem worth of attention is the effect of leaves from trees fallen on the rails. These form a hard layer on the rails, reducing friction and leading to station overruns and signals passed at danger [56]. A recent paper provides evidence derived from line tests of how humid weather conditions and the presence of fallen leaves influence wheel/rail contact conditions [57].

Moreover, the presence of water at wheel/rail interface may lead to accelerated surface damage due to rolling contact fatigue as water will get trapped in small surface cracks and will promote accelerated propagation of the cracks [58].

9.5.4 Effects of extreme wind

Strong wind is well known as a significant danger to the running safety of railway vehicles, as it may lead to overturning. The problem is dealt with at normative level by the European standard EN14067 where a combination of wind tunnel tests on a scaled model of the vehicle and of Multi-Body Systems simulation is proposed. A summary of problems and approaches to deal with railway vehicle aerodynamics and particularly with the effect of crosswinds is proposed in [59,60].

It is also worth of mention the problem of a train passing over a tall bridge. In this case, wind speed can be particularly high, due to the boundary layer effect being rapidly reduced with the height over the ground. If the expected maximum wind speed at the site is incompatible with the safe running of trains, wind barriers are used to mitigate the effect of crosswind. However, for flexible bridges such as cable-stayed or suspension bridges, wind screens negatively affect the aerodynamics of the bridge deck, leading to a design conflict where the maximization of protection of trains from the wind collides with the aerodynamic design of the bridge itself [78].

9.5.5 Effect of an earthquake

Earthquakes have an obvious effect on the dynamics of a railway vehicle running over a track and suitable modeling approaches have been developed [63] to analyze this running condition with the final aim of verifying the running safety of the vehicle and of designing mitigation measures both for the prederailment and postderailment phase. In these models, the effect of the earthquake is represented by a motion applied to the subgrade of the track, considering the stochastic properties of the seismic phenomenon.

Mitigation measures typically consist of designing a stopper to constrain the vehicle in the postderailment phase of its motion [64]. Recently, the use of brake disks as postderailment stoppers has also been proposed [65].

An alternative measure to mitigate the effect of an earthquake is to use a network of seismographic sensors to provide early warning of an incoming earthquake so that the train can be slowed down or even stopped before the earthquake reaches the train [79].

9.5.6 Effect of sand contamination

A final environmental effect worth being mentioned in this section is the contamination of ballast produced by the cumulation of sand over the track, representing a special case of ballast fouling. Sand grains tend to fill the interstices between ballast stones and also affect the friction between stones, resulting in increased stiffness of the layers supporting the sleepers. In this condition, increased dynamic effects caused by train-track interaction are to be expected and a considerable number of derailments have been reported in sand contaminated tracks [66]. To analyze the effect of sand contamination, a standard mathematical model of train-track interaction can be used, but the parameters of the track model need to be tuned based on measurements, to reflect the effect of ballast fouling. Tests to determine the change in ballast stiffness caused by sand contamination are described e.g. in [66,67].

9.6 Conclusions

This chapter has presented an overview of train and track interactions with focus on some representative improvements and extensions developed in the train/vehicle-track coupled dynamics model. Then, the theoretical framework for modeling of train/vehicle-track coupled systems has been introduced briefly with some new insights. To elucidate its engineering applications, the influences of long-term infrastructure degradation and various extreme weather conditions on train-track interactions have been illustrated and reviewed in detail.

Acknowledgments

This work was supported by the National Natural Science Foundation of China (Grant No. 11790283, No.51978587), the International Joint Laboratory on Railway Engineering System Dynamics, the 111 project (Grant No. B16041) and the EU H2020-MSCA-RISE-RISEN project (Grant No. 691135). The authors would like to thank their colleagues from the Southwest Jiaotong University and Politecnico di Milano for their help in preparing this chapter.

References

[1] W. Zhai, Vehicle–Track Coupled Dynamics: Theory and Applications, Springer, Singapore, 2020.
[2] V.K. Garg, R.V. Dukkipati, Dynamics of Railway Vehicle Systems, Academic Press Canada, Ontario, 1984.
[3] A.H. Wickens, Fundamentals of Rail Vehicle Dynamics: Guidance and Stability, Swets & Zeitlinger Publishers, Lisse, The Netherlands, 2003.
[4] K. Knothe, Gleisdynamik, Ernst & Sohn, Berlin, 2001.

[5] S.L. Lian, Track Dynamics, Tongji University Press, Shanghai, 2003 (in Chinese).
[6] W. Zhai, X. Sun, A detailed model for investigating vertical interaction between railway vehicle and track, Veh. Syst. Dyn. 23 (Suppl) (1994) 603–615.
[7] W. Zhai, C. Cai, S. Guo, Coupling model of vertical and lateral vehicle/track interactions, Veh. Syst. Dyn. 26 (1) (1996) 61–79.
[8] W. Zhai, K. Wang, C. Cai, Fundamentals of vehicle–track coupled dynamics, Veh. Syst. Dyn. 47 (11) (2009) 1349–1376.
[9] W. Zhai, Two simple fast integration methods for large-scale dynamic problems in engineering, Int. J. Numer. Methods Eng. 39 (1996) 4199–4214.
[10] T. Arvidsson, R. Karoumi, Train-bridge interaction—a review and discussion of key model parameters, Int. J. Rail Transp. 2 (3) (2014) 147–186.
[11] W. Zhai, Z. Han, Z. Chen, et al., Train-track-bridge dynamic interaction: a state-of-the-art review, Veh. Syst. Dyn. 57 (7) (2019) 984–1027.
[12] T. Arvidsson, A. Andersson, R. Karoumi, Train running safety on non-ballasted bridges, Int. J. Rail Transp. 7 (1) (2019) 1–22.
[13] S. Zhou, X. Zhang, H. Di, C. He, Metro train-track-tunnel-soil vertical dynamic interactions—semi-analytical approach, Veh. Syst. Dyn. 56 (12) (2019) 1945–1968.
[14] S. Alfi, S. Bruni, Mathematical modelling of train-turnout interaction, Veh. Syst. Dyn. 47 (5) (2009) 551–574.
[15] Z. Ren, Wheel/Rail Multi-Point Contacts and Vehicle-Turnout System Dynamic Interactions, Science Press, Beijing, 2014 (in Chinese).
[16] Z. Ren, S. Sun, W. Zhai, Study on lateral dynamic characteristics of vehicle/turnout system, Veh. Syst. Dyn. 43 (4) (2005) 285–303.
[17] L. Xu, Z. Yu, C. Shi, A matrix coupled model for vehicle-slab track-subgrade interactions at 3-D space, Soil Dyn. Earthq. Eng. 128 (2020), 105894.
[18] P. Liu, W. Zhai, K. Wang, Establishment and verification of three-dimensional dynamic model for heavy-haul train–track coupled system, Veh. Syst. Dyn. 54 (11) (2016) 1511–1537.
[19] Z. Chen, W. Zhai, K. Wang, Dynamic investigation of a locomotive with effect of gear transmission under tractive conditions, J. Sound Vib. 408 (2017) 220–233.
[20] T. Zhang, Z. Chen, W. Zhai, et al., Establishment and validation of a locomotive-track coupled spatial dynamics model considering dynamic effect of gear transmission, Mech. Syst. Signal Process. 119 (2019) 328–345.
[21] J. Luo, S. Zhu, W. Zhai, An advanced train-slab track spatially coupled dynamics model: theoretical methodologies and numerical applications, J. Sound Vib. 501 (2020), 116059.
[22] J. Yang, S. Zhu, W. Zhai, A novel dynamics model for railway ballastless track with medium-thick slabs, Appl. Math. Model. 78 (2020) 907–931.
[23] S. Zhu, C. Cai, P.D. Spanos, A nonlinear and fractional derivative viscoelastic model for rail pads in the dynamic analysis of coupled vehicle-slab track systems, J. Sound Vib. 335 (2015) 304–320.
[24] L. Ling, Q. Zhang, X. Xiao, et al., Integration of car-body flexibility into train-track coupling system dynamics analysis, Veh. Syst. Dyn. 56 (4) (2018) 485–505.
[25] J. Han, S. Zhong, X. Xiao, Z. Wen, G. Zhao, X. Jin, High-speed wheel/rail contact determining method with rotating flexible wheelset and validation under wheel polygon excitation, Veh. Syst. Dyn. 56 (8) (2018) 1233–1249.
[26] G. Kouroussis, O. Verlinden, C. Conti, On the interest of integrating vehicle dynamics for the ground propagation of vibrations: the case of urban railway traffic, Veh. Syst. Dyn. 48 (12) (2010) 1553–1571.
[27] W. Zhai, Z. He, X. Song, Prediction of high-speed train induced ground vibration based on train-track-ground system model, Earthq. Eng. Eng. Vib. 9 (4) (2010) 545–554.

[28] G. Kouroussis, D.P. Connolly, O. Verlinden, Railway-induced ground vibrations—a review of vehicle effects, Int. J. Rail Transp. 2 (2) (2014) 69–110.
[29] S. Zhu, J. Wang, C. Cai, et al., Development of a vibration attenuation track at low frequencies for urban rail transit, Comput. Aided Civ. Inf. Eng. 32 (2017) 713–726.
[30] J. Yang, S. Zhu, W. Zhai, et al., Prediction and mitigation of train-induced vibrations of large-scale building constructed on subway tunnel, Sci. Total Environ. 668 (2019) 485–499.
[31] X. Zhang, W. Zhai, Z. Chen, et al., Characteristic and mechanism of structural acoustic radiation for box girder bridge in urban rail transit, Sci. Total Environ. 627 (2018) 1303–1314.
[32] L. Xu, W. Zhai, A new model for temporal-spatial stochastic analysis of vehicle-track coupled systems, Veh. Syst. Dyn. 55 (3) (2017) 427–448.
[33] Z. Wen, X. Jin, X. Xiao, et al., Effect of a scratch on curved rail on initiation and evolution of plastic deformation induced rail corrugation, Int. J. Solids Struct. 45 (2008) 2077–2096.
[34] Y. Sun, Y. Guo, W. Zhai, Prediction of rail non-uniform wear—influence of track random irregularity, Wear 420 (2019) 235–244.
[35] X. Liu, W. Zhai, Analysis of vertical dynamic wheel/rail interaction caused by polygonal wheels on high-speed trains, Wear 314 (1–2) (2014) 282–290.
[36] F. Braghin, R. Lewis, R.S. Dwyer-Joyce, S. Bruni, A mathematical model to predict railway wheel profile evolution due to wear, Wear 261 (11–12) (2006) 1253–1264.
[37] T.X. Wu, D.J. Thompson, On the impact noise generation due to a wheel passing over rail joints, J. Sound Vib. 267 (3) (2003) 485–496.
[38] Z. Xu, W. Zhai, Mechanism of wheel/rail noise for rail transit, Noise Vib. Control 1 (2006) 52–54 (in Chinese).
[39] Y. Sun, Y. Guo, W. Zhai, A robust non-Hertzian contact method for wheel–rail normal contact analysis, Veh. Syst. Dyn. 56 (12) (2018) 1899–1921.
[40] Y. Guo, W. Zhai, Y. Sun, A mechanical model of vehicle–slab track coupled system with differential subgrade settlement, Struct. Eng. Mech. 66 (1) (2018) 15–25.
[41] Y. Sun, Y. Guo, Z. Chen, et al., Effect of differential ballast settlement on dynamic response of vehicle–track coupled systems, Int. J. Struct. Stab. Dyn. 18 (07) (2018) 1850091.
[42] Z. Chen, W. Zhai, Q. Yin, Analysis of structural stresses of tracks and vehicle dynamic responses in train–track–bridge system with pier settlement, Proc. Inst. Mech. Eng. F J. Rail Rapid Transit 232 (2) (2018) 421–434.
[43] A. Lundqvist, T. Dahlberg, Load impact on railway track due to unsupported sleepers, Proc. Inst. Mech. Eng. F J. Rail Rapid Transit 219 (2) (2005) 67–77.
[44] S. Zhu, C. Cai, Interface damage and its effect on vibrations of slab track under temperature and vehicle dynamic loads, Int. J. Non Linear Mech. 58 (2014) 222–232.
[45] S. Zhu, C. Cai, Stress intensity factors evolution for through-transverse crack in slab track system under vehicle dynamic load, Eng. Fail. Anal. 46 (46) (2014) 219–237.
[46] S. Zhu, Q. Fu, C. Cai, et al., Damage evolution and dynamic response of cement asphalt mortar layer of slab track under vehicle dynamic load, Sci. China Technol. Sci. 57 (10) (2014) 1883–1894.
[47] J. Luo, S. Zhu, W. Zhai, An efficient model for vehicle-slab track coupled dynamic analysis considering multiple slab cracks, Constr. Build. Mater. 215 (2019) 557–568.
[48] J. Luo, S. Zhu, W. Zhai, Theoretical modelling of a vehicle-slab track coupled dynamics system considering longitudinal vibrations and interface interactions, Veh. Syst. Dyn. (2020), https://doi.org/10.1080/00423114.2020.1751860.
[49] Z. Chen, W. Zhai, K. Wang, Vibration feature evolution of locomotive with tooth root crack propagation of gear transmission system, Mech. Syst. Signal Process. 115 (2019) 29–44.
[50] X. Xiao, X. Jin, Z. Wen, et al., Effect of tangent track buckle on vehicle derailment, Multibody Sys. Dyn. 25 (1) (2011) 1–41.

[51] X. Jin, X. Xiao, L. Ling, et al., Study on safety boundary for high-speed train running in severe environments, Int. J. Rail Transp. 1 (1–2) (2013) 87–108.
[52] L. Ling, X. Xiao, Y. Cao, L. Wu, Z. Wen, X. Jin, Numerical simulation of dynamical derailment of high-speed train using a 3D train-track mode, in: Civil-Comp Proceedings, vol. 104, 2014.
[53] S. Zhu, C. Cai, W. Zhai, Interface damage assessment of railway slab track based on reliability techniques and vehicle-track interactions, J. Transp. Eng. 142 (10) (2016) 04016041.
[54] Y. Lyu, E. Bergseth, U. Olofsson, The effect of Subzero temperature and snow on the tribology of wheel-rail contact, in: Civil-Comp Proceedings, vol. 110, 2016.
[55] J. Wang, G. Gao, Y. Zhang, K. He, J. Zhang, Anti-snow performance of snow shields designed for brake calipers of a high-speed train, Proc. Inst. Mech. Eng. F J. Rail Rapid Transit 233 (2) (2019) 121–140.
[56] K. Ishizaka, S.R. Lewis, R. Lewis, The low adhesion problem due to leaf contamination in the wheel/rail contact: bonding and low adhesion mechanisms, Wear 378–379 (2017) 183–197.
[57] A. Meierhofer, G. Trummer, C. Bernsteiner, et al., Vehicle tests showing how the weather in autumn influences the wheel–rail traction characteristics, Proc. Inst. Mech. Eng. F J. Rail Rapid Transit 234 (4) (2020) 426–435.
[58] D.I. Fletcher, P. Hyde, A. Kapoor, Investigating fluid penetration of rolling contact fatigue cracks in rails using a newly developed full-scale test facility, Proc. Inst. Mech. Eng. F J. Rail Rapid Transit 221 (1) (2007) 35–44.
[59] C. Baker, F. Cheli, A. Orellano, N. Paradot, C. Proppe, D. Rocchi, Cross-wind effects on road and rail vehicles, Veh. Syst. Dyn. 47 (8) (2009) 983–1022.
[60] F. Cheli, F. Ripamonti, D. Rocchi, G. Tomasini, Aerodynamic behaviour investigation of the new EMUV250 train to cross wind, J. Wind Eng. Ind. Aerodyn. 98 (4–5) (2010) 189–201.
[61] W. Zhai, J. Yang, Z. Li, et al., Dynamics of high-speed train in crosswinds based on an air-train-track interaction model, Wind Struct. 20 (2) (2015) 143–168.
[62] I.A. Ishak, M.S. Mat Ali, M.F. Mohd Yakub, et al., Effect of crosswinds on aerodynamic characteristics around a generic train model, Int. J. Rail Transp. 7 (1) (2019) 23–54.
[63] L. Xu, W. Zhai, Stochastic analysis model for vehicle-track coupled systems subject to earthquakes and track random irregularities, J. Sound Vib. 407 (2017) 209–225.
[64] H. Sunami, T. Morimura, Y. Terumichi, M. Adachi, Model for analysis of bogie frame motion under derailment conditions based on full-scale running tests, Multibody Sys. Dyn. 27 (3) (2012) 321–349.
[65] X. Wu, M. Chi, H. Gao, et al., Post-derailment dynamic behavior of railway vehicles travelling on a railway bridge during an earthquake, Proc. Inst. Mech. Eng. F J. Rail Rapid Transit 230 (2) (2016) 418–439.
[66] A.R.T. Kian, J.A. Zakeri, J. Sadeghi, Experimental investigation of effects of sand contamination on strain modulus of railway ballast, Geomech. Eng. 14 (6) (2018) 563–570.
[67] D. Ionescu, D.T. Fedele, M.R. Trounce, J. Petrolito, Deformation and degradation characteristics of sand-contaminated railway ballast, in: Civil-Comp Proceedings, 2016.
[68] W. Zhai, K. Wang, J. Lin, Modelling and experiment of railway ballast vibrations, J. Sound Vib. 270 (4–5) (2004) 673–683.
[69] B. Wang, Subgrade and Track Engineering of High-Speed Railway, Tongji University Press, Shanghai, 2015 (in Chinese).
[70] R. Chen, J. Chen, X. Zhao, et al., Cumulative settlement of track subgrade in high-speed railway under varying water levels, Int. J. Rail Transp. 2 (4) (2014) 205–220.

[71] B. Olivier, D.P. Connolly, P.A. Costa, et al., The effect of embankment on high speed rail ground vibrations, Int. J. Rail Transp. 4 (4) (2016) 229–246.

[72] J. Luo, Z. Zeng, A novel algorithm for longitudinal track-bridge interactions considering loading history and using a verified mechanical model of fasteners, Eng. Struct. 183 (2019) 52–68.

[73] P.P. Camanho, C.G. Davila, Mixed-Mode Decohesion Finite Elements for the Simulation of Delamination in Composite Materials, NASA/TM-2002-2117372002, 2002, pp. 1–37.

[74] F. Braghin, S. Bruni, G. Diana, Experimental and numerical investigation on the derailment of a railway wheelset with solid axle, Veh. Syst. Dyn. 44 (4) (2006) 305–325.

[75] N. Wilson, H. Wu, A. Klopp, et al., Railway vehicle derailment and prevention, in: Handbook of Railway Vehicle Dynamics, second ed., Taylor & Francis, 2020, pp. 373–414.

[76] G.P. Pucillo, Thermal buckling and post-buckling behaviour of continuous welded rail track, Veh. Syst. Dyn. 54 (12) (2016) 1785–1807.

[77] O. Mirza, S. Kaewunruen, C. Dinha, E. Pervanic, Numerical investigation into thermal load responses of railway transom bridge, Eng. Fail. Anal. 60 (2016) 280–295.

[78] G. Diana, D. Rocchi, M. Belloli, Wind tunnel: a fundamental tool for long-span bridge design, Struct. Infrastruct. Eng. 11 (4) (2015) 533–555.

[79] S. Yamamoto, N. Iwata, Development of seismograph equipped with improved algorithm for earthquake early warning, Q. Rep. RTRI 58 (1) (2017) 65–69.

Approaches for weigh-in-motion and wheel defect detection of railway vehicles

10

Araliya Mosleh, Pedro Aires Montenegro, Pedro Alves Costa, and Rui Calçada
CONSTRUCT—LESE, Faculty of Engineering (FEUP), University of Porto, Porto, Portugal

10.1 Introduction

In recent years, the safety of rolling stock and its maintenance's economic consequences have been a significant concern for the railway industry. Excessive loads on the infrastructure lead to increased wear of the track, which enhances the maintenance costs. On the other hand, with the warning system, the maintenance team can use the sensor information to plan and perform wheel replacement before serious damages to the wheels. In this context, a system for weighing in motion and detecting wheel defects of rolling stock is of great importance [1,2].

The installation of dynamic weighing systems and wheel defect detections in railway lines evaluates the actual loads transmitted by the train into the track, as well as characterizes the rolling stock conditions. Such a monitoring system enables infrastructure managers to determine whether the loads applied from train to rail comply with fundamental standards. Additional charges may apply in some cases where the load is often exceeded. This study is critical for unbalanced freight trains, as they damage infrastructure on the one hand and increase the risk of derailment [3].

In recent decades, two approaches have been proposed to determine the dynamic load and detect wheel defect, namely onboard and wayside measurements. The first approach is seldom used due to cost and maintenance issues, as it requires the installation of sensors on all wheels [4,5]. The second method, which is applied in this study infers the loads imposed by the train from the dynamic response of infrastructure [6,7]. Moreover, wayside measurement methods are also an ideal solution to detect wheel defects, as the condition of all wheels is assessed during the passage of a train [8–11].

The development of efficient weigh-in-motion (WIM) and wheel defect detection systems with high-precision estimation methods of track measurements are one of the main topics that attract the attention of the railway industry and scientific researchers [12]. Two independent WIM systems were tested by O'Brien et al. [13], and the accuracy of the results were compared. An approach to determine axle and wheel loads of railway vehicles at high vehicle speeds was developed by Allotta et al. [14]. The proposed algorithm is able to estimate loads of the wheel by considering the vertical forces on the sleepers. Different monitoring schemes and experimental setups have been proposed for weighing in motion.

Zhou et al. [15] developed a vehicle-track system coupling dynamic model and analyzed the sensitivity of different sensor layouts for different wheel flat conditions. This algorithm determines the exact moment of wheel flat impact occurrence by jointly analyzing multisensor signals. Liu and Ni [16] developed a Fiber Bragg Grating (FGB)-based track-side wheel condition monitoring method to detect wheel tread defects. The track-side system with more than 20 Fiber Bragg Grating sensors installed at the rail allows measuring the bending moments of rail under a defective wheel. An algorithm is developed to identify potential wheel tread defects for using the online-monitored rail responses. The results show that by utilizing the proposed method to process the monitoring data, all the defects were identified and the results were in good agreement with those from the static inspection of the wheelsets.

In order to estimate train weight and detect wheel defects, postprocessing is needed, because the load measured during the passage of the train contains dynamic components caused by the inertial effects developed by the train movements. Moreover, the probability of reliably detecting a defect using conventional power spectral analysis (FFT) is low [17]. In addition, parameters such as train speed, track unevenness profile, and noise in the signal play a relevant role in estimating a train weight and detecting a defective wheel.

This research study aims to develop postprocessing algorithms to be used in a track-side system aiming the evaluation of the train weigh in motion and the detection of wheel flats. A wide range of numerical simulations based on a train-track interaction model has been performed in order to obtain synthetic data to test the aforementioned algorithms. From the shear evaluated in the rail, both the static and dynamic wheel axle loads are obtained. Moreover, a postprocessing methodology based on envelope spectrum analysis was also carried out to distinguish healthy from defective wheels based on the same shear data. One of the novelties of the presented paper is to implement this methodology in the detection of wheel flats. Finally, the aspects and parameters that affect the accuracy of the proposed methods were also analyzed.

10.2 Propose approaches to obtain WIM and wheel defect detection

10.2.1 Proposed method to obtain static loads from WIM system

From the analysis presented in the previous researches [7] it is obvious that the average value of the load assessed with the passage of the wheelset in the six pairs of strain gauges is closer to the static value when the running speed is lower and the quality of the track is higher. Moreover, the unsprung masses of the vehicle also play a relevant role. Since the vehicle has a large mass, the dynamic loads generated are also more extensive due to the track unevenness. Besides, over time, the track unevenness is not constant. Even if the stretch is classified in the same quality class, changes are expected in the track unevenness profile. Hence the static loads cannot be obtained in a deterministic framework, and should assess a statistical procedure with an established interval to get a certain level of confidence. To take into account all these complex effects, one of the main novelty of this research study is to propose an approach to calculate the static loads from weigh-in-motion system. The following framework (presented in Fig. 10.1) is proposed to estimate the static load interval, with a certain level of confidence, for each wagon's wheelsets. To estimate the wheel static load following steps should be considered.

(i) Using the experimental data from the passage of a train, compute the train speed and the average value of loads of each wheelset (P_{d_w}) presented in Fig. 10.1(i). The unsprung masses for each wheelset of the vehicle should also be identified (M_w). It should be highlighted that the present approach was developed within the hypothesis that the mass (M_w) of each wheelset that composes a train can be obtained from a database of rolling stock properties.

(ii) Obtain the dynamic loads, generated by the passage of one single wheelset system for 100 unevenness profiles (Q_i). Note that the irregularity profiles in this step should be compatible with the geometrical quality class of the track and for the speed assessed in step (i). Moreover, in this research, the mass (M_m) and the load (Q_m) of an arbitrary wheelset are considered as 2003 kg and 195,000 N respectively.

(iii) Generate the distribution of the dynamic loads obtained from the passage of one single wheelset system for 100 unevenness profiles (Q_i). Compute the mean value ($Q_{i\text{-mean}}$) and standard deviation (Std$_{Qi}$) for distributed loads for 100 dynamic load functions (Q_i). The mean value and standard deviation are shown in Fig. 10.1(iii).

(iv) Calculate the effect of dynamic load (P_{d_m}) for one known wheelset passing through 100 unevenness profiles. P_{d_m} can be obtained by subtracting the static load (Q_m) from the mean value of the distribution of dynamic loads ($Q_{i\text{-mean}}$).

(v) Therefore, when a wagon passes along the track, the static load transmitted by each wheelset to the track (Q_w) can be estimated from the load measured by the WIM system (P_{d_w}). The static load is then estimated by removing the dynamic component of the load in the value of P_{d_w}. This step is performed by subtracting the P_{d_w} from the dynamic load component evaluated for the passage of a known wheelset (P_{d_m}) affected by a correction factor that corresponds to the ratio between the masses of the wagon wheelset (M_w) and the mass of the known wheelset (M_m).

(vi) Define the interval of confidence and compute the range of the static load as:

$$Q_{w\_\min} = P_{d\_w} - \frac{M_w}{M_m} \times \left(\left|P_{d\_m} - \text{Std}_{Qi} \times \text{coef.}\right|\right)$$
$$Q_{w\_\max} = P_{d\_w} + \frac{M_w}{M_m} \times \left(\left|P_{d\_m} - \text{Std}_{Qi} \times \text{coef.}\right|\right)$$
(10.1)

10.2.2 Methodology for flat detection with envelope spectrum analysis

The envelope spectrum is commonly used in vibration monitoring, particularly for diagnosing rolling element bearings [5,8,18]. The present paper aims to implement this methodology in the detection of wheel flats. The idea is to extract the envelope of the time signal before calculating the spectrum. This relates to amplitude demodulation and is useful in cases where the vibration phenomenon is modulated above a

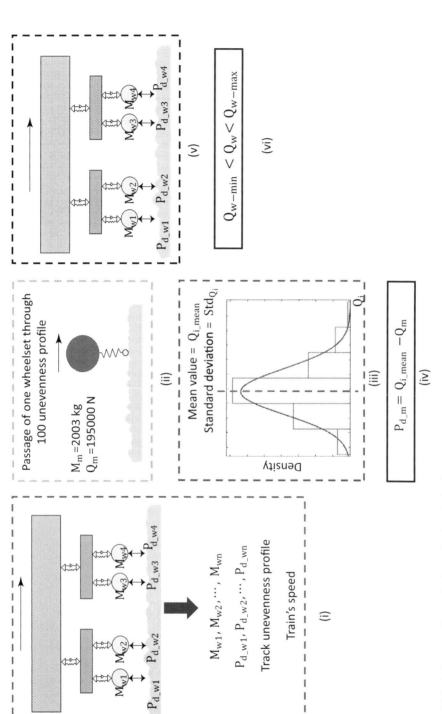

Fig. 10.1 A framework to calculate wheel static load.

specific frequency. The envelope spectrum is the spectrum of the envelope signal, and, evidently, has a much lower frequency than the original signal. Generally, in order to calculate the envelope spectrum, it is necessary to bandpass-filter the raw signal around the frequencies of interest [19] or perform complex demodulation [20]. In this research, the former method is used to obtain the envelope spectrum. The main purpose of the envelope spectrum technique is to eliminate the influence of disturbance and highlight the fault feature to detect the defective wheel [21]. To perform the envelope spectrum, the following steps should be taken into account [22]:

(i) Consider the time-series signal $X(t)$, as the shear indirectly obtained through the recorded strains obtained with strain gauge i, which is presented in Fig. 10.3. The sampling frequency (F_s) is considered as 2000 Hz.
(ii) Perform complex demodulation of the signal, multiplying the signal by the following factor

$$Y(t) = X(t)\exp(2\pi i f_0 t) \qquad (10.2)$$

where, f_0 is the center frequency of the band.

(i) Smooth the series $Y(t)$ by using a low-pass filter.
(ii) Compute the envelope signal as twice the absolute value of the analytic signal. Moreover, the DC bias from the envelope should be removed.
(iii) Compute the envelope spectrum using the FFT.
(iv) The presence of wheel flats is assessed due to a visual criterion in which the responses obtained by the envelope spectrum method in each strain gauge of the system are compared. When a healthy wheel passes over the system, the envelope spectrum amplitude is the same in all strain gauges and the responses are coincident. However, in the presence of wheel flat, a significant lag in the amplitude of the responses of the several strain gauges is clearly observed, resulting in a wheel flat detection.

The whole process is divided into two main blocks, as presented in Fig. 10.2. The first block shows the process of evaluating the envelope spectrum in each position of the strain gauge, while the second block aims to detect a defective wheel that may exist in the rolling stock based on the criterion defined in step (vi).

10.3 Numerical modeling

10.3.1 Overview of wayside system

The objective of the present study is to present postprocessing algorithms to obtain weigh-in-motion and detect wheel flats based on track-side monitoring. Thus, a numerical simulation was performed to produce synthetic data to test the aforementioned algorithms. For this reason, 12 positions distributed along the length of the equivalent perimeter of the wheel are considered in rail, allowing the measurement of rail shear evaluated by six pairs of strain gauges located between sleepers. A scheme of the 12 positions where the shear force in the rail is analyzed and the conventional wayside installation to indirect measure that shear is shown in Fig. 10.3.

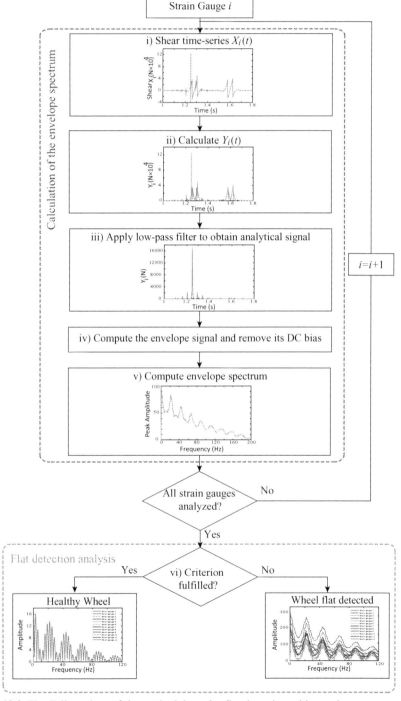

Fig. 10.2 The Fellow chart of the methodology for flat detection with envelope spectrum analysis.

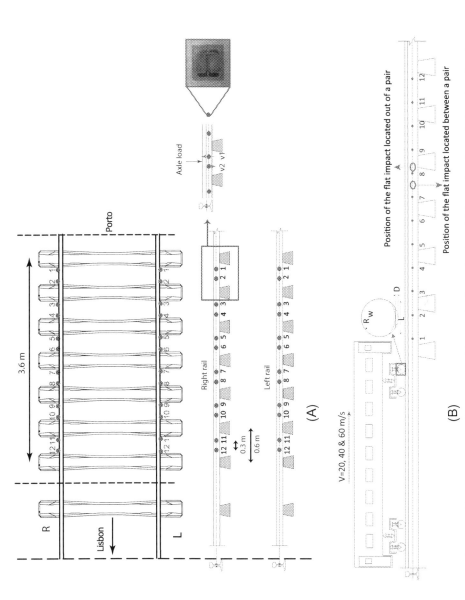

Fig. 10.3 Overview of the wayside system: (A) strain gauges positions, (B) schematic of a vehicle passing through six pairs of strain gauges and wheel flat shape.

This system presented here relies on the indirect evaluation of shear forces in the rail through the recorded strains obtained with the strain gauges. The difference between the shear acquired in the two consecutive sections is equivalent to the applied external load. Accordingly, the dynamic axle loads resulting from the passage of the train wheels equals the shear differences between V_1 and V_2 obtained from the strains assessed in Sections 10.1 and 10.2. The equation is shown as follows:

$$V_1 - V_2 = Q \tag{10.3}$$

where Q is wheel axle load. In this study, two approaches are presented to estimate the weigh-in-motion and identify defective wheels. Numerical modeling to generate a set of synthetic data is described in the following sections.

10.3.2 Framework of the vehicle-track coupling system

10.3.2.1 Track numerical model

The numerical model of the track, which is schematically presented in Fig. 10.4, was developed with the software ANSYS [23]. The numerical modeling is based on a three-layer model, simulating the ballast, sleeper and rail, connected through spring-dampers simulating the ballast and pads/fastner pads [24–26]. The foundation, ballast, and pads are modeled using linear spring-dampers, while the rails and sleepers are modeled with beam finite elements. The track properties, including ballast stiffness, pads/fasteners stiffness and damping, were adopted from the literature, while a simplified approach considered the foundation properties based on a Winkler model (presented in detail in [7]). The mechanical properties of the track are shown in Table 10.1. It should be mentioned that the cross-section considered for both rails of the track is modeled with section properties equivalent to two UIC60 [27] rails.

10.3.2.2 Vehicle numerical model

The train model was developed in ANSYS [23] using rigid beam finite elements to model the different components of the vehicle, namely the carbody, the bogies and the wheelsets. Mass elements were used to simulate the properties of those components

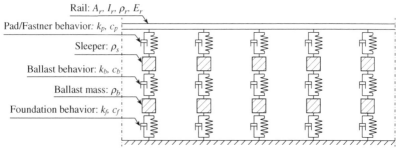

Fig. 10.4 A dynamic model of the railway track.

Table 10.1 Mechanical properties of the track.

Parameter		Value	
Rail	A_r (m^2)	7.67e-4	[28]
	ρ_r (kg/m^3)	7850	[28]
	I_r (m^4)	30.38e-6	[28]
	ν_r	0.28	[28]
	E_r (N/m^2)	210e9	[28]
Rail pad, longitudinal	K_p (N/m)	20E+06	[25]
	C_p (N s/m)	50E+03	[25]
Rail pad, lateral	K_p (N/m)	20E+06	[25]
	C_p (N s/m)	50E+03	[25]
Rail pad, vertical	K_p (N/m)	500E+06	[29]
	C_p (N s/m)	200E+03	[29]
Sleeper	ρ_s (kg/m^3)	2590	
	ν_s	0.2	
	E_s (N/m^2)	40.9e9	
Ballast, longitudinal	$K_{b,x}$ (N/m)	9000E+03	[30]
	$C_{b,x}$ (N s/m)	15E+03	[31]
Ballast, lateral	$K_{b,y}$ (N/m)	2250E+03	[32]
	$C_{p,y}$ (N s/m)	15E+03	[31]
Ballast, vertical	$K_{p,z}$ (N/m)	30E+06	[32]
	$C_{p,z}$ (N s/m)	15E+03	[31]
Foundation, longitudinal	$K_{f,x}$ (N/m)	20E+06	[33]
	$C_{f,x}$ (N s/m)	5.01E+02	[33]
Foundation, lateral	$K_{f,y}$ (N/m)	20E+06	[33]
	$C_{f,y}$ (N s/m)	5.01E+02	[33]
Foundation, vertical	$K_{f,z}$ (N/m)	20E+06	[33]
	$C_{f,z}$ (N s/m)	5.01E+02	[33]

in terms of mass and rotary inertia and spring-dampers to model the suspensions. The dynamic model of the train is shown in Fig. 10.5, in which k, c, m, and I represent stiffness, damping, concentrated mass and rotary inertia, respectively, while subscripts c_b, b, and w denote the carbody, the bogies, and the wheelsets, respectively.

Finally, the longitudinal, transversal and vertical distances are denoted by a, b, and h, respectively, while s indicates the gauge and R_0 is the nominal wheel radius. The train's geometrical and mechanical properties, which have been derived from a more detailed model of the Alfa Pendular developed by Ribeiro et al. [34] and calibrated by experimental data, are presented in Table 10.2. Moreover, the geometrical and mechanical properties of one wagon of the freight train are presented in Table 10.3.

10.3.2.3 Track unevenness profile

Synthetic unevenness profiles could be simulated by a stationary stochastic process which is described by a spectral density function (PSD). The Portuguese Railway Administration performs, the measurement of the geometric indicators of the track,

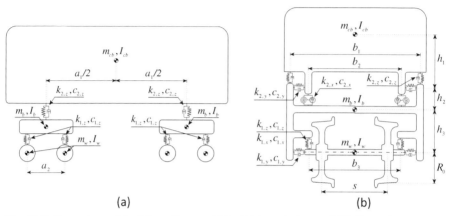

Fig. 10.5 Dynamic model of the railway vehicle (A) lateral view and (B) front view.

Table 10.2 Geometrical and mechanical properties of the Alfa Pendular train.

Parameter	Value
Car body mass, m_{cb}	35,640 kg
Car body roll moment of inertia, $I_{cb,x}$	55,120 kg m^2
Car body pitch moment of inertia, $I_{cb,y}$	1,475,000 kg m^2
Car body yaw moment of inertia, $I_{cb,z}$	1,477,000 kg m^2
Bogie mass, m_b	2829 kg
Bogie roll moment of inertia, $I_{b,x}$	2700 kg m^2
Bogie pitch moment of inertia, $I_{b,y}$	1931.49 kg m^2
Bogie yaw moment of inertia, $I_{b,z}$	3878.76 kg m^2
Wheelset mass, m_w	1711 kg
Wheelset roll moment of inertia, $I_{w,x}$	733.4303 kg m^2
Wheelset yaw moment of inertia, $I_{w,z}$	733.4303 kg m^2
Stiffness of the primary longitudinal suspension, $k_{1,x}$	44,981,000 N/m
Stiffness of the primary transversal suspension, $k_{1,y}$	30,948,200 N/m
Stiffness of the primary vertical suspension, $k_{1,z}$	1,652,820 N/m
Damping of the primary vertical suspension, $c_{1,z}$	16,739 N s/m
Stiffness of the secondary longitudinal suspension, $k_{2,x}$	4,905,000 N/m
Stiffness of the secondary transversal suspension, $k_{2,y}$	2,500,000 N/m
Stiffness of the secondary vertical suspension, $k_{2,z}$	734,832 N/m
Damping of the secondary longitudinal suspension, $c_{2,x}$	400,000 N s/m
Damping of the secondary transversal suspension, $c_{2,y}$	17,500 N s/m
Damping of the secondary vertical suspension, $c_{2,z}$	35,000 N s/m
The static load transmitted by each wheel	64,000 N
Longitudinal distance between bogies, a_1	19 m
Longitudinal distance between wheelsets, a_2	2.7 m
Transversal distance between vertical secondary suspensions, b_1	2.144 m
Transversal distance between longitudinal secondary suspensions, b_2	2.846 m
Transversal distance between primary suspensions, b_3	2.144 m
Vertical distance between carbody center and secondary suspension, h_1	0.936 m

Approaches for weigh-in-motion and wheel defect detection   193

Table 10.2 Continued

Parameter	Value
Vertical distance between bogie center and secondary suspension, h_2	0.142 m
Vertical distance between bogie center and wheelset center, h_3	0.065 m
Nominal rolling radius, R_0	0.43 m
Gauge, s	1.67 m

Adapted from D. Ribeiro, R. Calçada, R. Delgado, M. Brehm, V. Zabel, Finite-element model calibration of a railway vehicle based on experimental modal parameters, Veh. Syst. Dyn. 51 (2013) 821–856.

Table 10.3 Geometrical and mechanical properties of the freight wagon.

Parameter	Value
Car body mass	93,640 kg
Body pitch inertia	4,410,000 kg m^2
Bogie mass	1880 kg
Bogie pitch inertia	1720 kg m^2
Secondary vertical stiffness per bogie	6,200,000 N/m
Secondary vertical damping per bogie	1,000,000 N s/m
Primary vertical stiffness per axle	13,000,000 N/m
Primary vertical damping per axle	90,000 N s/m
Bogie center	8.078 m
Bogie wheelbase	1.75 m
Wheelset mass	1150 kg
Wheel diameter	0.84 m
linearized wheel-rail contact spring stiffness	2,950,000,000
The static load transmitted by each wheelset	255,000 N/wheelset

every 6 months. Based on this information, PSD curves are generated due to the real data and then used to simulate artificial unevenness profiles with the same power spectral density. The generated PSD of unevenness profile for wavelengths between 3 m and 70 is shown in Fig. 10.6.

A PSD curve based on the general equation from Federal Railroad Administration (FRA) is adjusted by the least-square method, and the track coefficient is calculated as 1.02 (m^3/rad). Therefore the unevenness profiles are generated by considering wavelengths ranging from 1 to 30 m as follow:

$$S(K_1) = \frac{10^{-7} A K_3^2 \left(K_1^2 + K_2^2\right)}{K_1^4 \left(K_1^2 + K_3^2\right)} \tag{10.4}$$

where K_2 and K_3 are constant values as 0.1464 and 0.8168 rad/m respectively. K is wave number which dependants on the cyclic spatial frequency of irregularity, define as $2\pi/\lambda$. Since λ is taken between 1 and 30 m, therefore K is varying between $2\pi/30$ and

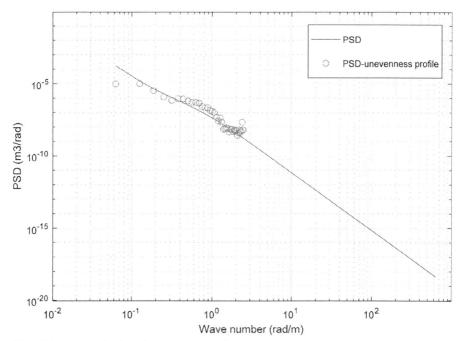

Fig. 10.6 Generated PSD of unevenness profile.

2π. Parameter A is calculated as 1.02 (m³/rad). More details about the unevenness profile measurement are provided by Mosleh et al. [7]. As an example, Fig. 10.7 depicts a 220 m stretch—the total length of the simulation—of the vertical irregularity profile corresponding to the right rail. The sample of the track irregularity profile shown in Fig. 10.7 is used for all analyses in this research study.

10.3.2.4 Wheel flat geometry

In this study, the vertical profile deviation of the wheel flat is defined as:

$$Z = -\frac{D}{2}\left(1 - \cos\frac{2\pi x}{L}\right) \cdot H(x - (2\pi R_w - L)), \quad 0 \leq x \leq 2\pi R_w \tag{10.5}$$

where, D is flat depth, L is flat length, R_w is the radius of the wheel, and H represents the Heaviside function. The vertical profile deviations of the flats are added to the rail unevenness profile as a periodic artificial irregularity in order to consider their effects in the vehicle-track dynamic interaction analysis (see Section 10.3.2.5). The schematic of a vehicle with a wheel flat passing through 12 positions of strain gauges is illustrated in Fig. 10.3B. When a defective wheel rotates, a periodic impulse is imposed by the wheel on the track with a specific frequency. Once the wagon speed (V) and perimeter of the wheel (R_W) are identified, the frequency of the periodic impulse referring to the flat impact frequency can be calculated as follows:

Approaches for weigh-in-motion and wheel defect detection 195

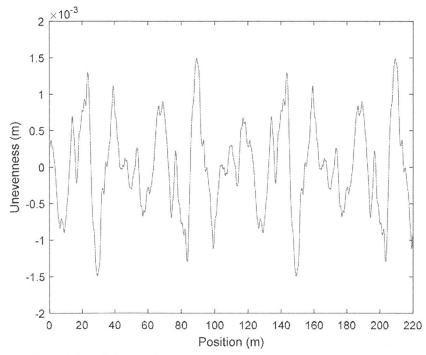

Fig. 10.7 Track irregularity sample.

$$f_f = \frac{V}{(2\pi R_W)} \tag{10.6}$$

10.3.2.5 Vehicle-track interaction method

In the present study, the direct method developed by [35,36] is used to solve the train-track interaction coupling system. This method is based on the Lagrange Multipliers contact technique, where a set of constraint equations are imposed on the system to establish the relationship between the displacements of the vehicle's wheels with the nodal displacements of the track. These equations, in which the track unevenness profiles and the periodic wheel flat geometry are included (see Sections 10.3.2.3 and 10.3.2.4, respectively), are added to the governing equilibrium equations of motion of the two subsystems, forming a single system that defines the interaction between them. Later, Montenegro, Neves [37] extended the formulation to take into account the lateral interaction by considering the actual geometry of the wheel and rail and the contact model between them. Based on this work, the train-structure coupling system can be mathematically expressed as:

$$\begin{bmatrix} \overline{\mathbf{K}} & \mathbf{D} \\ \overline{\mathbf{H}} & \mathbf{0} \end{bmatrix} \begin{bmatrix} \Delta \mathbf{a}_F^{i+1} \\ \Delta \mathbf{X}^{i+1} \end{bmatrix} = \begin{bmatrix} \mathbf{\psi}\left(\mathbf{a}^{t+\Delta t, i}, \mathbf{X}^{t+\Delta t, i}\right) \\ \overline{\mathbf{r}} \end{bmatrix} \quad (10.7)$$

where $\overline{\mathbf{D}}$ is a transformation matrix that relates the contact forces in the local coordinate system of the track with the nodal forces in the global coordinate system, $\overline{\mathbf{K}}$ is the effective stiffness matrix of the train-structure system, and $\overline{\mathbf{H}}$ is also a transformation matrix that relates the displacements of the nodes of the rails in the global coordinate system to the displacements of the contact nodes in the vehicle (see [37] for details). Vectors $\Delta \mathbf{a}$ and $\Delta \mathbf{X}$ are the incremental nodal displacements and contact forces, respectively, $\mathbf{\psi}$ is the residual force vector and $\overline{\mathbf{r}}$ the vector of irregularities, both rail unevenness and periodic wheel flat geometry (see Sections 10.3.2.3 and 10.3.2.4, respectively), that may exist in the contact interface. Superscript $t + \Delta t$ refers to the current time step, while i and $i + 1$ indicate the previous and current Newton iteration, respectively. The normal contact forces are computed through the nonlinear Hertz contact theory [38], while the tangential contact forces are established through the Kalker's USETAB algorithm [39]. The formulation of the train-track interaction tool used in this paper (presented in detail in [37]) has been programmed in MATLAB [40], which imports the structural matrices of both the track and the train, previously modeled in ANSYS [23].

10.4 Results and discussion

10.4.1 Estimation of static loads using a WIM system

In order to analyze the dynamic interaction behavior between track and train, the numerical model previously described in Section 10.3 is adopted. A passage of a freight wagon is considered to obtain weigh in motion. The geometrical and material properties of the wagon is described in Table 10.3. The load assessed with the WIM system is composed of two components: the quasistatic component, which corresponds to the distribution of the dead weight by the different wheelsets, and the dynamic component, which is due to the inertial dynamic effects induced by the train due to several sources, as for instance the track unevenness. The load imposed by the rolling stock onto the track depends on the track unevenness, which gives rise to dynamic loads added to the quasistatic ones. It should be highlighted that this increment of interaction loads increases the track degradation process.

10.4.1.1 Statistical evaluation of the wheel loads

In order to better discern the variation of the Q load due to speed, a parametric study is presented here. To calculate the static load, 100 unevenness profiles were generated with Eq. (10.4). The mean value and standard deviation of the load assessed at each wheel were computed for running speeds of 10 and 40 m/s. The load for each one of the wagon wheelsets is obtained for each of the 100 unevenness profiles. When the

Approaches for weigh-in-motion and wheel defect detection 197

wagon passes through the 12 strain gauges (six pairs), the average of the four-wheel set loads is computed. Hence, 100 loads are calculated for each wheelset for 100 unevenness profile, for which the distributions for a speed equal to 10 and 40 m/s are plotted. From the results presented in Fig. 10.8, it is possible to infer that the train speed and track quality affect the mean value and standard deviation. Since the track has a good quality, the mean value is near the static load. In this study, the static value for a freight wagon to obtain weigh-in-motion is 255,000 N/wheelset, which is presented in Table 10.4. Moreover, the standard deviation increases when the speed of the train enhances.

For freight trains, not only the static axle load is a relevant parameter, but also the weight of the train has essential information for railway administrations. Therefore, when the wagon is passing over the 12 strain gauges (six pairs), the static load of the wagon is calculated as the sum of the average load of each wheelset. Regarding the wagon weight assessment, the mean value and standard deviation for speed equal

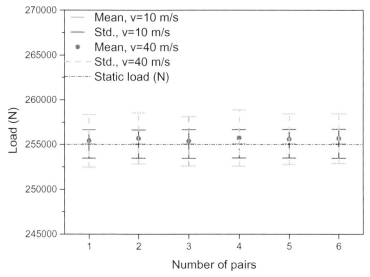

Fig. 10.8 Mean value and standard deviation for speed = 10 and 40 m/s for one wagon passage of freight train (load per wheelset).

Table 10.4 Mean value and standard deviation for speed equal to 10 and 40 m/s for one wagon passage of freight train considering six pairs.

Velocity (m/s)	Mean value of distribution (N)	Standard deviation (N)	Distribution
10	1,020,290	5340	Normal
40	1,022,310	7921	Normal

to 10 and 40 m/s for one wagon are illustrated in Table 10.4. As shown in this table, the mean value of data distribution is near the static weight of the freight wagon (1,020,000 N).

With the results presented here, it can be concluded that when a significantly large number of strain gauge pairs are installed in the track, the average value of the Q load measured on the passage of each wheelset would give a reasonable estimation of the static load. However, the resources are limited and Q load can only be assessed in few sections (pairs of strain gauges). Therefore, when calculating static load with WIM, a correction must be taken into account. Accordingly, due to all these complex effects, the static load cannot be estimated in a deterministic framework.

10.4.1.2 Estimation of the static loads

From the analysis presented in the previous section, it is clear that the average value of the load assessed with the passage of the wheelset in the six pairs of strain gauges is closer to the static value when the running speed is lower and the quality of the track is higher. Moreover, the unsprung masses of the vehicle also play a relevant role. Since the vehicle mass is large, the dynamic loads generated are also more extensive due to the track unevenness. Besides, over time, the track unevenness is not constant. Hence as explained before, the static loads cannot be obtained in a deterministic framework, and should assess in a statistical procedure with an established interval to get a certain level of confidence.

Fig. 10.9A shows the load distribution for the passage of the single mass system, running at a speed of 40 m/s. For this distribution, the mean value ($Q_{i\text{-mean}}$) is 195,659 N and the standard deviation (Std_{Qi}) is 551 N. The static load (Q_m) imposed by the system is 195,000 N, which means that there is a difference $P_{d\_m} = 659$ N between the mean value and the static value. Therefore, when the wagon (presented

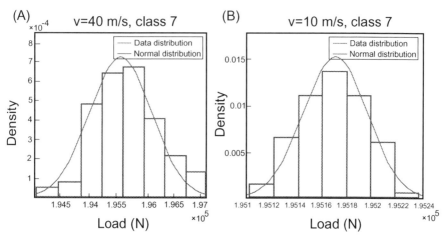

Fig. 10.9 Load distribution for one moving mass for class 7 subjected to 100 unevenness profiles due to (A) $V = 40$ m/s, (B) $V = 10$ m/s.

Approaches for weigh-in-motion and wheel defect detection

in Fig. 10.3) running at a speed of 40 m/s along the track, the dynamic load measured for one wheelset (P_{d_w}) can be obtained as 255,160 N. Hence, the static load for the wheelset (Q_w) can be calculated from the dynamic load measured for the wheelset of the wagon ($P_{d\_w} = 255,160$ N), subtract from the effect of the dynamic load for one known wheelset ($P_{d\_m} = 659$ N), by considering the wheel masses for wagon ($M_w = 1150$ kg) and one single mass ($M_m = 2003$ kg). The upper and lower limits to obtain the static load for one wheel, with speed $= 40$ m/s for 95% reliable results, are calculated as follow:

$$Q_{W\_min} = 255,160 - \frac{1150}{2003} \times (|659 - 551 \times 1.96|) = 254,920 \text{ N}$$

$$Q_{w\_max} = 255,160 + \frac{1150}{2003} \times (|659 + 551 \times 1.96|) = 256,160 \text{ N}$$

(10.8)

Therefore, as it can be observed for a static load, the statistical load can be obtained for each wheelset of the wagon. Fig. 10.9B shows the load distribution for the passage of the single mass with speed $= 10$ m/s. The mean value ($Q_{i\text{-mean}}$) and the standard deviation (Std_{Qi}) are 195,168 and 26.28 N respectively. Therefore, P_{d_m} is $Q_{i\text{-mean}} - Q_m = 168$ N. Hence, Q_{W_min} and Q_{W_max} for speed $= 10$ m/s are obtained as follows:

$$Q_{W\_min} = 254,932 - \frac{1150}{2003} \times (|168 - 26 \times 1.96|) = 254,865 \text{ N}$$

$$Q_{w\_max} = 254,932 + \frac{1150}{2003} \times (|168 + 26 \times 1.96|) = 255,056 \text{ N}$$

(10.9)

Steps (ii and iii) of the proposed approach (presented in Fig. 10.1) to obtain the mean value ($Q_{i\text{-mean}}$) and standard deviation (Std_{Qi}) for the distributed load (Q_i), generated by the passage of one single wheelset system for 100 unevenness profiles, is just performed once for an arbitrary value of M_m (here suggested as 2003 kg). This step can take some hours, but it is just performed once to obtain the mean value and standard deviation of dynamic loads from the passage of one known single wheelset along 100 unevenness profiles. Apart from that, the computational effort demanded by the method is relatively reduced, taking only a few seconds.

10.4.2 An approach to detect wheel flat

Previous researches have been shown that the probability of detecting a reliable defect using conventional power spectral analysis (FFT) is low [17]. Therefore, to detect a defective wheel, the envelope spectrum technique is considered. The present section aims to demonstrate the robustness of the proposed methodology described in Section 10.2.2 to detect wheel flats. The numerical model described in Section 10.3 is adopted to analyze the dynamic interaction behavior between track and train. The train adopted in this section is the Alfa Pendular train that runs over the system presented in Fig. 10.3. The geometrical and mechanical properties of the train are shown in Table 10.2. Moreover, the proposed methodology is tested for several flat defect scenarios (described in Section 10.3.2.4), starting with a flat length L of 150 mm, depth D is considered equal to 0.5 mm (scenario 1), 1 mm (scenario 2), and 1.5 mm (scenario 3). Subsequently, for scenario 4, $L = 50$ mm and $D = 1$ mm. For all scenarios, the speeds of the wagon are 20, 40, and 60 m/s. Shear of the track for rounded and defective wheels is obtained for 12 strain gauge positions.

10.4.2.1 Sensitivity of the proposed approach to the flat geometry

Fig. 10.10 presents the envelope spectrum analysis for the 12 positions of strain gauges presented in Fig. 10.3, considering four different flat geometries. The train speed is set as 60 m/s. The results show that there is a significant lag in the amplitude of the responses of the several strain gauges in the presence of a defective wheel, resulting in a wheel flat detection. Moreover, as presented in all figures, the increasing trend of the envelope spectrum's amplitude is evident. The maximum amplitude corresponds to the strain gauges nearest to the flat's impact. Afterward, the amplitude decreases again. Moreover, in the absence of severe defects (scenario 4), the amplitudes of the envelope spectrum are lower for 12 positions. This effect is evident in

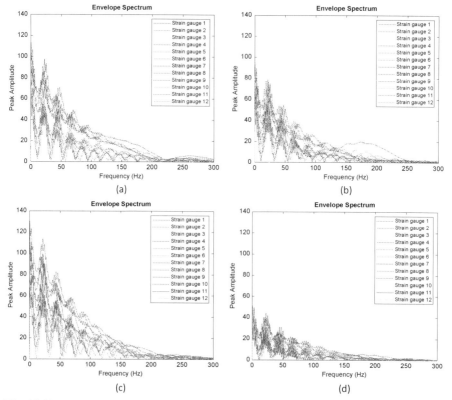

Fig. 10.10 Envelope spectrum analysis for 12 strain gauges considering flat geometry for (A) $L = 150, D = 0.5$ mm; (B) $L = 150, D = 1$ mm; (C) $L = 150, D = 1.5$ mm; (D) $L = 50, D = 1$.

Fig. 10.10D. The amplitudes of the envelope spectrum are lower for 12 positions. However, when there is a severe defect, the amplitude of the envelope spectrum for different strain gauges increases.

10.4.2.2 Sensitivity of the proposed approach to the random position of the defect with respect to the sensors

Fig. 10.11 illustrates the envelope spectrum analysis for the 12 positions of strain gauges with the flat's impact occurring between a pair of strain gauges (between strain gauges 7 and 8 presented in Fig. 10.3B). The front wheel of the wagon has a 150 mm flat length and a 1.5 mm flat depth (scenario 1), and the impact of the wheel flat occurred between strain gauges 7 and 8. The train speed is set as 20 m/s. As presented in this figure, this method does not depend on the wheel flat position in relation to the strain gauges. In the presence of a defective wheel, a significant lag in the amplitude of the responses of the several strain gauges is clearly observed, resulting in the detection of a wheel flat. The strain gauges nearest to the flat occurrence can always detect the wheel flat. The strain gauge nearest to the flat impact shows lower attenuation and, therefore, higher energy. Consequently, in Fig. 10.11, strain gauges 7 and 8 have higher amplitudes.

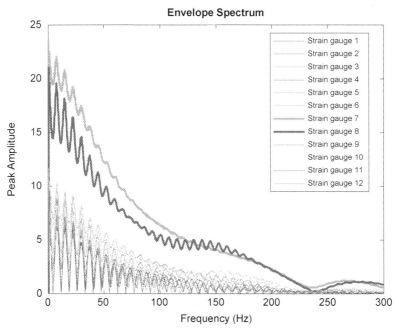

Fig. 10.11 Envelope spectrum analysis for 12 strain gauges where the impact of the flat occurs between a pair, $L = 150$, $D = 1.5$, $V = 20$ m/s.

10.4.2.3 Sensitivity of the proposed approach to the rail unevenness profile

To check the accuracy of the proposed method to detect a defective wheel, one vehicle's passage along a track considering the track unevenness profile is also analyzed. The irregularity profile previously presented in Section 10.3.2.3 is used in the numerical simulation to test the accuracy of the proposed method. According to Eq. (10.6), for a train speed of 60 m/s, the flat impact frequency is 22.17 Hz.

Fig. 10.12 presents the envelope spectrum obtained for the 12 strain gauges considering the unevenness profile of the track. Again, and considering the unevenness of the track, it is evident that the strain gauge closest to the flat impact location is the one with a higher amplitude, confirming the presence of the flat. As shown in Fig. 10.12B when the healthy wheel passes over the track, the amplitude of the envelope spectrum is the same for all strain gauges. However, it is clear that the amplitude of the envelope spectrum is different for all 12 positions in the presence of a flat (Fig. 10.12A).

10.4.2.4 Sensitivity of the proposed approach to the noise of the signal

Another factor that plays an essential role in the results is the noise of the signal, which perturbs the shear assessment of the rail. To estimate the sensitivity of the envelope spectrum analysis method to the intensity of the noise, an artificial noise based on the maximum shear of the signal is generated and added to the main signal. Hence, new signals are generated considering different noise levels as 2.5% and 10% of the maximum shear of the initial signal. Fig. 10.13 presents the envelope spectrum obtained for the 12 positions of strain gauges considering the unevenness of the track and the noisy signal, to analyze the accuracy of the method to detect a defective wheel.

As depicted in Fig. 10.13, the flat impact frequency is obtained at 22.12 Hz. Once again, by considering a vehicle speed of 60 m/s, the wheel rotational frequency

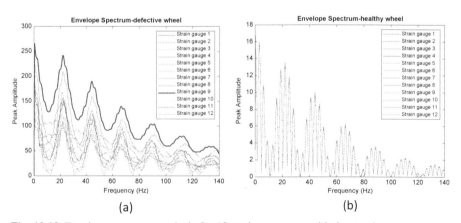

Fig. 10.12 Envelope spectrum analysis for 12 strain gauges considering track unevenness profile for (A) defective wheel, (B) healthy wheel.

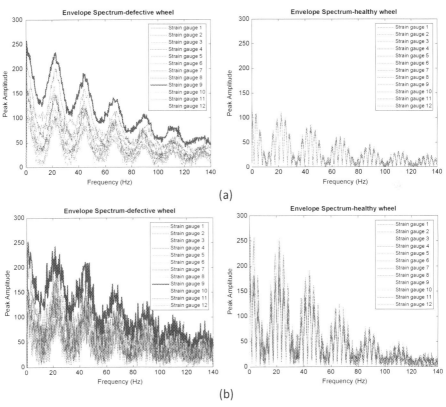

Fig. 10.13 Envelope spectrum analysis for 12 strain gauges considering noise in the signal for a defective and healthy wheel (A) signal with 2.5% noise, (B) signal with 10% noise.

(22.17 Hz) is in good agreement with the flat impact frequency detected by the envelope spectrum (22.12 Hz). Different strain gauges show different frequency ranges and strain gauge 9 has a higher amplitude due to its proximity to the wheel flat. Then the amplitude of the envelope spectrum decreases again. By comparing the results of the defective wheel and the healthy one, it is obvious that the amplitude of the envelope spectrum for different strain gauges is lagged, thus fulfilling the criterion defined in Section 10.2.2 to detect wheel flats. However, in the case of healthy wheels, the amplitudes of the envelope spectra for all strain gauges are very similar and no lag appears between the amplitudes.

10.5 Conclusion

The present study aims to implement two approaches to detect wheel flat impacts accurately and weigh the train in motion based on on-side monitoring. A wide range of numerical simulations based on a train-track interaction model has been performed for different train speeds and different types of flat geometries. Afterward, based on the trends identified through the comprehensive study, two approaches were performed to obtain weigh-in-motion and wheel defect detection. Based on the results, the following conclusions can be drawn:

- The results presented show that when a significantly large number of strain gauge pairs are installed in the track, the average value of the Q load measured on the passage of each wheelset would give a reasonable estimation of the static load. However, the resources are limited and Q load can only be assessed in few sections (pairs of strain gauges). Therefore, when calculating static load with WIM, correction must be taken into account. Accordingly, the static load cannot be estimated in a deterministic framework due to all these complex effects. Therefore, this study proposes a procedure for estimating the interval of the static load, with a certain level of confidence and for each of the vehicle wheelsets.
- For the final development of the proposed WIM approach, experimental validation is necessary and essential. For this purpose, the passage of trains, previously weighted, must be performed. For the application of the proposed methodology, track unevenness should be measured along the stretch of installation of the WIM system. Moreover, the mass of the wheelsets must be known. After that, the experimental validation will be carried out by comparing the static load of each wheelset and the predicted load-interval obtained through the application of the present proposal.
- An envelope spectrum analysis is applied as an effective to detect a defective wheel. From the results and comments, it is possible to conclude that the wheel flat cannot be detected by simply performing a FFT of the shear time-series since the frequency responses evaluated in all strain gauges are coincident for both the healthy and the defective wheels. However, by applying the proposed methodology, a clear lag between the amplitude of the envelope spectrum is observed only for the defective wheel, showing that a flat has been detected.
- There is no indication regarding the presence of flats in the envelope spectrum analysis approach for a healthy wheel. The envelope spectrum of the signal obtained by the 12 positions of strain gauges does not reveal a significant lag when compared with a defective wheel.
- All strain gauges can detect the defective wheel when the wagon passes through the track. The only difference lies in the intensity of the shear recorded by the different strain gauges, being strain gauge 9, the one with the highest shear intensity. This is due to the fact that strain gauge 9 is located immediately after the flat. The results have shown a considerable fluctuation in the strain signal, especially when the wheels have more severe defects.

From the results presented in this research, it is possible to visually detect whether a wheel passing over a system is healthy or defective. Moreover, it is necessary to define an indicator that automatically distinguishes a healthy wheel from a defective one for the final development of the proposed approach. Besides, the application of the proposed method will be studied and examined in field conditions and presented in the forthcoming publication.

Acknowledgments

This work was financially supported by: Project POCI-01-0145-FEDER-007457—CONSTRUCT—Institute of R&D In Structures and Construction funded by FEDER funds through COMPETE2020—Programa Operacional Competitividade e Internacionalização (POCI)—and by national funds through FCT—Fundação para a Ciência e a Tecnologia. The paper reflects research developed in the ambit of the project PEDDIR DEMO, NORTE-01-0247-FEDER-006397, founded by Agência Nacional de Inovação S.A., program P2020|COMPETE—Projetos Demonstradores em Copromoção. The authors are also sincerely grateful to the European Commission for the financial sponsorship of H2020 MARIE SKŁODOWSKA-CURIE RISE Project, Grant No. 691135 "RISEN: Rail Infrastructure Systems Engineering Network," and project IN2TRACK2—Research into enhanced track and switch and crossing system 2, funded by European funds through the H2020 (SHIFT2RAIL Innovation Programme).

References

[1] A. Meixedo, A. Gonçalves, R. Calçada, J. Gabriel, H. Fonseca, R. Martins, Weighing in Motion and Wheel Defect Detection of Rolling Stock, IEEE, Portugal, 2015.
[2] A. Meixedo, A. Gonçalves, R. Calcada, J. Gabriel, H. Fonseca, R. Martins, On-line monitoring system for tracks, in: 3rd Experiment International Conference, University of Azores, Ponta Delgada, Sao Miguel, Azores Island, 2015, pp. 133–134.
[3] M. Molodova, Z. Li, A. Nunez, R. Dollevoet, Axle box acceleration: measurement and simulation for detection of short track defects, Wear 271 (2011) 349–356.
[4] H. Kanehara, T. Fujioka, Measuring rail/wheel contact points of running railway vehicles, Wear 253 (2002) 275–283.
[5] M. Entezami, C. Roberts, P. Weston, E. Stewart, A. Amini, M. Papaelias, Perspectives on railway axle bearing condition monitoring, Proc. Inst. Mech. Eng. F J. Rail Rapid Transit 234 (2019) 17–31.
[6] A. Mosleh, A. Meixedo, P. Costa, R. Calçada, Trackside monitoring solution for weighing in motion of rolling stock, in: TESTE2019—2nd Conference on Testing and Experimentations in Civil Engineering—Proceedings, 19–21 February, Porto, Portugal, 2019, https://doi.org/10.5281/zenodo.33553542019.
[7] A. Mosleh, P. Costa, R. Calçada, A new strategy to estimate static loads for the dynamic weighing in motion of railway vehicles, Proc. Inst. Mech. Eng. F J. Rail Rapid Transit 234 (2020) 183–200.
[8] A. Amini, M. Entezami, Z. Huang, H. Rowshandel, M. Papaelias, Wayside detection of faults in railway axle bearings using time spectral kurtosis analysis on high-frequency acoustic emission signals, Adv. Mech. Eng. 8 (2016) 1–9.
[9] G. Kouroussis, D. Kinet, V. Moeyaert, J. Dupuy, C. Caucheteur, Railway structure monitoring solutions using fibre Bragg grating sensors, Int. J. Rail Transp. 4 (2016) 135–150.
[10] G. Alexandrou, G. Kouroussis, O. Verlinden, A comprehensive prediction model for vehicle/track/soil dynamic response due to wheel flats, Proc. Inst. Mech. Eng. F J. Rail Rapid Transit 230 (2016) 1088–1104.
[11] A. Mosleh, P. Costa, R. Calçada, Development of a low-cost trackside system for weighing in motion and wheel defects detection, Int. J. Railway Res. 7 (2020) 1–9.
[12] P. Adimo, L. Marini, E. Meli, L. Pugi, A. Rindi, Development of a dynamical weigh in motion system for railway applications, Meccanica 51 (2016) 2509–2533, https://doi.org/10.1007/s11012-016-0378-2.
[13] E. OBrien, L. Zhang, H. Zhao, D. Hajializadeh, Probabilistic bridge weigh-in-motion, Can. J. Civ. Eng. 45 (2018) 667–675.
[14] B. Allotta, P. DAdamio, L. Marini, E. Meli, L. Pugi, A. Rindi, A new strategy for dynamic weighing in motion of railway vehicles, IEEE Trans. Intell. Transp. Syst. 16 (2015) 3520–3533.
[15] C. Zhou, L. Gao, H. Xiao, B. Hou, Railway wheel flat recognition and precise positioning method based on multisensor arrays, Appl. Sci. 10 (2020) 1–23.

[16] X. Liu, Y. Ni, Wheel tread defect detection for high-speed trains using FBG-based online monitoring techniques, Smart Struct. Syst. 21 (2018) 687–694.

[17] A. Amini, M. Entezami, M. Papaelias, Onboard detection of railway axle bearing defects using envelope analysis of high frequency acoustic emission signals, J. Case Stud. Nondestruct. Test. Eval. 6 (2016) 8–16.

[18] A. Mosleh, P.A. Montenegro, P. Costa, R. Calçada, An approach for wheel flat detection of railway train wheels using envelope spectrum analysis, Struct. Infrastruct. Eng. (2020), https://doi.org/10.1080/15732479.2020.1832536.

[19] J.S. Bendat, The Hilbert Transform and Applications to Correlation Measurements, Bruel & Kjaer, 1991.

[20] T. Hasan, Complex demodulation: some theory and applications, Handb. Stat. 3 (1983) 125–156.

[21] J.W. Tukey, Discussion emphasizing the connection between analyses of variance and spectrum analysis, Technometrics 3 (1961) 1–29.

[22] R. Randall, Vibration-Based Condition Monitoring, John Wiley & Sons, Ltd, 2011. ISBN: 978-0-470-74785-82011.

[23] ANSYS®, Release 19.2, Academic Research, ANSYS Inc, Canonsburg, PA, USA, 2018.

[24] W. Zhai, H. Xia, C. Cai, M. Gao, X. Li, X. Guo, et al., High-speed train–track–bridge dynamic interactions—part I: theoretical model and numerical simulation, Int. J. Rail Transp. 1 (2013) 3–24.

[25] W. Zhai, K. Wang, C. Cai, Fundamentals of vehicle-track coupled dynamics, Veh. Syst. Dyn. 47 (2009) 1349–1376.

[26] Y.B. Yang, J.D. Yau, Y.S. Wu, Vehicle–Bridge Interaction Dynamics With Applications to High-Speed Railways, World Scientific, Singapore, 2004.

[27] European Committee for Standardization (CEN), Railway Applications—Measurement of Vertical Forces on Wheels and Wheelsets—Part 1: On-Track Measurement Sites for Vehicles in Service, EN 15654-1, 2018.

[28] European Standard, Railway Applications Railway Applications—Track-Rail-Part 1: Vignole Railway 46 kg/m and Above, peEN 13674-1, final draft, Brussels, 2002.

[29] ERRI D 214/RP 5, Rail Bridges for Speeds >200 km/h: Numerical Investigation of the Effect of Track Irregularities at Bridge Resonance, European Rail Research Institute, Utrecht, 1999.

[30] UIC 774-3-R, Track/Bridge Interaction—Recommendations for Calculation, second ed., International Union of Railways (UIC), Paris, 2001.

[31] Y.S. Wu, Y.B. Yang, Steady-state response and riding comfort of trains moving over a series of simply supported bridges, Eng. Struct. 25 (2003) 251–265.

[32] ERRI D 202/RP 11, Improved Knowledge of Forces in CWR Track (Including Switches): Parametric Study and Sensivity Analysis of CWERRI, European Rail Research Institute, Utrecht, 1999.

[33] L. Auersch, Dynamic interaction of various beams with the underlying soil—finite and infinite, half-space and Winkler models, Eur. J. Mech. A. Solids 27 (2008) 933–958.

[34] D. Ribeiro, R. Calçada, R. Delgado, M. Brehm, V. Zabel, Finite-element model calibration of a railway vehicle based on experimental modal parameters, Veh. Syst. Dyn. 51 (2013) 821–856.

[35] P.A. Montenegro, S.G.M. Neves, A.F.M. Azevedo, R. Calçada, M. Papadrakakis NDL, A nonlinear vehicle-structure interaction methodology with wheel-rail detachment and reattachment, in: V. Plevris (Ed.), COMPDYN 2013—4th ECCOMAS Thematic Conference on Computational Methods in Structural Dynamics and Earthquake Engineering, Kos, Greece, 2013.

[36] S.G.M. Neves, P.A. Montenegro, A.F.M. Azevedo, R. Calçada, A direct method for analyzing the nonlinear vehicle–structure interaction, Eng. Struct. 69 (2014) 83–89.
[37] P.A. Montenegro, S.G.M. Neves, R. Calçada, M. Tanabe, M. Sogabe, Wheel-rail contact formulation for analyzing the lateral train-structure dynamic interaction, Comput. Struct. 152 (2015) 200–214.
[38] H. Hertz, Ueber die Berührung fester elastischer Körper [On the contact of elastic solids], J. Reine Angew. Math. 92 (1882) 156–171.
[39] J.J. Kalker, Book of tables for the Hertzian creep-force law, in: I. Zobory (Ed.), 2nd Mini Conference on Contact Mechanics and Wear of Wheel/Rail Systems, Budapest, Hungary, 1996.
[40] MATLAB®, Release R2018a, The MathWorks Inc., Natick, MA, USA, 2018.

Railway ground-borne vibrations: Comprehensive field test development and experimental validation of prediction tools

Aires Colaço, Alexandre Castanheira-Pinto, Pedro Alves Costa, and Rui Calçada
CONSTRUCT—LESE, Faculty of Engineering (FEUP), University of Porto, Porto, Portugal

11.1 Introduction

Prediction and mitigation of ground-borne vibrations are one of the largest environmental challenges for the railway exploitation in urban areas. Nowadays, in a final attempt to drastically reduce CO_2 emissions, there is a global shift where investment in rail transport takes precedence over other transportation options. Also, considering the imbalances in the demographic distribution observed, with a strong trend of concentration of the world population in urban areas, the management of urban mega centers, their mobility and the need to combat climate change are the major challenges of society, properly addressed in the 2030 Agenda for sustainable development of the United Nations.

The expansion and improvement of the railway network, associated with the high standards of comfort required by nowadays' society, demands the technical community to assess the environmental impact of the exploitation of railway infrastructures in the surroundings, especially in terms of the annoyance caused by railway-induced vibrations in the inhabitants living in nearby buildings. This impact is specially relevant in fully urban railway projects which are normally implemented in the underground space, putting the ground-borne vibrations as the major source of annoyance.

Taking into account the expressed concerns, considerable efforts have been enforced by the technical and scientific communities on the development of prediction tools for ground-borne vibrations. Such models comprise different degrees of approximation and complexity, extending from empirical and scoping approaches [1–5] up to detailed and complex numerical formulations [6–12]. The application of those approaches to practical engineering problems allowed to achieve a better understanding of the phenomenology of the problem from the theoretical point of view, with clear benefits in the design of mitigation countermeasures, technically and economically feasible [13–18].

Despite the achievements in the numerical field, the experimental research cannot be neglected. Experimental characterization of the track-ground-building system, as well as the measurement of ground-borne vibrations, is a need and actual topic, allowing to observe the main trends of the problem, to establish behavior patterns and to identify the main sources of uncertainty. Reliable studies were performed in this field, comprising the characterization and measurement of vibrations on the track-ground system [19–22] or comprising the receiver (building) [23–25]. However, the existence of a comprehensive experimental test site of the global system, from the source to the receiver, as well as the incorporation of mitigation measures is not found in the bibliography.

In that way, the description and characterization of a comprehensive experimental test site correspond to the main objective of the current work. The main data presented along the manuscript is available for download from www.fe.up.pt\~csf\DataCarregado.zip. This represents a relevant output for the scientific and technical communities dealing with induced vibration by railway traffic, since the experimental data reported can be used to validate prediction models. Additionally, an integrated numerical tool for the prediction of vibrations over all the system is described and experimentally validated.

11.2 Experimental characterization of the Carregado test site

11.2.1 General description

The development and continuous exploration of the test site allows to achieve a deeper understanding about the phenomena related with ground-borne vibrations and also to construct a database of experimental results that can be used by the technical and scientific communities. With this purpose, an experimental test site was implemented in a stretch of the Portuguese railway network ("Linha do Norte," that connects Lisbon to Oporto), near Carregado, in Portugal. Its location in the Portuguese territory is shown in Fig. 11.1.

On this test site, several experimental campaigns have occurred during last 10 years, where a full characterization of the site was performed and several measurements of

Fig. 11.1 Experimental test site of Carregado: (A) location; (B) detailed satellite photography.

Fig. 11.2 General view of the structure and the nearby railway track.

Fig. 11.3 Some of the trains circulating in the studied railway stretch: (A) Alfa Pendular train; (B) Suburban train; (C) Intercity train.

vibrations were recorded and analyzed. Additionally, a significant improvement of the test facility was reached with the construction of a small-scaled structure to better understand the aspects related with vibrations in buildings due to railway traffic. A general view of the structure and the nearby railway track is shown in Fig. 11.2.

Due to its proximity to Lisbon, the analyzed railway stretch has a very significant traffic volume, traveling several types of passenger and freight trains. Some of them are represented in Fig. 11.3, including the Alfa Pendular train, the fastest train operating in the Portuguese railway network.

In the following sections, an overview of the experimental activities performed over the last years is presented.

11.2.2 Characterization of the track

11.2.2.1 General description

The railway line, illustrated in Fig. 11.4, corresponds to a double ballasted track with a straight alignment along a substantial distance. The analysis of the renovation design project, accompanied by the results of some in situ tests, allowed the identification of ballast and subballast layers with a thickness of 0.35 and 0.30 m, respectively.

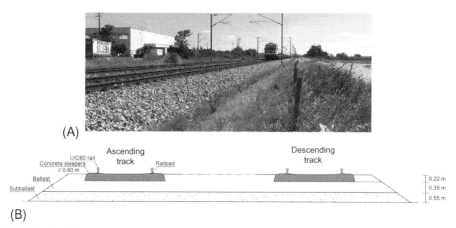

Fig. 11.4 Railway track at the test site: (A) general view; (B) cross-section.

Beneath the subballast, it was also possible to identify a layer of granular material, with a thickness around 0.25 m. The prestressed concrete sleepers, spaced 0.60 m, support continuous welded rails (UIC60).

11.2.2.2 Measurement of track unevenness

Characterization of the track unevenness is a mandatory step for understanding the train-track dynamic interaction problem, since it corresponds to one of the main sources of dynamic excitation. The spatial representation of the rail vertical unevenness profile of the descending track is presented in Fig. 11.5B, being the track unevenness measured by the inspection vehicle shown in Fig. 11.5A.

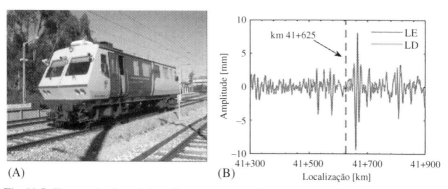

Fig. 11.5 Characterization of the rail unevenness profile: (A) Inspection vehicle EM 120; (B) Spatial representation (*red line* (*light gray* in print version): right side of the descending track; *blue line* (*dark gray* in print version): left side of the descending track).

11.2.2.3 Assessment of the mechanical properties of the track

The assessment of reliable properties of the track dynamic behavior is another challenging issue that deserves specific experimentation. Usually, the properties of the rail and sleepers are well defined. However, the same statement cannot be extended to other track components, such as the ballast or the railpads. In order to minimize the uncertainty associated with the properties of those elements, receptance tests are frequently performed. These tests are extremely useful, allowing to understand the dynamic behavior of the track, the definition of its properties and the calibration of the numerical models [26].

Fig. 11.6 shows the adopted setup for the receptance tests. As represented, the impact load (provided by an instrumented impulse hammer) is applied at the mid-span of the sleeper and the response is measured, simultaneously, by accelerometers installed on the sleeper ends and on the overlying rails. At the first stage, the tests were performed in the reference section located at km 41 + 625 and then repeated in different sections of the track (20 sleepers within a 40 m radius from the reference section) in order to assess the variability of the dynamic response along the track development.

Once the applied load and the response of the track are known, it is possible to calculate the receptance function given by:

$$H(\omega) = \frac{x_i^{av}(\omega)}{F_h^{av}(\omega)} \tag{11.1}$$

where $x_i^{av}(\omega)$ and $F_h^{av}(\omega)$ represent the values of the displacement at the observation point and of the impulse load, respectively, defined in the frequency domain. The

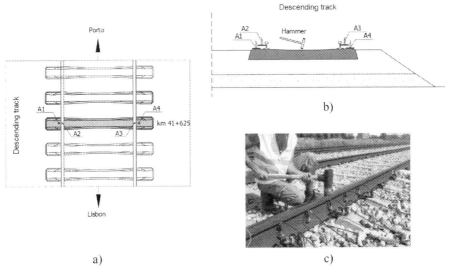

Fig. 11.6 Layout of the receptance tests (A) plan; (B) cross-section; (C) photographic record.

consideration of a large number of events makes reasonable the consideration of mean values, minimizing the effect of random noise.

This aspect can be confirmed by the computation of the coherence between the excitation and the response, where a value above 0.95 means a high similarity between the multiple events, and therefore the absence of random disturbances. The coherence function is given by:

$$\gamma^2(\omega) = \frac{S_{hi}^{av}(\omega)\overline{S}_{hi}^{av}(\omega)}{S_{hh}^{av}(\omega)\overline{S}_{ii}^{av}(\omega)} \tag{11.2}$$

where S_{jk} is the power spectrum and the overlying dash represents the conjugate variable.

Fig. 11.7 presents the receptance curves, based on the response measured at the end of the sleeper and on the rail, for all the tested sections. Only experimental data with coherence values greater than 0.95 were considered.

A global analysis of the receptance curves presented in Fig. 11.7 allows the identification of two characteristics resonance peaks: the first one appears at a frequency close to 110 Hz and corresponds to the overall vibration of the track, i.e., the vibration of rail and sleepers over the different granular layers of the track. The second peak

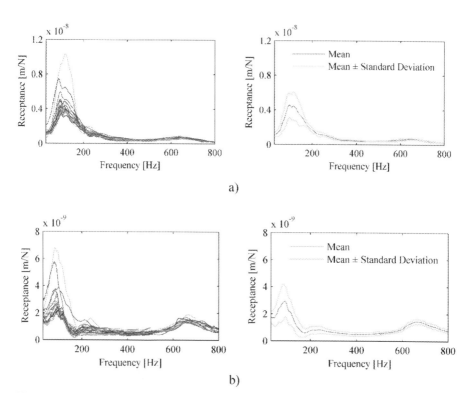

Fig. 11.7 Receptance curves corresponding to the different sections: (A) sleeper; (B) rail.

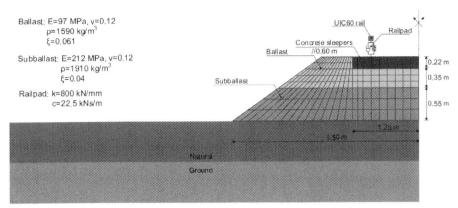

Fig. 11.8 Geomechanical properties of the railway track.

occurs at higher frequencies, around 650 Hz, and corresponds to the vibration of the rail over the rail pads. This peak, less evident in the sleeper receptance, becomes clearer in the rail cross receptance. A simplified analysis, assuming a SDOF system where the mass is given by the rail and the spring stiffness by the rail pad, enables to conclude that this resonance frequency is related to a railpad stiffness of 620 kN/mm. This value is in correspondence with the value previously pointed out by the railpad supplier for the rail pads adopted in this stretch (Vossloh Zw 687a).

As previously stated, several properties of the track elements are very difficult to characterize, namely the stiffness, the damping and the mass of the ballast and subballast. These properties can be evaluated by an inversion procedure, where the numerical model is calibrated in order to obtain a reasonable fit between the measured and computed receptances. A complete description of the optimization procedure can be found in Alves Costa et al. [19]. Fig. 11.8 summarizes the properties of the track after the inversion and updating procedure.

11.2.3 Ground characterization

11.2.3.1 General description

According to a geotechnical characterization conducted during the renovation of the railway track, the ground is characterized by the existence of an embankment consisting of clay-sandy material with a thickness of about 2.0 m. This formation is covering clay and clay-sandy alluvial formations with variable thickness. Table 11.1 shows the results from the SPT test (Standard Penetration Test), accompanied by a short lithological description of the different layers.

Beyond classical geotechnical tests, such CPT and SPT, also geophysical characterization tests were performed, namely Cross-hole and SASW tests. The resource to geophysical testing is a key aspect when dealing with ground-borne vibration problems, since the small-strain stiffness properties govern the wave propagation over the ground. Fig. 11.9 shows a plan with location of the tests, all of them very close to the railway track infrastructure.

Table 11.1 Information provided by the SPT test.

z (m)	Description		N (SPT)
1,5	Embankment clayey sand		4
3,0	Clayey sand with organic material		13
4,5	Clay with sand intercalations		0
6.0			21
7.5			14
9.0			20
10.5	Clay with limestone fragments		16
12.0			15
13.5			14

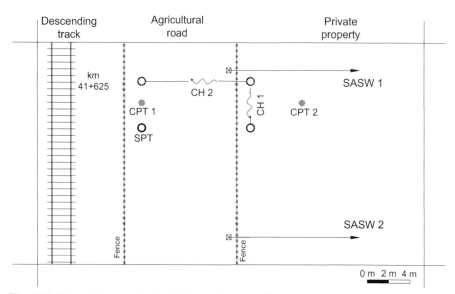

Fig. 11.9 Plan of the test site, including the location of the geotechnical tests.

The information provided by the classical geotechnical tests can be used from a qualitative point of view since they involve deformation levels well above those generated by railway traffic, whereby they are not here presented. A detailed description and interpretation of the results from SPT e CPT tests can be found in Alves Costa et al. [19] and Santos et al. [27]. As already mentioned, geophysical characterization tests are essential, allowing the estimation of the dynamic properties of the ground that are usually compatible with the strains induced by railway traffic.

11.2.3.2 Geophysical tests

Cross-Hole tests are currently used in the characterization of the ground when soil properties compatible with low strain levels are required. As shown in Fig. 11.9, two cross-hole tests were performed (CH1 and CH2), with a depth of up to 9.0 m and a spacing of 1.0 m in depth. The employed equipment (see Fig. 11.10) allows the generation of polarized shear waves, allowing the measurement of the P and S waves velocities.

In addition to the cross-hole tests, two SASW tests were performed. This non-intrusive geophysical test includes an experimental component, as well as an inversion procedure.

The experimental part consists of the application of an impulse excitation to the ground surface (the signal is measured by a strain gauge device placed between the falling mass and the ground) and the transient signal is recorded using accelerometers placed in a straight line starting from the source (Fig. 11.11). In order to maximize accuracy/resolution of the transfer function in the wavenumber domain, it is advised to acquire the response along a length as long as possible (e.g. 100 m) with a sensor spacing of 1 m. Obviously, a different array can be chosen. However, it has implications in terms of results resolution: if a shorter array length is considered it implies some loss of accuracy in the low-frequency range.

Using the previously described setup, P-wave profile can be easily assessed by a time-domain analysis of the first wave arrival of P-waves in each receiver location, as shown in Fig. 11.12.

a)   b)

Fig. 11.10 Cross-hole test: (A) source; (B) receiver.

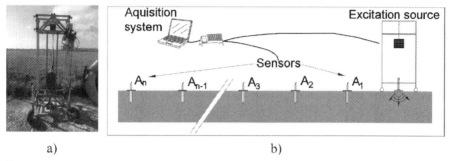

Fig. 11.11 SASW test: (A) falling weight device; (B) experimental setup.

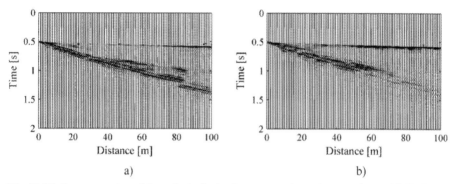

Fig. 11.12 Representation of the arrival of seismic waves to the various receivers: (A) SASW1; (B) SASW2.

Regarding S-waves and damping profiles, they are obtained by an inversion procedure used to fit the theoretical curves to the experimental ones. The optimization problem ends when a reasonable correspondence between experimental and theoretical curves is reached. Additional details about the mathematical formulation can be found in Degrande et al. [28]. The determination of the experimental attenuation curve is based on the half-power bandwidth method, proposed by Badsar et al. [29]. The comparison between experimental and theoretical curves for both tests is shown in Fig. 11.13.

From the reference geophysical tests, the dynamic properties of the soil, in terms of S and P-waves velocities, as well as hysteretic damping coefficients, were obtained as depicted in Fig. 11.14.

As evidenced in Fig. 11.14, some differences can be seen between results given by the SASW and the cross-hole tests. The heterogeneity of the ground and the region covered by each test can be pointed out as possible explanations to these differences.

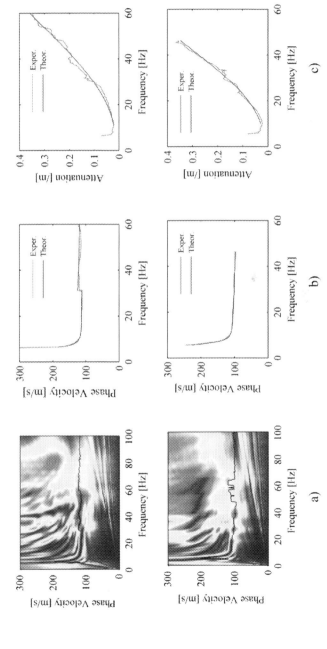

Fig. 11.13 SASW results: (A) frequency-wavenumber spectrum; (B) dispersion curves; (C) attenuation curves [top—SASW 1; bottom—SASW 2].

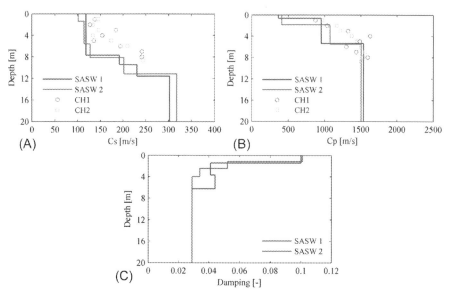

Fig. 11.14 Dynamic properties of the soil along the depth: (A) S-waves velocity profile; (B) P-waves velocity profile; (C) damping.

In fact, SASW tests involve a larger volume of the ground, while cross-hole gives information corresponding to a restricted area.

Another evident difference is related to the position of the water table: according to SASW tests it should be located at depths close to 6.0 m, while cross-hole suggest a more superficial location. These deductions are performed based on an analysis of the P-waves profiles. In fact, the depth of the water level presents a strong influence in the P wave velocity, once the dynamic response of the ground beneath the water level occurs without volumetric deformation, which explains the P-wave velocities around 1500 m/s presented in Fig. 11.14B. An explanation for the difference between results is related to the seasonal variation of the water level: SASW tests were developed in the end of the summer and cross-hole tests were performed in the end of the winter, when the phreatic level reaches the upper level.

11.2.4 Dynamic characterization of the structure

11.2.4.1 General description

The structure built next to the railway track consists in a three-story steel building (using structural IPE 100 steel profiles), with plan area of 2.1 × 2.1 m^2 and a total height of 3 m. All the steel profiles are linked by bolted connections. The slabs, constructed in MDF (medium-density fiberboard), have a nominal thickness of 30 mm. As evidenced in Fig. 11.15, each slab is supported by a set of 16 discrete points, uniformly distributed along its entire perimeter. These supports confer a rigid connection between the MDF panels and the horizontal IPE100 steel profiles.

Fig. 11.15 General view of the structure and details of the slab-beam connections and the column-footing connections.

Table 11.2 Material properties of the different elements of the structural system.

Material	Mass density (kg/m^3)	Poisson's ratio	Elasticity modulus (MPa)
IPE100	7850	0.30	210,000
MDF	700	0.25	4000

The prototype was designed to present a mechanical behavior similar to a regular concrete building in what concerns to the natural frequencies associated to the bending motion of the slabs (1st natural frequency around 15 to 20 Hz). The mechanical properties of the different structural elements are presented in Table 11.2.

The structure is supported by 4 reinforced concrete square spread footings, with 0.70 m of edge and a height of 0.35 m. The connection between the structure and the footings was performed by sleeve anchor bolts (see Fig. 11.15), conferring a rigid connection between both structural elements. Fig. 11.16 schematics the location of the structure in relation to the track.

11.2.4.2 Modal identification tests

To characterize the overall dynamic behavior of the structure, essential to the calibration process of the numerical model, an ambient vibration test was carried out. Globally, two different experimental setups were adopted to evaluate the modal parameters. The first one is focused on the experimental assessment of the mode shapes and natural frequencies of the structure associated to flexural and torsional motions of the global structure and the second one is used to investigate the modal

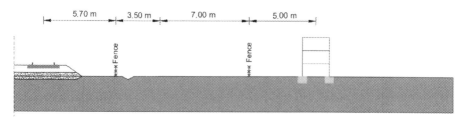

Fig. 11.16 Schematic representation of the structure placement at the test site: cross-section view.

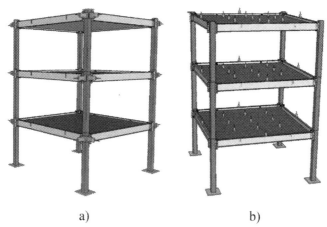

Fig. 11.17 Modal identification test: (A) measurement layout focused on the flexural and torsional motion of the structure; (B) measurement layout focused on the flexural motion of the slabs.

behavior of the slabs. For that purpose, the measurement layouts shown in Fig. 11.17 were adopted.

The collected acceleration time series from the two setups were processed through the modal identification algorithm Peak-Picking implemented in the commercial software ARTeMIS Extractor [30]. The obtained response spectra are presented in Fig. 11.18 for both layouts presented in Fig. 11.17.

The structural mode shapes of the first three natural frequencies associated to each layout are depicted in Figs. 11.19 and 11.20.

11.2.4.3 Experimental assessment of the dynamic stiffness of the footings

An additional experimental activity was developed to evaluate the dynamic stiffness curves of the footing-ground system. The use of such curves is relevant in the context of soil-structure interaction analysis. Thus, the experimental work developed corresponds to the application of a high number of impacts on the footing surface, through

Railway ground-borne vibrations

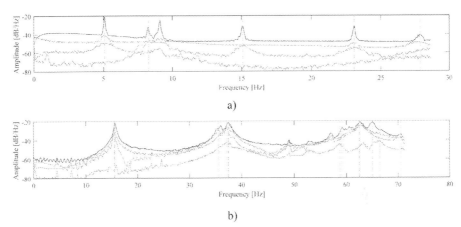

Fig. 11.18 Modal identification test: (A) response spectra associated to the flexural and torsional motion of the structure; (B) response spectra associated to the bending motion of the slabs.

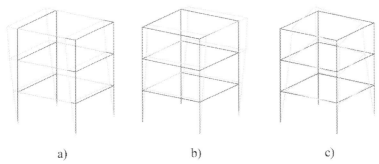

Fig. 11.19 Experimental mode shapes: (A) 1st mode: $f = 5.08$ Hz; (B) 2nd mode: $f = 8.27$ Hz; (C) 3rd mode: $f = 9.10$ Hz.

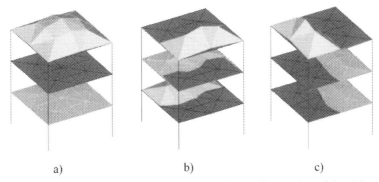

Fig. 11.20 Experimental mode shapes associated to the bending motion of the slabs: (A) 1st mode typology: $f = 15.41$ Hz; (B) 2nd mode typology: $f = 35.52$ Hz; (C) 3rd mode typology: $f = 37.23$ Hz.

Fig. 11.21 Experimental activities used to evaluate the dynamic stiffness curves of the footings.

an instrumented impact hammer, being the transient signal recorded using accelerometers also placed on the footing surface. The photographic record of the experimental activities can be observed in Fig. 11.21.

Once the applied impulse and the response at the receivers are known, it is then possible to calculate the frequency response function according to the following expression:

$$K_{\text{foot}}(\omega) = \frac{F(\omega)}{x(\omega)} \tag{11.3}$$

where $x(\omega)$ and $F(\omega)$ represent the values of the displacement and the impact force in the frequency domain, respectively. The use of multiple events makes reasonable the application of mean values, minimizing the effect of ambient noise.

The dynamic stiffness curves for the different footings (module and phase angle) are depicted in Fig. 11.22. The dynamic stiffness is a complex quantity, where the real and imaginary data represent, respectively, the footing response in phase and 90° out of phase relative to the force applied. The imaginary component of the dynamic

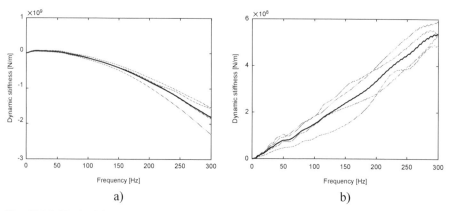

Fig. 11.22 Vertical dynamic stiffness of the footings: (A) real part; (B) imaginary part (*gray lines*: individual curves; *black line*: mean curve).

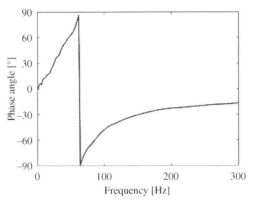

Fig. 11.23 Phase angle of the mean curve.

stiffness function is due to the presence of hysteretic and radiation damping in the soil. The black line represented on the figures corresponds to the mean curve, i.e., obtained by averaging the individual dynamic stiffness curve of all footings (gray lines).

As can be seen, there is some variability of the curves evaluated for each foundation that can be related to local heterogeneities of the soil. However, the mean curve could be a reasonable approximation for the present purpose. Bearing that in mind, and take into account the mean curve, the phase angle is computed considering the Eq. (11.4). The resultant curve is presented in Fig. 11.23, where it is possible to observe a resonant frequency of the footing over the ground around 55 Hz.

$$\theta(°) = \tan^{-1}\left(\frac{\text{imag}(H_{vv}(\omega))}{\text{real}(H_{vv}(\omega))}\right) \tag{11.4}$$

11.2.5 Testing of mitigation measures based on seismic metamaterial (phononic crystal) concept

One of the issues related with traffic ground-borne vibrations is associated with the efficiency of the mitigation measures. Actually, although the potentialities of the nowadays numerical tools, the experimental assessment of the mitigation measures is one key aspect of indubitable value. One of the advantages of the developed field test is the possibility of installing different mitigation measures along the propagation path and check the reliability of them based on experimental evidence.

In this context it should be highlighted that the mitigation of train-induced vibration is a widely explored topic in the scientific community [16,17,31], being the use of seismic metamaterials a newest concept to be explored. By definition, a metamaterial corresponds to a structure with higher attenuation characteristics than the ones enhanced by the material itself. For example, a periodic arrangement with stiffer inclusions is a metamaterial since the periodicity of the structure allows the development of the Bragg effect, creating a band-gap, i.e., a range of frequencies where the energy content is highly attenuated [32]. Despite the study of metamaterials, namely

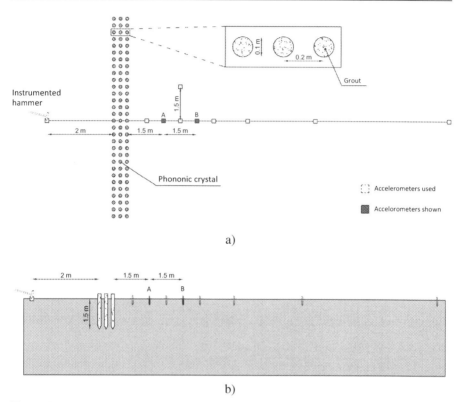

Fig. 11.24 Illustrative scheme of the experimental campaign: (A) plan view; (B) cross-section.

the phononic crystals, in acoustics is already advanced, the application in an elastodynamic context is reduced, being the experimental campaigns practically inexistent.

Having this in mind, a seismic metamaterial barrier was installed in the field test of Carregado. Fig. 11.24 shows a plan view, as well as a cross-section, of the experiment developed.

Fig 11.25 shows a picture before and after the phononic crystal barrier installation.

The experimental campaign was composed by two steps, one prior to the introduction of the phononic crystal on the ground and the other after construction. For each one of the steps, impacts were applied in the ground surface and the response was measured accordingly to the layout presented in Fig. 11.24. Based on the records of excitation and response signals, the transfer functions between the source and receiver are computed for both scenarios, i.e., before and after the construction of the seismic metamaterial barrier. Fig. 11.26 shows some examples of those results with the indication of the frequency range where Bragg's effect is expected to occur [17].

As can be seen by Fig. 11.26, the adoption of a phononic crystal barrier induces a complexity in the response pattern compared to the unreinforced scenario. It is important to note that although the phononic crystal barrier is designed to filter a specific range of frequencies, it does not mean that the remaining frequency spectrum will not

Railway ground-borne vibrations

a)   b)

Fig. 11.25 Experimental site overview: (A) before construction of phononic crystal barrier; (B) after construction.

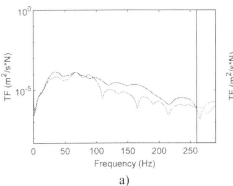 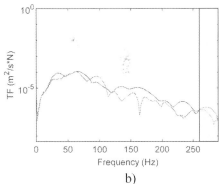

a)   b)

Fig. 11.26 Transfer function for: (A) Point A; (B) Point B (*blue line* (*dark gray* in print version)—without phononic crystal; *red line* (*light gray* in print version)—with phononic crystal; *light blue area* (*light gray* in print version)—Bragg's effect frequency range).

suffer changes. Actually, for frequencies larger than 100 Hz, there is a relevant attenuation of the transfer function of mitigated scenario when compared with the nonmitigated scenario. As can be seen, the mechanical behavior of a seismic metamaterial is very complex, representing a theme that remains largely unexplored.

11.2.6 Measurement of vibrations induced by railway traffic

The range of experiments presented above are related to the comprehensive characterization of the test site. Those steps are primordial for the identification of sources of error and uncertainty when proceeding to the numerical modeling of ground-borne vibrations induced by railway traffic.

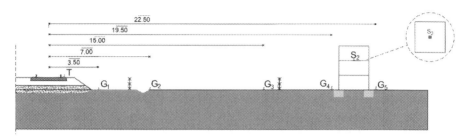

Fig. 11.27 Experimental setup adopted for the measurement of vibrations induced by traffic.

Beyond the characterization experimental activities, several campaigns for measurement of vibrations induced by railway traffic were developed on the test site over the last 10 years. In the present work, only some those results are summarized.

For the purpose of validation the prediction models, vertical dynamic responses of the track-ground-structure system were evaluated during the passage of several trains in the descending track. For that, unidirectional accelerometers were placed at the track, free-field and structure. The layout of the measurements points is illustrated in Fig. 11.27.

The results of these experiments are shown in a further section, where the experimental validation of a numerical model is presented.

Although the railway line, where the present field test is installed, is explored by different types of rolling stock, the results here presented are referred to the passage of train Alfa-Pendular. The Alfa-Pendular is the fastest train operating in Portugal, being a conventional train composed by six vehicles, as indicated in Fig. 11.28.

The main mechanical properties of the Alfa-Pendular train were provided by the constructor and they are summarized in Table 11.3. Additional information concerning the modeling of the rolling stock can be found in Alves Costa et al. [33] and Colaço et al. [34].

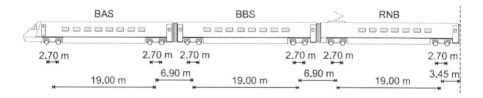

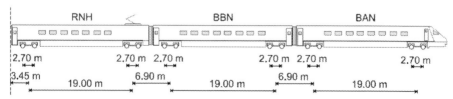

Fig. 11.28 Configuration of the Alfa Pendular train.

Table 11.3 Summary of the mechanical properties of the Alfa Pendular train.

Vehicle			BAS	BBS	RNB	RNH	BBN	BAN
Car body		Mass (kg)	36,901	37,810	36,924	38,524	38,510	37,301
		Rotational inertia (kg m^2)	2,083,600					
Secondary suspension		Stiffness (kN/m)	1320					
		Damping (kN s/m)	36					
Bogies	1	Mass (kg)	4932	4823	4712	4712	4712	4823
		Rotational inertia (kg m^2)	5150	5090	5000	5000	5090	5150
	2	Mass (kg)	4823	4823	4712	4712	4823	4923
		Rotational inertia (kg m^2)	5150	5090	5000	5000	5090	5150
Primary suspension		Stiffness (kN/m)	3200					
		Damping (kN s/m)	35					
Axles	1	Mass (kg)	1538	1538	1538	1538	1538	1538
		Load (kN)	129.8	131.4	128.8	132.6	133.2	130.2
	2	Mass (kg)	1884	1884	1538	1538	1884	1884
		Load (kN)	133.2	134.8	128.8	132.6	136.6	133.6
	3	Mass (kg)	1884	1884	1538	1538	1884	1884
		Load (kN)	132.6	134.8	128.8	132.6	136.6	134.2
	4	Mass (kg)	1538	1538	1538	1538	1538	1538
		Load (kN)	129.2	131.4	128.8	132.6	133.2	130.8

11.3 Numerical modeling of ground-borne vibrations

11.3.1 Overview

The numerical prediction of vibrations induced by railway traffic involves the simulation of a complex system, composed by distinct components: train, track, soil, and building. To address the global system different numerical techniques can be applied. In the present study, a modular numerical model, based on a substructuring approach, is selected. The proposed approach is divided into three main modules: one includes the track-ground structure, modeled by a 2.5D approach taking into account the three-dimensionality nature of the domain; the second concerns the simulation of the dynamic behavior of the train, which is simulated by a multibody formulation considering the main masses and suspensions of the vehicles; the last one comprises the modeling of the structure and its dynamic interaction with the soil. Fig. 11.29 summarizes the procedure for train-track interaction. In the following sections the main aspects and assumptions of each one of the modulus are described.

11.3.2 Track-ground system—2.5D FEM-BEM model

The computation of the 3D track-ground dynamic response induced by the train passage is performed using a numerical procedure based on the 2.5D coupling of finite (FEM) and boundary elements (BEM) formulation [7,19,35,36]. The 2.5D formulation assumes the linearity and the invariability of the domain (in the track development direction), being the equilibrium established in the wavenumber-frequency domain,

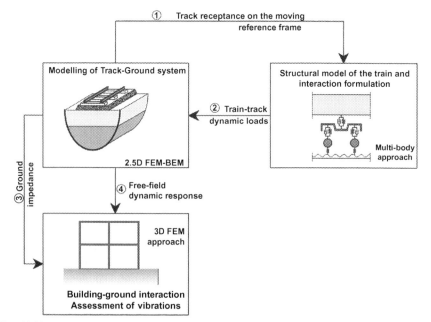

Fig. 11.29 Representative scheme of the numerical modeling approach.

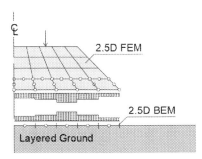

Fig. 11.30 2.5D FEM-BEM coupling.

employing Fourier expansions for space and time. Therefore, the 3D solution is obtained without the need of numerical discretization along the development direction of the track, resulting in a reduction of the computational effort when compared with fully discretized models.

Usually, a domain decomposition technique is used to solve the dynamic problem, being, in the current case, the track modeled by the 2.5D FEM and the layered ground simulated through 2.5D BEM, as shown in Fig. 11.30.

The coupling between both domains is done by a finite element formulation, comprising the transformation of the flexibility matrix that governs the dynamic behavior of the ground (BEM domain) into a dynamic stiffness matrix.

$$\left([K]_1^{\text{global}} + ik_1[K]_2^{\text{global}} + k_1^2[K]_3^{\text{global}} + k_1^4[K]_4^{\text{global}} - \omega^2[M]^{\text{global}} \right. \\ \left. + [K]_5^{\text{global}}(k_1,\omega)\right)\{u_n(k_1,\omega)\} = \{p_n(k_1,\omega)\} \tag{11.5}$$

where K_1^{global} to K_4^{global} are stiffness matrices of the FEM domain, M^{global} is the mass matrix; k_1 is the Fourier image of the coordinate x; ω is the frequency; u_n is the vector of the nodal displacements; p_n is the vector of the external forces. The matrix K_5^{global} is computed from the flexibility matrix that governs the dynamic behavior of the domain described by 2.5D BEM.

Detailed information about the deduction of the above-mentioned matrices, as well as the strategies for modeling particular aspects of the system, as for instance the sleepers and the rail, can be found in Refs. [19,37,38].

The nodal displacements and pressures along the coupling boundary FEM-BEM in the transformed domain are obtained solving the system of Eq. (11.5). Once these pressures and the Green's functions of the displacements are known, the computation of the free-field response is a simple step.

11.3.3 Train model and train-track interaction

In relation to the load applied by the train to the track, two components are considered: the static load (resulting from the movement of the dead loads—weight of the train) and the dynamic load (due to the train-track dynamic interaction). The latter component requires the simulation of the train-track dynamic interaction problem. Herein,

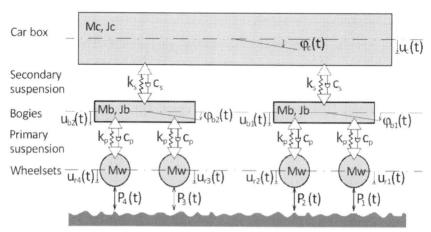

Fig. 11.31 Complete 2D vehicle model.

this problem is solved by a compliance procedure formulated in a frame of reference that moves with the train.

In the present study, the vehicle is simulated by a 2D multibody approach, as represented in Fig. 11.31. As a result, only the movements on the vertical plane of the train are taken into account, disregarding the dynamic loads induced by the movement of the train in other directions. So, assuming that the source of excitation corresponds to the track unevenness, the train-track interaction problem can be written in a matrix form on the frequency domain:

$$\{N(\Omega)\} = -\left([F] + [A] + [F^H]\right)^{-1}\{\Delta u(\Omega)\} \quad (11.6)$$

where $N(\Omega)$ gives the vector of the train-track interaction loads, being Ω the driven frequency (frequency of oscillation of the wheelset due to the unevenness with a wavelength k); F corresponds to the train compliance matrix at the contact points with the track; A is the compliance matrix of the track; F^H is a diagonal matrix where the terms are equal to $1/k_H$ (k_H is the linearized Hertzian stiffness); and $\Delta u(\Omega)$ is the complex displacement amplitude of the track unevenness. A detailed description of the mathematical formulation inherent to the evaluation of each one of these matrices can be found in Alves Costa et al. [19,33]. Nevertheless, it should be highlighted that the dynamic loads are assessed taking into account the motion of the train, which means that the response at a given frequency is a combination of responses induced by different excitation frequencies due to the Doppler effect in the ground.

11.3.4 Soil-structure interaction model

For the modeling of the structure, a 3D FEM approach is adopted. The general equation of motion, in the frequency domain, is given by the following conventional matrix equation:

$$\left(K^b + i\omega C^b - \omega^2 M^b\right)u^b = F^b \quad (11.7)$$

where K^b, C^b and M^b corresponds to stiffness, damping and mass matrices of the structure, respectively; u^b is the displacement vector, and F^b is the loading vector. The damping matrix is proportional to the stiffness and the mass matrices, according to the Rayleigh damping model.

Given the nature of the present problem, the soil-structure dynamic interaction that occurs when the building is excited by an incident wave field (here represented in terms of displacements u_0) needs to be considered. For that, and taken into account the inertial forces generated by the motion of the structural elements caused by this incident wave field, the displacement vector of the degrees of freedom (DOF) in connection with the ground is given by

$$u_s^b = \Delta u^b + u_0 \tag{11.8}$$

where Δu^b corresponds to the displacement increment of the DOF due to the inertial forces generated in the building.

For a given dynamic stiffness matrix of the soil, K_s, and assuming the compatibility of displacements between the ground and the structure and the equilibrium condition between both systems, the following relationships can be easily inferred:

$$F_s = K_s \Delta u^b \tag{11.9}$$

$$F^b = -F_s = -K_s \Delta u^b \tag{11.10}$$

where F_s is the vector that corresponds to the incremental loads applied by the structure to the ground and F^b is the vector of incremental loads generated in each DOF of the structure in connection to the ground.

Finally, the following relationship can easily be derived when introducing Eqs. (11.8), (11.10) into Eq. (11.7):

$$\begin{bmatrix} K_b^{ii} & K_b^{ij} \\ K_b^{ji} & K_b^{jj} + K_s \end{bmatrix} \begin{bmatrix} u_i \\ \Delta u^b \end{bmatrix} = - \begin{bmatrix} K_b^{ii} & K_b^{ij} \\ K_b^{ji} & K_b^{jj} \end{bmatrix} \begin{bmatrix} 0 \\ u_0 \end{bmatrix} \tag{11.11}$$

where the term K_b^{mn} corresponds to the dynamic stiffness matrix of the structure (the indices mn referring to interior (m) or interface (n) nodes of the FEM), which can easily be obtained from a commercial finite element program.

The computation of the dynamic stiffness of the soil, K_s, is not so straightforward, existing different approaches, comprising different degrees of approximation and complexity. In the present chapter, a simplified methodology, which consists of the use of a lumped-parameter model, is adopted.

In a general way, the lumped-parameter model represents the frequency-dependent soil-structure interaction of a massless rigid footing. The dynamic behavior of the ground is simulated by a range of spring-dashpots-masses, with lumped values representing the stiffness, the damping and the inertial effects. Different lumped-parameter models are available in the bibliography [39,40]. The present study adopts the Monkey-tail fundamental lumped model, with the configuration depicted in

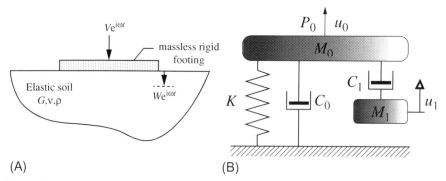

Fig. 11.32 Monkey-tail lumped-parameter model: (A) problem description; (B) structural system.
Adapted from L. Ibsen and M. Liingaard, Lumped-Parameter Models, Aalborg University, 2006, p. 36.

Fig. 11.32. This methodology has been previously validated by an experimental in situ campaign, as detailed exposed in Colaço et al. [41].

11.4 Experimental validation

11.4.1 General description

As emphasized before, the experimental test site developed and properly characterized allows the validation of numerical prediction tools. In this sense, the present section intends to present a general validation of the general numerical approach presented before. Thus, the experimental data recorded using the setup illustrated above is presented simultaneously with the numerical results achieved by the numerical modeling strategy presented above.

11.4.2 Dynamic response of the track-ground system

Starting with the evaluation of the track response, specifically in what concerns the sleeper response, Fig. 11.33 compares the experimental and numerical time series of the vertical velocity induced by the passage of the train Alfa-Pendular at 212 km/h.

A very good agreement between both results is evidenced in Fig. 11.33, where a clear distinction of the passage of each bogie is identified in the temporal record. Despite the high quality of the results, for frequencies above 75 Hz slight differences between numerical and experimental curves are visible. These differences can be justified by the fact that only the excitation induced by the track unevenness is taken into account in the numerical modeling, discarding other causes that influence the dynamic excitation mechanism, namely irregularities of the wheels or the inhomogeneities of the track support.

In terms of the dynamic response at the free-field, the time records of the vertical velocity for the observation points listed for different distances are depicted in Fig. 11.34. The numerical prediction is again overlapped with the experimental record. Some remarks can be pointed out:

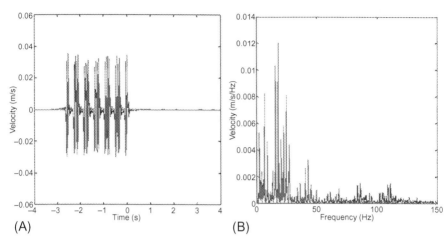

Fig. 11.33 Vertical vibration velocity at the sleeper induced by the train passage: (A) time record; (B) frequency content (*blue line* (*black* in print version)—experimental; *red line* (*dark gray* in print version)—numerical).

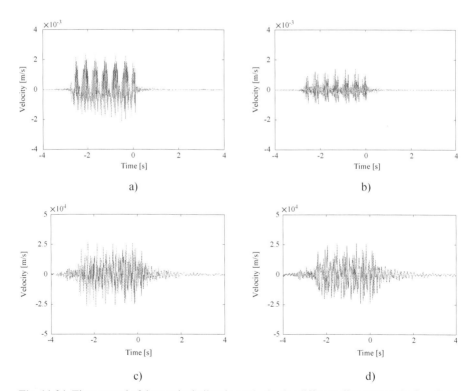

Fig. 11.34 Time record of the vertical vibration velocity for different distances at the free-field: (A) 3.5 m; (B) 7.0 m; (C) 15.0 m; (D) 22.5 m (*blue line* (*dark gray* in print version)—experimental; *red line* (*gray* in print version)—numerical).

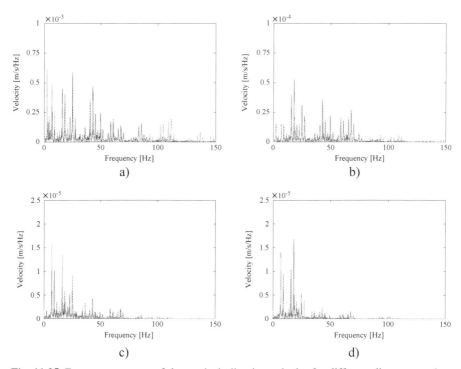

Fig. 11.35 Frequency content of the vertical vibration velocity for different distances at the free-field: (A) 3.5 m; (B) 7.0 m; (C) 15.0 m; (D) 22.5 m (*blue line* (*dark gray* in print version)—experimental; *red line* (*gray* in print version)—numerical).

- The passage of the axles and bogies in the observation points at the free-field becomes imperceptible with the increase of the distance to the track.
- Decrease of the amplitude of the vertical velocity with the increase of the receiver-source distance, while the duration of the event increases.
- A very good agreement between numerical and experimental results for the observation points closest to the track, companied by a loss of accuracy with the increase of the distance.

In what concerns to a frequency domain analysis, a comparison between the numerical and the experimental vertical velocity at the free-field is provided by Fig. 11.35. As expected, it was not possible to achieve a perfect match. However, the numerical model leads to a very reasonable assessment of the main characteristics of the response. In fact, the dominant frequencies of the numerical response agree well with the experimental ones. Moreover, the main trends of the system are well estimated by the numerical model, with a clear emphasis to the evolution of the amplitude with the increase of the distance source-receiver.

11.4.3 Dynamic response of the structure

Regarding the structure response, the experimental activities were performed in a different period compared to the previous results. Thus, and during a new passage of the Alfa-Pendular train, circulating on the descending track at a speed of 220 km/h, the

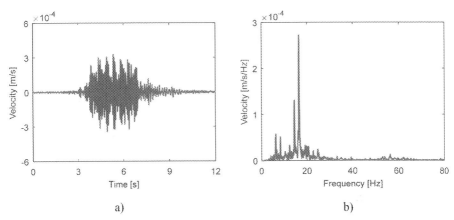

Fig. 11.36 Vertical vibration velocity at the observation point G4 due to the passage of the Alfa-Pendular train at speed of 220 km/h: (A) time record; (B) frequency content.

vibration field was recorded on the center of the slab of the 2nd floor and at the free-field, near the structure's base (observation points G4 to G5), accordingly to the layout illustrated in Fig. 11.27. In this way, and as an example, Fig. 11.36 presents the vertical vibration velocity for observation point G4.

Taking into account the excitation records evaluated for observation points G4 and G5, the dynamic response of the structure is computed following the procedure explained in Section 11.3.4. Fig. 11.37 illustrates the comparison between experimental and numerical results for the observation point S2, located in the center of the 2nd floor of the building.

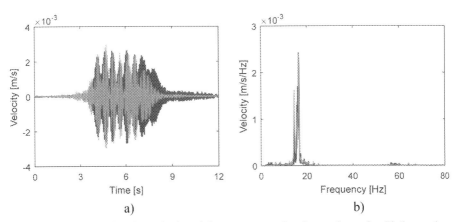

Fig. 11.37 Vertical vibration velocity of the structure at the observation point S2 due to the passage of the Alfa-Pendular train at speed of 220 km/h: (A) time record; (B) frequency content (*blue line* (*dark gray* in print version)—experimental; *red line* (*gray* in print version)—numerical).

The comparison between numerical and experimental results illustrated in Fig. 11.37 shows that the proposed model allows to achieve a realistic prediction of the vibration levels inside the building. This is valid for both frequency and time domain analyses. Notwithstanding this general comment, it is evident some loss of accuracy on the predicted values at the final instants of the vibration record, after the passage of the train, where the numerical results present a quicker decay of the response of the structure in free vibration conditions. A plausible reason for the obtained results can be attributed to a less accurate consideration of the damping conditions of the structure. However, peak vibration velocity and frequency content of the vibration field are well represented by the proposed numerical approach.

11.5 Conclusions

This chapter summarizes the main outputs of the exploration of a railway ground-borne vibration field test over the last 10 years. The field test has been implemented in the Portuguese railway network and comprises a comprehensive approach to the problem, from the source to the receiver. In order to obtain a deeper discernment of the problem, several sets of experiments were developed, having in mind a full characterization of each part of the problem. Thus, rolling stock was characterized, the track was subject to test in order to extract the main mechanical and geometrical properties and a comprehensive geotechnical ground characterization was developed.

Moreover, special attention is given to the dynamic response of structures when excited by the ground vibrations induced by railway traffic. For that purpose, a small-scale structure was constructed in the vicinity of the railway infrastructure.

The collected experimental data allowed to better understand the problem and provide an experimental database that can be used by the technical and scientific communities on the experimental validation of prediction tools.

Regarding the prediction tools, a substructure and the efficient numerical tool was presented. This tool is constituted by several modulus and is able to simulate the whole system, since the generation of vibrations up the effects of railway traffic on the dynamic response of nearby structures. The comparison of the numerical results with the measurements performed in the field test allowed to find a good agreement and, therefore, to check the reliability of the proposed numerical approach on the prediction of railway traffic vibrations. Actually, the proposed approach is a useful tool not only for the prediction of railway vibrations but also on the design of mitigation measures.

The studies presented in this chapter show the state of art in terms of development of experimental field test related with railway traffic ground-borne vibrations and on the development of reliable, efficient, and robust numerical prediction tools. The combination of experimental evidence with advanced numerical tools will allow the development of environmentally friendly railway infrastructures prepared to face the challenges of this century.

Acknowledgments

This work was financially supported by: Programmatic funding—UIDP/04708/2020 and Base Funding—UIDB/04708/2020 of the CONSTRUCT—Instituto de I&D em Estruturas e Construções—funded by national funds through the FCT/MCTES (PIDDAC); Project PTDC/ECI-EGC/3352/2021, Project PTDC/ECI-EGC/29577/2017—POCI-01-0145-FEDER-029577 and project PTDC/ECM-COM/1364/2014—POCI-01-0145-FEDER-016783—both funded by FEDER funds through COMPETE2020—Programa Operacional Competitividade e Internacionalização (POCI) and by national funds (PIDDAC) through FCT/MCTES. The authors are also sincerely grateful to European Commission for the financial sponsorship of H2020 MARIE SKŁODOWSKA-CURIE RISE Project, Grant No. 691135 "RISEN: Rail Infrastructure Systems Engineering Network"; this work was partially carried out under the framework of In2Track2, a research project of Shift2Rail.

References

[1] A. Quagliata, M. Ahearn, E. Boeker, C. Roof, L. Meister, H. Singleton, Transit Noise and Vibration Impact Assessment Manual, Federal Transit Administration—Department of Transportation, Office of Planning and Environment, Washington, DC, 2018.
[2] D.P. Connolly, G. Kouroussis, A. Giannopoulos, O. Verlinden, P.K. Woodward, M.C. Forde, Assessment of railway vibrations using an efficient scoping model, Soil Dyn. Earthq. Eng. 58 (2014) 37–47.
[3] D.P. Connolly, G. Kouroussis, P.K. Woodward, A. Giannopoulos, O. Verlinden, M.C. Forde, Scoping prediction of re-radiated ground-borne noise and vibration near high speed rail lines with variable soils, Soil Dyn. Earthq. Eng. 66 (2014) 78–88.
[4] H. Verbraken, G. Lombaert, G. Degrande, Verification of an empirical prediction method for railway induced vibrations by means of numerical simulations, J. Sound Vib. 330 (8) (2011) 1692–1703.
[5] D. López-Mendoza, A. Romero, D.P. Connolly, P. Galvín, Scoping assessment of building vibration induced by railway traffic, Soil Dyn. Earthq. Eng. 93 (2017) 147–161.
[6] P. Lopes, P. Alves Costa, M. Ferraz, R. Calçada, A. Silva Cardoso, Numerical modelling of vibrations induced by railway traffic in tunnels: from the source to the nearby buildings, Soil Dyn. Earthq. Eng. 61–62 (2014) 269–285.
[7] P. Galvín, S. François, M. Schevenels, E. Bongini, G. Degrande, G. Lombaert, A 2.5D coupled FE-BE model for the prediction of railway induced vibrations, Soil Dyn. Earthq. Eng. 30 (12) (2010) 1500–1512.
[8] S. François, L. Pyl, H.R. Masoumi, G. Degrande, The influence of dynamic soil–structure interaction on traffic induced vibrations in buildings, Soil Dyn. Earthq. Eng. 27 (7) (2007) 655–674.
[9] H. Chebli, D. Clouteau, L. Schmitt, Dynamic response of high-speed ballasted railway tracks: 3D periodic model and in situ measurements, Soil Dyn. Earthq. Eng. 28 (2) (2008) 118–131.

[10] J.N. Varandas, P. Hölscher, M.A.G. Silva, Dynamic behaviour of railway tracks on transitions zones, Comput. Struct. 89 (13–14) (2011) 1468–1479.

[11] P. Fiala, G. Degrande, F. Augusztinovicz, Numerical modelling of ground-borne noise and vibration in buildings due to surface rail traffic, J. Sound Vib. 301 (3–5) (2007) 718–738.

[12] P. Galvín, A. Romero, J. Domínguez, Fully three-dimensional analysis of high-speed train-track-soil-structure dynamic interaction, J. Sound Vib. 329 (2010) 5147–5163.

[13] M. Buonsanti, F. Cirianni, G. Leonardi, A. Santini, F. Scopelliti, Mitigation of railway traffic induced vibrations: the influence of barriers in elastic half-space, Adv. Acoust. Vib. (2009) 956263. https://doi.org/10.1155/2009/956263.

[14] P. Lopes, P. Alves Costa, R. Calçada, A. Silva Cardoso, Mitigation of vibrations induced by railway traffic in tunnels through floating slab systems: numerical study, in: A. Cunha (Ed.), Eurodyn 2014, European Association for Structural Dynamics (EASD), 2014, pp. 871–878. Porto.

[15] A. Colaço, P.A. Costa, P. Amado-Mendes, L. Godinho, R. Calçada, Mitigation of vibrations and re-radiated noise in buildings generated by railway traffic: a parametric study, Procedia Eng. 199 (2017) 2627–2632, https://doi.org/10.1016/j.proeng.2017.09.401.

[16] J. Barbosa, P. Alves Costa, R. Calçada, Abatement of railway induced vibrations: numerical comparison of trench solutions, Eng. Anal. Bound. Elem. 55 (2015) 122–139.

[17] A. Castanheira-Pinto, P. Alves-Costa, L. Godinho, P. Amado-Mendes, On the application of continuous buried periodic inclusions on the filtering of traffic vibrations: a numerical study, Soil Dyn. Earthq. Eng. 113 (2018) 391–405.

[18] D.J. Thompson, J. Jiang, M.G.R. Toward, M.F.M. Hussein, E. Ntotsios, A. Dijckmans, P. Coulier, G. Lombaert, G. Degrande, Reducing railway-induced ground-borne vibration by using open trenches and soft-filled barriers, Soil Dyn. Earthq. Eng. 88 (2016) 45–59.

[19] P. Alves Costa, R. Calçada, A. Silva Cardoso, Track–ground vibrations induced by railway traffic: in-situ measurements and validation of a 2.5D FEM-BEM model, Soil Dyn. Earthq. Eng. 32 (1) (2012) 111–128.

[20] D.P. Connolly, P. Alves Costa, G. Kouroussis, P. Galvin, P.K. Woodward, O. Laghrouche, Large scale international testing of railway ground vibrations across Europe, Soil Dyn. Earthq. Eng. 71 (2015) 1–12.

[21] G. Lombaert, G. Degrande, J. Kogut, S. François, The experimental validation of a numerical model for the prediction of railway induced vibrations, J. Sound Vib. 297 (2006) 512–535.

[22] P. Galvín, J. Domínguez, Experimental and numerical analyses of vibrations induced by high-speed trains on Córdoba-Málaga line, Soil Dyn. Earthq. Eng. 29 (4) (2009) 641–657.

[23] M. Sanayei, P. Maurya, J.A. Moore, Measurement of building foundation and ground-borne vibrations due to surface trains and subways, Eng. Struct. 53 (2013) 102–111.

[24] C. Zou, Y. Wang, J.A. Moore, M. Sanayei, Train-induced field vibration measurements of ground and over-track buildings, Sci. Total Environ. 575 (2017) 1339–1351.

[25] M. Crispino, M. D'Apuzzo, Measurement and prediction of traffic-induced vibrations in a heritage building, J. Sound Vib. 246 (2) (2001) 319–335.

[26] A. Mann, DYNATRACK: A Survey of Dynamic Railway Track Properties and Their Quality, Technische Universiteit Delft, Delft, 2002.

[27] N.C. dos Santos, A. Colaço, P.A. Costa, R. Calçada, Experimental analysis of track-ground vibrations on a stretch of the Portuguese railway network, Soil Dyn. Earthq. Eng. 90 (2016) 358–380.

[28] G. Degrande, S.A. Badsar, G. Lombaert, M. Schevenels, A. Teughels, Application of the coupled local minimizers method to the optimization problem in the spectral analysis of surface waves method, J. Geotech. Geoenviron. 134 (10) (2008) 1541–1553.

[29] S. Badsar, M. Schevenels, W. Haegeman, G. Degrande, Determination of the material damping ratio in the soil from SASW tests using the half-power bandwidth method, Geophys. J. Int. 182 (2010) 1493–1508.
[30] SVS, Structural Vibration Solution—ARTeMIS Extractor Pro. Release 5.3, ed, 2012.
[31] P. Alves Costa, R. Calçada, A. Silva Cardoso, Ballast mats for the reduction of railway traffic vibrations. Numerical study, Soil Dyn. Earthq. Eng. 42 (2012) 137–150.
[32] S. Brûlé, E.H. Javelaud, S. Enoch, S. Guenneau, Flat lens effect on seismic waves propagation in the subsoil, Sci. Rep. 7 (1) (2017) 1–9.
[33] P. Alves Costa, R. Calçada, A. Cardoso, Influence of train dynamic modelling strategy on the prediction of track-ground vibrations induced by railway traffic, Proc. Inst. Mech. Eng. F J. Rail Rapid Transit. 226 (4) (2012) 434–450.
[34] A. Colaço, P.A. Costa, D.P. Connolly, The influence of train properties on railway ground vibrations, Struct. Infrastruct. Eng. 12 (5) (2016) 517–534.
[35] R. Calçada, R. Delgado, A. Cardoso, P. Alves Costa, N. Santos, C. Ribeiro, B. Coelho, Train-track-ground interaction on high speed lines, in: Noise and Vibration on High-Speed Railways, FEUP, Porto, 2008.
[36] S. François, M. Schevenels, P. Galvín, G. Lombaert, G. Degrande, A 2.5D coupled FE–BE methodology for the dynamic interaction between longitudinally invariant structures and a layered halfspace, Comput. Methods Appl. Mech. Eng. 199 (23–24) (2010) 1536–1548.
[37] L. Gavric, Computation of propagative waves in free rail using a finite element technique, J. Sound Vib. 185 (3) (1995) 531–543.
[38] S. François, M. Schevenels, G. Degrande, J. Borgions, B. Thyssen, A 2.5D finite element-boundary element model for vibration isolating screens, in: P. Sas, B. Bergen (Eds.), KATHOLIEKE UNIV LEUVEN, DEPT WERKTUIGKUNDE, 2008, pp. 2765–2776. Leuven.
[39] J.P. Wolf, Foundation Vibration Analysis Using Simple Physical Models, Prentice Hall, Englewood Cliffs, 1994.
[40] L. Ibsen, M. Liingaard, Lumped-Parameter Models, Aalborg University, Aalborg, 2006, p. 36.
[41] A. Colaço, P. Alves Costa, A. Castanheira-Pinto, P. Amado-Mendes, R. Calçada, Experimental validation of a simplified soil-structure interaction approach for the prediction of vibrations in buildings due to railway traffic, Soil Dyn. Earthq. Eng. 141 (2021), https://doi.org/10.1016/j.soildyn.2020.106499.

Lateral resistance of different sleepers for the resilience of CWR tracks

Guoqing Jing and Peyman Aela
Beijing Jiaotong University, Beijing, China

12.1 Introduction

The Ballasted track is the most traditional type of railway track widely utilized in heavy-haul and high-speed railway tracks. The predominant feature of the ballast layer is to provide vertical, lateral, and longitudinal resistance to keep track role [1]. The construction of continuously welded rails (CWR) in ballasted tracks leads to the high compressive force along the rails due to the high-temperature variations or passing trains [1–6]. In order to prevent railway track buckling, the stability of the railway track is a continuing concern within CWR tracks in terms of vertical, longitudinal, and lateral resistance as well as the influence of temperature variation.

The compressive force is uniformly distributed along the rail track before buckling. However, the thermal force increment or dynamic loading induced by trains led to track buckling when the axial force is higher than the longitudinal resistance of the track (Fig. 12.1). In the first case, buckling occurs with the increase of the temperature above the neutral temperature (T_B). In this regard, track safe operating temperature (T_0) is calculated as follows:

$$T_B = T_N + dT_B \tag{12.1}$$

$$\begin{cases} T_0 = T_N + dT_s, & \text{DMS} \geq 10°C \\ T_0 = T_N + dT_B - 10, & \text{DMS} \leq 10°C \end{cases} \tag{12.2}$$

where T_B, buckling temperature; T_N, actual neutral temperature; dT_B, temperature increment above neutral temperature; T_0, safe operating temperature; DMS, dynamic margin of safety $= dT_B - dT_S$; dT_S, safe temperature increment above neutral temperature.

Temperature increment leads to the high axial compression along the rail and consequential movements of the rail track toward the outside [8]. This axial force is calculated:

$$N = \alpha EA \quad \Delta t° \tag{12.3}$$

Rail Infrastructure Resilience. https://doi.org/10.1016/B978-0-12-821042-0.00017-4
Copyright © 2022 Elsevier Inc. All rights reserved.

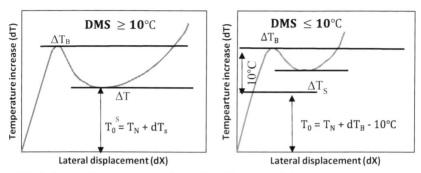

Fig. 12.1 Safe operating temperature limits for different conditions [7].

where α is the steel expansion coefficient, E is the steel elasticity modulus, and A is the rail section area [9]. To overcome this issue, white painting of the rails by different coating systems was a solution to reduce the rail temperature down to 10°C.

The impact of train loads on buckling has been investigated using experimental and theoretical techniques (e.g., Kish et al. [10]; Poulton [11]; Van [12]). Misalignments, lateral resistance, sleeper/ballast interaction, longitudinal and torsional resistance, sleeper spacing, curvature, and static and dynamic loading resulting from passing trains are the crucial parameters of track buckling associated with ballast layer characteristics [2,3,13,14]. The shape of lateral buckling in CWR tracks is a sinusoidal waveform caused by the following energies accumulated in a railway track:

- Strain energy caused by longitudinal force
- Strain energy generated by rail bending
- Internal energy caused by ballast resistance

In this way, buckling strength is calculated by the application of the principle of virtual work to these energies as follows:

$$P_l = P + \left\{ \frac{\gamma^2 r^2}{P} + \frac{\alpha r}{P^3 (P)^{1/2}} \left[\left(g - \varepsilon \frac{P}{R} \right)^2 + k \left(g - \varepsilon \frac{P}{R} \right) \frac{P}{R} \right] \right\}^{1/2} - \frac{\gamma r}{(P)^{1/2}} \quad (12.4)$$

where is P_l, buckling strength; P, longitudinal rail force balanced after buckling; g, longitudinal ballast resistance; r, lateral ballast resistance.

Generally, the lateral resistance of ballasted tracks is governed by different factors illustrated in Fig. 12.2. As shown in Fig. 12.3, the lateral displacement of the track is provided by rail/fastening, fastening/sleeper, and sleeper/ballast interaction so that the contribution of rail, fastening system, and ballast is 10%, 30%, and 60%, respectively [15]. Since the ballast layer has the high contribution to the sleeper lateral resistance, several methods have been employed to measure this parameter, including Track Lateral Pull Test (TLPT), Single Tie Push Test (STPT), Discrete Cut Panel Pull Test (DCPPT), Continuous Dynamic Measurement (DTS) [10]. In this regard, empirical formulas for railway track lateral forces and sleeper lateral resistance were recommended by researchers [16]. The STPT is the most convenient method to determine

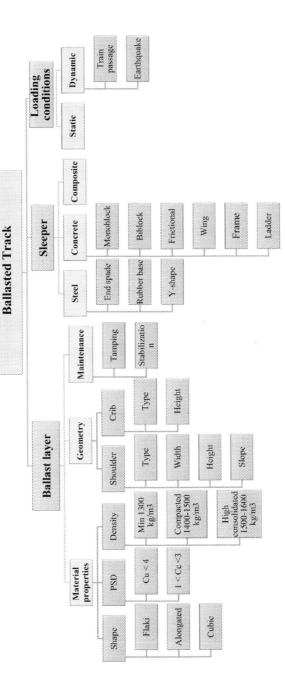

Fig. 12.2 Effective factors relevant to the lateral resistance.

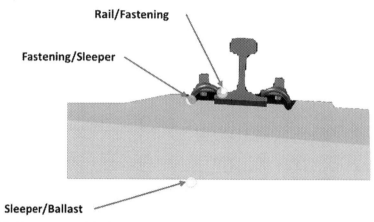

Fig. 12.3 Components of the lateral resistance of the track.

the sleeper lateral resistance, whose procedure was presented in AREMA standard [17]. This chapter focuses on the contribution of the components of the ballast layer, sleeper types, and shapes to the lateral resistance of railway tracks.

12.2 Fastening/sleeper resistance

The rail fastening system consists of rail anchors, rail tie plates, chairs, fasteners, spikes, screws, and bolts to fix rails to railway sleepers [18]. According to British standards, the minimum resistance force of fasteners is 7 and 9 kN for high-speed rail and heavy haul tracks, respectively. Fig. 12.4 illustrated the load distribution

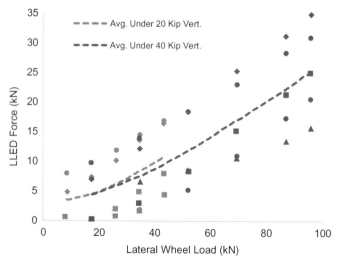

Fig. 12.4 Relationship between LLED force and lateral wheel load for a rail seat.

through the rail section. The lateral bearing force on the shoulder was measured by the Lateral Load Evaluation Device (LLED). The results indicated that the growth in the lateral wheel load from 20 to 100 kN leads to a 20 kN increase in LLED force, while the value of the vertical loading has a low impact on the lateral bearing restraint forces [19]. In addition, the increase of the train speed causes a major reduction in lateral resistance of fastening systems [20]. To enrich the performance of rail fasteners, 32% of the lateral stiffness increment leads to a 100% increase in lateral load transmission [21].

12.3 The effect of ballast specifications on the lateral resistance

Ballast specifications could be evaluated from two aspects of ballast particle properties (e.g., type, shape, density, and particle size distribution) and geometry of the ballast layer (e.g., dimensions of shoulder ballast and crib ballast). In the following, each parameter is discussed in detail.

12.3.1 Ballast material properties

Since the mechanical properties of ballast aggregates impact ballast crushing, the breakage of particles results in the lower lateral resistance of the ballast layer [22]. For instance, friction between ballast aggregates remarkably influences the lateral resistance of sleepers, while the friction between ballast and sleeper has an ignorable influence on the ultimate force between sleepers and ballast aggregates. Therefore, the hardness and toughness of aggregates should be considered for ballasting. Table 12.1 shows the allowable value of properties for different ballast materials according to different standards [23].

Table 12.1 Ballast material properties [23–25].

Ballast property	Australia	AREMA	Canada	China
Aggregate crushing value	<25%			<22%
LAA	<25%	<30%	<20%	<27%
Flakiness index	<30%	<5%		<20%
Misshapen particles	<30%		<25%	
Sodium sulfate soundness		<5%	<5%	
Magnesium sulfate soundness			<10%	
Soft and friable pieces		<5%	<5%	
Fines (<no. 200 sieve)		<1%	<1%	
Clay lumps		<0.5%	<0.5%	
Bulk unit weight (kg/m^3)	>1200	>1120		
Particle specific gravity	>2.5	>2.6	>2.6	>2.55

From the perspective of environmental issues, steel slag is a practical solution as a substitution of mineral stone aggregates. The high resistance to abrasion, lateral movement on curves, the variation of temperature, and low sulfate soundness losses (ASTM C88) are advantages of steel slag ballast, as stated by the National Slag Association [26–28]. The recommended size of steel slag aggregates was proposed by American Railway Engineering Association (AREA) Specification for Prepared Slag, Stone and Gravel Ballast; Federal Specification SS-S-449 [29]. In order to examine the lateral stability of the steel slag layer, a series of STPTs was performed by Esmaeili et al. [30] to replace limestone ballast with steel slag aggregates. The results show that there was a 27% increase in the lateral resistance of the track constructed by steel slag aggregates. In fact, the higher roughness of steel slag particles is the main factor to improve the lateral stability of ballasted tracks.

12.3.2 Ballast particle size

The density of the ballast layer is directly related to the size of particles due to changes in the porosity of the ballast layer. Sun [31] confirmed that the increase of particle size caused more rounded and fewer columnar particles by the three-dimensional laser scanning method. To minimize the deformation under high-frequency cyclic loading, the uniformity coefficient of particles was recommended in the range of $1.8 \leq Cu \leq 2.0$ for the ballast to reach the optimum particle size distribution. In another research, the optimum ballast size was presented by Profillidis [32] for different shapes and sizes of ballast particles. It is obvious that the cube-shaped crush quarry material has a high impact on lateral resistance increment. In contrast, the use of gravel ballast is not recommended due to the lower lateral resistance, which was already reported by Mulhall et al. [33].

12.3.3 Influence of the ballast profile

Sleeper lateral resistance has a direct relationship with the geometry of ballast components. Thus, the depth of the ballast bed, shoulder ballast width and height, shoulder slope angle, and crib ballast height should be taken into consideration. According to the results of previous studies [1,10,15,17,34–37], bearing lateral force is the main function of the ballast bed and crib ballast, particularly under vertical loading conditions.

However, Irazábal [38] point out that the portion of ballast components to the lateral resistance changes throughout the displacement of the sleeper so that the frictional force between ballast bed and sleeper bottom surface was dropped significantly after 10 mm displacement, whereas the contribution of shoulder ballast increased by 20% due to the rise of passive pressure between sleeper end and ballast (Fig. 12.5). In addition, there was a 10% increase in the frictional force between sleeper sides and crib ballast after 10 mm sleeper movement. However, in order to prevent the ballast flight from the ballast surface, China's high-speed ballast track standard [39] recommended the reduction of ballast about 50 mm down to the sleeper top surface.

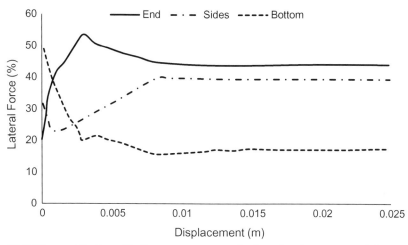

Fig. 12.5 The contribution of ballast components to the lateral resistance toward the displacement.

From the perspective of ballast depth, there was a 120% increase in the lateral resistance of the ballast panel when the ballast depth increased from 100 to 200 mm, as reported by Zeng [40]. Minimum ballast depth for different conditions was proposed by Selig and Waters [1], as shown in Table 12.2. On the other hand, DEM simulations conducted by Khatibi et al. [34] have shown that there is no justification for the increment of ballast depth higher than 300 mm owing to the lower efficiency of the ballast track. Thus, the ballast depth of 300 mm has been recommended as the optimum depth of the ballast layer.

As shown in Table 12.3, the influence of shoulder width is related to the type of sleeper on the ballast layer. For instance, the increment of shoulder width from 300 to 500 has only 5% improvement in the resistance of the ladder sleeper panel, while that was about 17.5% and 25.5% increment for mono-block and steel sleeper panels, respectively. In contrast, changes in shoulder ballast height remarkably enriched the interaction of shoulder ballast and sleeper ends for different types of sleepers so that the resistance increased by 44.2% for mono-block sleeper panel.

Table 12.2 Proposed ballast depths by Selig and Waters [1].

Function	Ballast depth (mm)
To provide resiliency to track on a subgrade less stiff than the ballast	150–300
To provide enough ballast for storage of fouling material	300–375
To accommodate tamping	Around 225
For drainage to keep water surface well below the ties	300–375
To reduce stress on underlying subballast or subgrade	150–450

Table 12.3 Lateral resistance increment for different types of sleepers [41–43].

Sleeper type	Mono-block	Bi-block	Winged	Ladder	Steel
Shoulder width increment from 300 to 500 mm	14.6%	17.5%	12.8%	5%	25.5%
Shoulder height increment from 0 to 150 mm	44.2%	49.3%	28.5%	24.3%	19.7%

Crib ballast is another component of the ballast layer that has frictional contact with sleeper sides. Le Pen and Powrie [36] reported that the crib ballast has 37%–50% contribution to the lateral resistance of sleepers. On the other hand, concerning the longitudinal resistance of ballast, the more crib ballast, the higher stability of the ballast layer. However, the ballast flight phenomenon causes major problems in terms of safety issues for high-speed railway tracks. Therefore, there should be a reduction in the height of the shoulder and crib ballast down to the sleeper top surface. According to the France and China standards, the crib ballast height should be 5 cm lower than the sleeper top surface [44].

12.4 Influence of the sleeper type and shape on lateral resistance

Sleeper type and shape lead to changes in the track lateral resistance due to the different weight and texture of sleepers as well as contact force with ballast layer components. A summary of sleepers' functions and features was already listed by Fischer [45]. In general, concrete sleepers are the most conventional type of sleeper due to the long service life and high stability in comparison with other types. Previous studies indicated that the lateral resistance of mono-block and timber sleepers is 1.8–2.7 and 1–1.3 times higher than that of steel sleepers under static loading [32,46]. On the other hand, the results of the pendulum loading test device (PLTD) conducted by Esmaeili et al. [47] confirmed that the ballast-sleeper interaction force of mono-block concrete sleeper is about 1.8 and 4.7 times higher than timber and steel sleepers, respectively.

Various alternatives were recommended in order to improve the performance of different types of sleepers. In this regard, increasing sleeper length or height, installation of safety caps, the attachment of end-plate to sleeper ends, and cutting grooves in the sleeper base were countermeasures to improve lateral resistance of timber sleepers. Modification of sleeper ends to spade shape [48], attachment of web steel stiffeners or rubber layer to the sleeper bottom surface [43], and application of Y-shape steel sleepers were recommended alternatives to improve lateral resistance of steel sleepers [49].

12.4.1 Steel sleeper

As illustrated in Fig. 12.6, to examine the influence of rubber layer and web steel stiffeners, a series of STPTs were performed on six steel sleepers reinforced by small and large bumped rubber layers (SRL, LRL) or web steel stiffeners (WSS1, WSS2, WSS3). Results show that there is an ignorable difference between the lateral resistance of the conventional steel sleeper and steel sleeper reinforced by a large bumped rubber layer owing to the ballast movement through the large gap between rubber bumps. In contrast, the attachment of the small bumped rubber layer led to a 25% increase in the lateral resistance of steel sleepers. On the other hand, the more web stiffeners, the higher lateral resistance of steel sleepers, so that the lateral resistance of WSS1 and WSS3 were 25% and 70% higher than conventional steel sleepers, respectively. Consequently, reinforcement of steel sleepers with web steel stiffeners is more efficient in terms of lateral resistance increment.

12.4.2 Winged sleeper

The proposed ideas for the improvement of concrete sleeper resistance are mostly in relevance with the shape of sleepers. The modification of the sleeper bottom surface

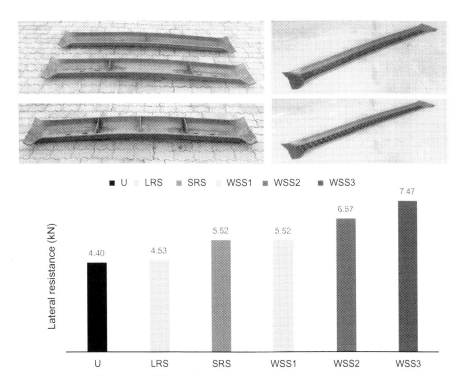

Fig. 12.6 Steel sleepers used in experimental tests [43].

Fig. 12.7 Winged sleepers shapes and dimensions.

influences the contact area between the sleeper and ballast particles as well as the lateral resistance of the sleeper. As reported earlier, the lateral resistance of a frictional concrete sleeper is approximately 64% higher than a mono-block sleeper [50,51]. The application of long sleepers is an approach implemented in the German railway tracks. Consequently, there was 15%–20% increase in the transverse resistance of ballasted tracks [32]. In another study, Pieringer [52] confirmed that the replacement of the B90.2 sleeper with BBS 4 led to a 40% increase in the lateral resistance of the track due to the higher sleeper base area.

The widen of sleepers in the rail seats is an innovative idea proposed by Hayano [53,54] to increase the interlocking of the crib ballast with sleepers. Based on experiments, the lateral resistance of 40 mm-winged sleepers was 67% higher than the typical type. To assess the performance of winged sleepers, a series of STPT tests were conducted on two types of winged sleepers named end-winged sleepers (EW) and middle-winged (MW) sleepers (Fig. 12.7). According to the results, EW sleepers had higher resistance to sleeper movement. With the consideration of ballast track maintenance, sleeper wings with 13 and 14 cm in width and height were recommended for the construction of winged sleepers. As already mentioned, the increase of shoulder ballast width and height could have a significant impact on the lateral resistance of sleepers which is obvious in Fig. 12.8.

12.4.3 Bi-block and mono-block sleepers

In another research, the application of bi-block sleepers was recommended as the substitution of mono-block sleepers [55]. Results show that there was a 23% lateral resistance increment when bi-block sleepers were under operation. Additionally, there is a 30% reduction in the total weight of the track when mono-block sleepers are replaced with bi-block sleepers [56]. However, an increase in the ballast depth is required in the case of using bi-block sleepers [32]. Fig. 12.9 displays STPTs for mono-block and bi-block sleepers performed on a ballast layer with the density, height, and shoulder

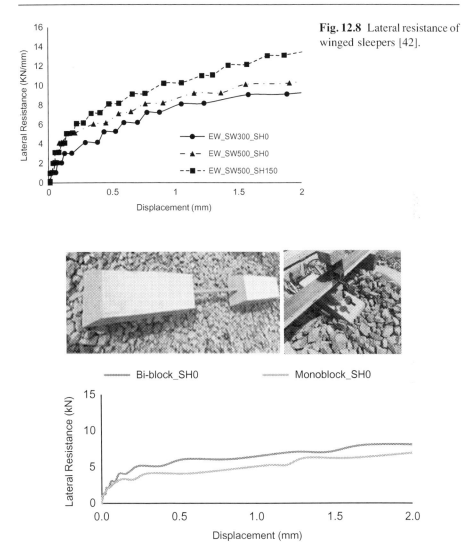

Fig. 12.8 Lateral resistance of winged sleepers [42].

Fig. 12.9 STPTs on mono-block and bi-block sleepers [57].

width of 1450 kg/m^3, 350 mm, and 500 mm, respectively. Based on results, bi-block sleeper resistance was 16% higher than the resistance of the mono-block sleeper after 2 mm displacement due to the better interlocking between crib ballast and bi-block sleeper.

12.4.4 Ladder sleeper

In order to improve the stability of railway tracks, the application of ladder sleepers was proposed by Moses et al. [58]. Also, the application of ladder sleepers is an

Fig. 12.10 STPT results and instrumentation for a ladder sleeper [41].

appropriate solution for high-speed railway tracks owing to the reduction in the ground vibration caused by passing trains [59]. Fig. 12.10 shows the structural detail of a ladder sleeper. The lower width of the ladder sleeper leads to less occupation of the land. The increase in the interaction between crib ballast and sleeper sides and higher weight of the sleeper is the advantage of ladder sleepers in terms of ballast lateral resistance. The results of STPTs carried out by Jing [41] indicated that the lateral resistance of panels with 25%, 50%, and 100% crib ballast height increased by 84%, 116%, 137% in comparison with the track with 0% crib ballast, respectively.

12.4.5 Anchor-reinforced sleeper

A novel concept suggested by Hill [60] as buckling prevention plates was the installation of steel anchors to the sleeper ends. Yamamoto et al. [13] suggested in another study the attachment of these plates at sleeper ends. However, the rise in plate height has caused the lateral resistance to increase; due to the maintenance process, the distance between the sleeper base and the ballast bed rises.

For concrete sleepers, the installation of sleeper anchors has recently been suggested to enhance the lateral resistance of sleepers under operation. Earlier, for timber sleepers, sleeper anchors have been installed to increase the lateral track resistance by up to 50% [61]. There are two types of sleeper anchors to improve the interlock between the ballast bed and the sleeper bottom surface or the crib ballast and sleeper sides. One of the advantages of sleeper anchors is related to the maintenance process due to the quick removal and assembling of anchors.

Moreover, by reducing the height of the ballast top surface on high-speed railway lines [62], sleeper anchors compensate for the resistance loss due to the lack of crib and shoulder ballast around sleepers. To date, companies have developed various shapes of sleeper anchors to be placed in the sleeper center and sleeper ends of concrete sleepers. In case the radius of curves is in the range of 351–500 m, alternate sleepers should be reinforced by anchors, while for curves with a radius of 250–300 m, each sleeper is required to be reinforced by anchors. To mount the anchor on each sleeper, the following steps need to be considered (Fig. 12.11):

- Removing aggregates of ballast about the sleeper.
- Attachment of sleeper anchors in the middle or sleeper ends.
- Tightening the anchor appropriately.
- Reballasting around sleeper anchors.

Fig. 12.11 Attachment of sleeper anchors to operated sleepers.

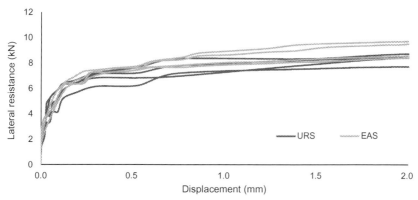

Fig. 12.12 STPTs results of anchor reinforced and unreinforced sleepers.

As Zarembski [61] has already reported, there is a fairly slight increase in lateral track resistance when anchors are used on each sleeper instead of an alternative sleeper. Consequently, in this analysis, the use of sleeper anchors was introduced on the alternate sleeper.

12.4.5.1 Results

Since a mono-block sleeper was reinforced by anchors, STPT results of the anchor-reinforced sleeper were compared with the mono-block sleeper. STPTs have been carried out four times for both unreinforced and reinforced sleepers. For each test condition, the average value of the lateral force after 2 mm sleeper movement was specified as a single sleeper's lateral resistance. According to the field test results given in Fig. 12.12, the lateral resistance for the EAS sleeper increased by 10%.

12.4.6 Bottom-textured sleeper (frictional sleeper)

Texturing the sleeper bottom surface leads to an increase in the lateral resistance without disruption or delay during the railway track tamping process, and the lateral resistance of the frictional sleeper can increase by 64%, as reported in Refs. [51,63]. The frictional sleeper is provided by texturing the bottom surface (Fig. 12.13). Many studies evaluated the portion of the base, crib, and shoulder ballast on the lateral resistance [36,65]. All the studies specified that the ballast bed significantly contributes to lateral resistance. Overall, frictional sleepers can increase the lateral resistance by approximately 32.0% compared with conventional concrete sleepers.

12.4.7 Nailed sleeper

The improvement of lateral track resistance could be achieved by better interaction of ballast bed and sleeper bottom. In this regard, using steel bars for constructing nailed

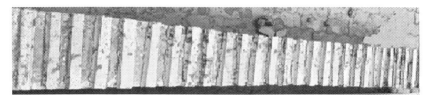

Fig. 12.13 Bottom texture sleeper [64].

sleepers is a new idea to provide high-lateral resistance sleepers, which have the following advantages [66]:

- Considerable increase in the lateral track resistance
- Without disturbing tamping operations
- Higher longitudinal resistance of the track

Fig. 12.14A demonstrates the structural detail of a nailed sleeper. The two holes are used for inserting nails with a constant diameter of 40 mm with different lengths (100–400 mm). Fig. 12.14B shows the relationship between lateral resistance and sleeper displacement. The lateral resistance increased by 53.7% when the mono-block sleeper was replaced with the nailed sleeper. The nail length has a substantial effect on the lateral resistance, as shown in Fig. 12.14B. In comparison with the mono-block

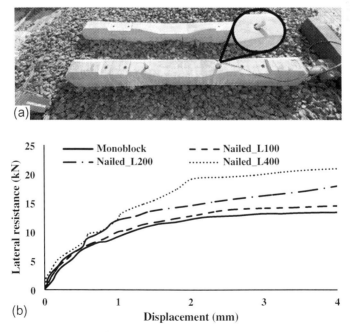

Fig. 12.14 (A) Construction of a nailed sleeper using steel bars. (B) Nailed sleeper lateral resistance [67].

sleeper, the lateral resistance increased by 6.1%, 19.1%, and 53.7% for 100, 200, and 400 mm-nail length sleepers, respectively.

12.4.8 FFU synthetic sleeper

Composite sleepers are other alternatives used recently in railway tracks instead of existing sleepers [68]. Typically, composite sleepers are made of Engineered Polymer Composite (EPC) or Engineered Wood Product (EWP) [69]. The advantages of composite sleepers are comprised of the low weight of the sleeper, high resistance against corrosion, and good insulation against electricity and temperature. However, the high construction cost, insufficient lateral resistance, and permanent deformation caused by creep are the most important challenges for applying this type of sleeper [68,70,71]. Fiber-reinforced polymer composite sleeper (FFU) is a new type of sleeper developed in Japan. To date, there have been no investigations on how the interactions of sleeper/ballast components affect the lateral resistance of FFU sleepers and on the impact of the shoulder height and width on the lateral resistance. In addition, the use of sleeper wings could be another method to improve the lateral stability of composite sleepers, having already been implemented for concrete sleepers [42,53].

12.4.8.1 FFU synthetic sleeper characteristics

FFU synthetic sleepers are produced by reinforcing synthetic resin foam with glass fibers with lengths of 20–80 mm and diameters of 20–35 nm so that the synthetic resin foam is made by a foaming chemical reaction on hard ethyl carbamate resin. The material properties and FFU sleepers are shown in Table 12.4. As depicted in Table 12.5, the mechanical characteristics of FFU sleepers were controlled by KEBOS [74].

To increase the sleeper/ballast interlocking capacity, optimized FFU sleepers might be fabricated by adding strip blocks or wings of the same material to the FFU sleeper. In the following, the characteristics of each type are described.

Optimized type A
Two forms of optimization in the structure are shown in Fig. 12.11. The extruded blocks are set to the outside to increase the disturbing ballast aggregates of the crib

Table 12.4 FFU synthetic sleeper properties [72].

Material	Mass percentage (%)
Long glass fiber	18.5
Hard urethane resin foam	80
Trimeric cyanamide tinting material	0.3
Releasing agent	0.3
Coupling agent	0.4
Whipping agent	0.5

Table 12.5 FFU synthetic sleeper characteristics [73].

Property			Unit	Beech New	FFU synthetic wood New	10 years	15 years	30 years	Standard
Density			kg/m^3	750	740	740	740	740	JIS Z 2101
Flexural strength			kN/cm^2	8	14.2	12.5	13.1	11.7	JIS Z 2101
Elastic modulus			kN/cm^2	710	810	800	816	816	JIS Z 2101
Compressive strength			kN/cm^2	4	5.8	6.6	6.3	6.0	JIS Z 2101
Shear strength			kN/cm^2	1.2	1.0	0.95	0.96	0.93	JIS Z 2101
Hardness			kN/cm^2	1.7	2.8	2.5	2.7	2.4	JIS Z 2101
Impact flexural strength	+20C		J/cm^2	20	41	–	–	–	JIS Z 2101
	−20C		J/cm^2	8	41	–	–	–	JIS Z 2101
Water absorption			mg/cm^2	137	3.3	–	–	–	JIS Z 2101
Insulation resistance		Dry	Ω	6.6*10^7	1.6*10^{13}	2.1*10^{12}	3.6*10^{12}	8.2*10^{11}	JIS K 6852
		Wet	Ω	5.9*10^4	1.4*10^8	5.9*10^{10}	1.9*10^9	–	JIS K 6852
Pull-out force dog rail			kN	25	28	28	23	22	RTRI
Pull-out force screw			kN	43	65	–	–	–	RTRI

ballast. Due to the lower ballast confinement between sleepers, there is a low average contact force between the crib ballast and sleeper sides. The added blocks are used to increase the confinement pressure on crib aggregates and increase the number of contacts. The added blocks are 60 mm * 60 mm * 260 mm in length, width, and height, respectively, and are made of the FFU material.

Optimized type B

Other forms of FFU sleepers consisted of strip blocks with dimensions of 10 mm * 10 mm * 240 mm on the sleeper base for type B1, and strip blocks with dimensions of 10 mm * 10 mm * 200 mm on lateral sides for type B2, and a 100 mm spacing between blocks was used for both sleeper types.

12.4.8.2 Test method and plan

According to the UIC International Standard for Lateral Track Resistance (UIC 2019), the STPT method was implemented to evaluate the lateral resistances of FFU sleepers. Fig. 12.12 shows the instrumentation of the panel for the measurement of lateral force-displacement. First, the ballast bed was poured into four layers so that each layer was compressed five times by a vibrating compactor. In the second step, after sleepers were placed, the crib ballast was poured into three layers compacted by a vibrating compactor. The spacing of the sleepers was 600 mm for all tests. Afterward, a lateral force was applied by a hydraulic jack fixed by steel rods at a rate of 0.05 kN/s. The lateral displacement of sleepers was measured by the attachment of two linear variable differential transformers (LVDTs) to the sleeper end. In the end, the lateral displacement of each sleeper was equal to the mean of the values obtained from the LVDTs. The data logger INV3018A was used to record the lateral resistance force, which corresponded to a 2 mm displacement of the sleeper. The experiment was repeated three times under each test condition, and finally, the average number was considered the sleeper lateral resistance.

12.4.8.3 Results

Fig. 12.13 shows the lateral resistance of sleepers for three sets of sleepers with different test conditions at a horizontal displacement of $d = 2$ mm. In the following section, the STPT results are discussed further regarding shoulder height and shoulder-width variations, as shown in Table 12.3.

The influence of shoulder height and width

As depicted in Table 12.6, along with the increase in shoulder ballast width from 300 to 500 mm and shoulder ballast height from 0 to 150 mm, the lateral resistances of sleepers all increased. The data for lateral resistance showed an ascending trend throughout the increase in the shoulder width and height so that changes in shoulder width from 300 to 500 mm lead to a 27.6% lateral resistance increase for the composite sleeper. On the other hand, the lateral resistance increased approximately 10.5% when the shoulder height was increased to 150 mm for panels with a shoulder width

Table 12.6 Shoulder width and height effects on lateral resistance of the prototype FFU sleeper.

Shoulder width-Shoulder height (mm)	Lateral resistance of different cases (kN)				
	Test No. 1	Test No. 2	Test No. 3	Avg	SD
SW300-SH0	5.86	5.12	6.95	**5.97**	0.92
SW400-SH0	6.23	7.82	7.13	**7.06**	1.29
SW500-SH0	7.86	7.52	7.48	**7.62**	0.79
SW300-SH150	6.08	8.15	5.78	**6.67**	1.50
SW400-SH150	6.23	7.63	9.23	**7.71**	0.21
SW500-SH150	7.68	7.45	10.13	**8.42**	1.48

Avg, average value; *SD*, standard deviation.

of 500 mm. It should be mentioned that the lateral resistance of the concrete sleeper was much higher than that of the composite sleeper under the same conditions, so the resistance of the concrete sleeper with shoulder widths and heights of 500 and 150 mm was approximately 21.7% higher than that of the prototype composite sleeper.

Lateral resistance of modified FFU sleepers

As illustrated in Fig. 12.13, with the application of wing-shaped FFU sleepers, types A1 and A2, the lateral resistance increased compared with that of the prototype composite sleeper. According to the data, sleeper type A2 had a better performance than type A1, but the increment was not significant for the same ballast condition. The lateral resistance of FFU sleeper types B1 and B2 with strip blocks along sleeper sides and the bottom surface was higher than those of all other types of composite sleepers used in this research. According to the results, sleeper type B2 exhibited an average resistance increase of 18.5%. However, no types of composite sleepers achieved the lateral resistance of the concrete sleeper in the same ballast layer condition. In addition, the increase in shoulder width and height increased the lateral resistance. For instance, the increase in shoulder width of sleeper types A2 and B2 from 300 to 500 mm resulted in 25.4% and 28.1% resistance increases, while the increase in shoulder height from 0 to 150 mm resulted in 12.5% and 8.2% resistance increases, respectively. Therefore, variations in shoulder width had a higher impact on the lateral resistance of sleepers than variations in shoulder height, which was already proven by previous studies [36,75].

12.5 Numerical assessment of sleeper lateral resistance

Recently, researchers tend to apply the Discrete Element Method (DEM) for the simulation of ballast particles. In this regard, ballast/sleeper interaction, settlement, breakage, and stress distribution of ballast particles were examined by DEM codes

[38,76–82]. The shape particles, the contact force between elements, friction coefficient are important parameters of DEM modeling. Irazábal [38] recommended 10–20 pebbles for the simulation of ballast particles in the macroscopic scale. As recommended by previous studies [34,83], the linear elastic model was applied for the simulation of contact force between particles in the following models. Subsequently, the geometry of the sleeper was imported as an STL file and created by wall command. The stages of STPT simulations were already described in previous studies [41,42]. As shown in Fig. 12.14, the shape of the sleeper influences the stress distribution through the ballast layer. Due to the continuous shape of the ladder sleeper, there is uniform stress distribution between ballast components and sleeper facets.

12.6 Ballast components contribution to lateral resistance for different sleepers

Figs. 12.15 and 12.16 show the contribution of ballast components to the lateral resistance of different types of sleepers. As already mentioned, the ladder sleeper has higher stability against lateral displacement, so that the lateral resistance is 1.4 times higher than a mono-block sleeper. It can be attributed to the high contribution of crib ballast between ladder sleeper beams. However, high construction cost and maintenance difficulties are the main disadvantages of this sleeper type. Thereafter, nailed sleeper and winged sleeper have high lateral resistance. For the nailed sleeper, the ballast bed and sleeper bottom interaction (62% contribution) significantly improved the lateral resistance, while the lateral resistance of the winged sleeper increased due to the interaction between crib ballast and sleeper sides (74% contribution). On the other hand, modifications of steel sleeper increase the contribution of ballast bed due to the interlocking between web steel stiffeners and ballast aggregates so that the lateral resistance WSS3 sleeper was about 80% higher than the conventional steel sleeper. For FFU sleepers, the use of strip block along the sleeper sides increased lateral resistance by 23.7% (Figs. 12.17–12.19).

12.7 Conclusion

The incidence of track buckling is provoked due to the expansion of the rails. The lateral resistance of ballasted tracks, which is controlled by various components, is an important factor in eliminating lateral track buckling. In this regard, the sleeper/ballast interaction was examined by means of laboratory and discrete element methods considering the impacts of track profile dimensions and sleeper type on the lateral track resistance. Firstly, the lateral track resistance for panels with different sizes of shoulder ballast and crib ballast was investigated using Single Tie Push Tests (STPTs). Thereafter, the portion of ballast components to the lateral track resistance was examined using DEM software. In this regard, the calibration of ballast properties is vital before the simulation of a track panel. Therefore, several experiments, including repose angle test, direct shear test, and ballast box tests, were conducted to

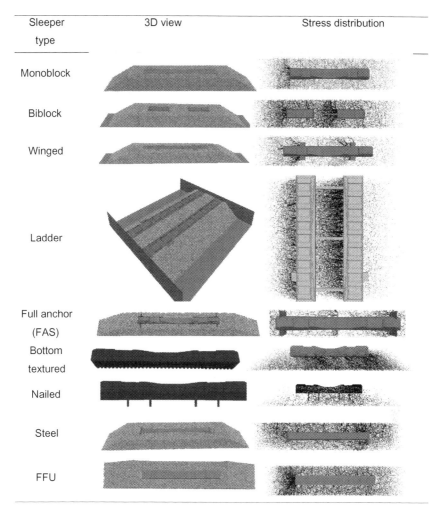

Fig. 12.15 3D modeling of STPTs for different sleepers [41–43,57].

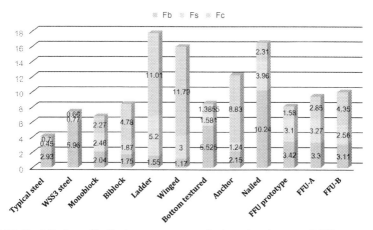

Fig. 12.16 Contribution of ballast components to the lateral resistance of different types of sleepers.

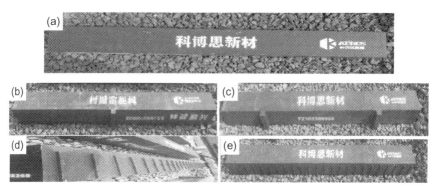

Fig. 12.17 Different FFU sleepers, (A) prototype, (B) type A1, (C) type A2, (D) type B1, and (E) type B2.

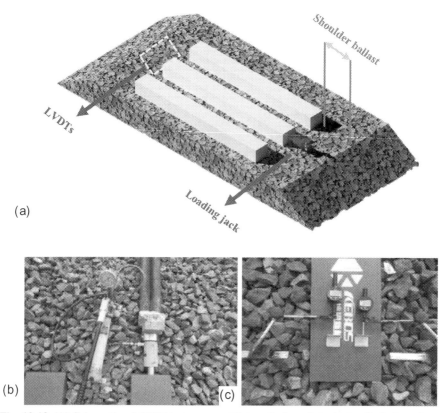

Fig. 12.18 (A) Schematic of STPT tests set up, (B) installation of the hydraulic jack, and (C) attachment of LVDTs.

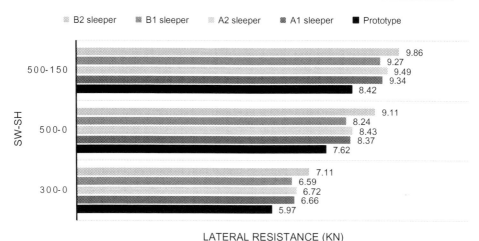

Fig. 12.19 Lateral resistance of the prototype composite sleeper.

calibrate friction coefficients, Young's modulus, Poisson's ratio, and restitution coefficient of particles. Afterward, the ballast panel was created using calibrated particles for STPTs simulation. Railway fasteners are other railway components that were considered in the present study. Several LTPTs were carried out to evaluate the impact of fasteners' loss on the lateral track resistance. The outcomes of this study are

1. The lateral track resistance for reinforced sleepers improved when the shoulder ballast height increased from 0 to 150 mm, such that the panels U-SH150, SRS-SH150, and WSS3-SH150 enriched the lateral resistance by 19.9%, 16.5%, and 11.3%, respectively, compared to track panels U-SH0, SRS-SH0, and WSS3-SH0. DEM results indicated that in the unreinforced sleeper track, the ballast bed offered 72% and 64% resistance to lateral loading when the shoulder height was 0 and 150 mm, respectively, while in the reinforced sleeper, welded steel plates with 45% contribution played a major role in the provision of lateral resistance. Strengthening steel sleepers with under sleeper plates improves the contact between the ballast bed and the bottom surface of the sleeper.
2. Lateral resistance for single mono-block and bi-block sleepers surrounded by shoulder ballast with 500 mm in width and 150 mm in height was approximately 1.45 and 1.5 times of a panel with 0 mm shoulder ballast height. On the other hand, with an increase in shoulder width from 300 to 500 mm, there is a 15% increase in the lateral resistance of the mono-block sleeper, while for the track panel of the bi-block sleeper, lateral resistance increased by around 20%. From the sleeper-type viewpoint, the lateral resistance of the bi-block sleeper was 15% larger than the mono-block. The crib ballast primarily provides the bi-block sleeper track's lateral resistance, and the lateral track resistance has increased by 50% in comparison with the mono-block sleeper track.
3. The lateral resistance of end-winged, middle-winged, and bumped sleepers on panels with 500 mm shoulder width and 150 mm shoulder height has increased by 28.6%, 27.3%, and 32.4%, respectively, compared to panels with 0 mm shoulder height. Using the discrete element method, the wings with 13 cm in width and 14 cm in length are proposed as the optimum size of sleeper wings.

4. The ladder track's lateral resistance was relatively stable when the shoulder width increased from 200 to 500 mm, which is relative to the influence of the shoulder ballast width on the mono-block sleeper resistance, with a minor increase from 17.75 to 18.65 kN/m Due to the presence of crib ballast between ladder track beams and resultant passive force, the lateral resistance of panels with 25%, 50%, and 100% crib ballast increased by 8%, 116%, and 137% compared to the track with 0% crib ballast, respectively. The presence of crib ballast serves as a passive force against the sleeper movement. Thus, the crib ballast significantly improves the lateral track resistance for ladder sleepers (about 63%) compared to mono-block sleepers in the unloaded state.
5. The lateral resistance of the end-anchor sleeper (EAS) was 30% and 21% greater than the middle-anchor sleeper (MAS) sleeper when the shoulder ballast heights were 0 and 150 mm, respectively. The full-anchor sleeper's (FAS) high resistance illustrates the considerable influence of anchors on lateral track resistance so that lateral resistance of FAS on a panel with 0 mm shoulder ballast height was about 45% and 100% higher than unreinforced sleeper on panels with shoulder ballast height of 150 and 0 mm, respectively. The implementation of FAS is also the most efficient method for meeting the specifications of high-speed railway ballast lines, and 200 mm is proposed as the optimal anchor spade length.

In summary, The lateral stability of ballasted track constructed by ladder sleepers was the maximum value in comparison with other sleeper types. However, an economic analysis is required to evaluate the application of ladder sleepers as the substitution of conventional sleepers. Since there is a passive pressure between ladder sleeper beams and crib ballast, the compaction of this zone would be beneficial for the provision of higher lateral resistance, which is recommended for future studies.

References

[1] E.T. Selig, J.M. Waters, Track Geotechnology and Substructure Management, Thomas Telford, 1994.
[2] A. Kish, G. Samavedam, D. Wormley, Fundamentals of Track Lateral Shift for High-Speed Rail Applications, National Technical Information Service, 2004.
[3] S. Ahmed, M. Kumar, A. Chattopadhy, A comparative study of track buckling parameters on continuous welded rail, in: Proceedings of the International Conference on Mechanical Engineering, 2009.
[4] A. Kish, G. Samavedam, Dynamic buckling of continuous welded rail track: theory, tests, and safety concepts, Transp. Res. Rec. 1289 (1991) 23–38.
[5] A.D. Kerr, Railroad Track Mechanics and Technology: Proceedings of a Symposium Held at Princeton University, April 21–23, 1975, Elsevier, 2014.
[6] S. Miura, Lateral track stability: theory and practice in Japan, in: Lateral Track Stability 1991: Proceeding of a Conference, St. Louis, Missouri, 1991.
[7] G. Jing, P. Aela, Review of the lateral resistance of ballasted tracks, Proc. Inst. Mech. Eng. Part F: J. Rail Rapid Transit 234 (8) (2020) 807–820.
[8] J.A. Zakeri, M. Barati, S. Mohammadzadeh, New definition of neutral temperature in continuous welded railway track curves, Period. Polytech. Chem. Eng. 62 (1) (2018) 143–147.
[9] Constantin, White Rails, 2016.
[10] A. Kish, On the Fundamentals of Track Lateral Resistance, American Railway Engineering and Maintenance of Way Association, 2011.

[11] A. Poulton, The Effect of Resurfacing on the Lateral Resistance of Narrow Gauge Low Profile Concrete Sleepers, USQ Project, 2016.
[12] M. Van, Buckling analysis of continuous welded rail track, Heron 41 (3) (1996) 9.
[13] S. Yamamoto, et al., Lateral resistance characteristics of ballasted tracks subjected to angular folding at structure boundaries, J. Jpn. Soc. Civ. Eng. Ser. E1 (Pavement Eng.) 72 (1) (2016) 21–30.
[14] F. Arbabi, M. Khalighi, Stability of railroad tracks under the effects of temperature change and earthquake, J Seismol and Earthq. Eng. 12 (3) (2010) 119.
[15] A. De Iorio, et al., Transverse strength of railway tracks: part 2. Test system for ballast resistance in line measurement, Frat. Integrita Strutt. 30 (2014) 578–592.
[16] M. Esmaeili, S.A.S. Hosseini, M. Sharavi, Experimental assessment of dynamic lateral resistance of railway concrete sleeper, Soil Dyn. Earthq. Eng. 82 (2016) 40–54.
[17] AREMA "American Railway Engineering and Maintenance of Way Association", Ties, Part 2, in: Manual for Railway Engineering, vol. 1, AREMA, 2014 (Chapter 30).
[18] 13481-1, B.S.B.E., Railway Applications. Track. Performance Requirements for Fastening Systems. Definitions, 2012.
[19] B.A. Williams, et al., Quantification of the lateral forces in concrete sleeper fastening systems, Proc. Inst. Mech. Eng. Part F: J. Rail Rapid Transit 230 (7) (2016) 1714–1721.
[20] B. Williams, et al., Analysis of the lateral load path in concrete crosstie fastening systems, in: 2014 Joint Rail Conference, American Society of Mechanical Engineers, 2014.
[21] B.W. Williams, et al., Lateral force measurement in concrete crosstie fastening systems, in: Proceedings of Transportation Research Board 93rd Annual Conference, TRB 14-3112, Washington, DC, 2014.
[22] J. Sadeghi, J.A. Zakeri, M.E.M. Najar, Developing track ballast characteristic guideline in order to evaluate its performance, Int. J. Rail. 9 (2) (2016) 27–35.
[23] AREMA "American Railway Engineering and Maintenance of Way Association", Roadway & ballast, Part 2, in: Manual for Railway Engineering, vol. 1, 2015 (Chapter 1).
[24] B. Indraratna, W. Salim, C. Rujikiatkamjorn, Advanced Rail Geotechnology–Ballasted Track, CRC Press, 2011.
[25] J. Xiao, et al., Evolution of longitudinal resistance performance of granular ballast track with durable dynamic reciprocated changes, Adv. Mater. Sci. Eng. 2018 (2018), 3189434.
[26] A.E.R.N.P.V. Hasheminezhad, Investigation of abrasion resistance of steel furnace slag as ballast in railway tracks, J. Comput. Civ. Eng. 1 (1) (2016).
[27] C88M-18, A.C, Standard Test Method for Soundness of Aggregates by Use of Sodium Sulfate or Magnesium Sulfate, ASTM International, West Conshohocken, PA, 2018.
[28] N.S. Association, Steel furnace slag, an ideal railroad ballast, Technical Report, NSA 173–3, Alexandria, Virginia, 2007.
[29] Specification, F, Slag and stone, crushed, grave, crushed and crushed (for railroad ballast), SS-S-449, 1966.
[30] M. Esmaeili, R. Nouri, K. Yousefian, Experimental comparison of the lateral resistance of tracks with steel slag ballast and limestone ballast materials, Proc. Inst. Mech. Eng. Part F: J. Rail Rapid Transit 231 (2) (2015) 175–184.
[31] Y. Sun, Effect of Particle Angularity and Size Distribution on the Deformation and Degradation of Ballast Under Cyclic Loading, Doctor of Philosophy Thesis, School of Civil, Mining and Environmental Engineering, University of Wollongong, 2017.
[32] V. Profillidis, Railway Management and Engineering, Routledge, 2014.

[33] C. Mulhall, et al., Large-scale testing of tie lateral resistance in two ballast materials, in: Third International Conference on Railway Technology: Research, Development and Maintenance, Cagliari, Sardinia, Italy, 2016.

[34] F. Khatibi, M. Esmaeili, S. Mohammadzadeh, DEM analysis of railway track lateral resistance, Soils Found. 57 (4) (2017) 587–602.

[35] J. Zakeri, A. Bakhtiary, Comparing lateral resistance to different types of sleeper in ballasted railway tracks, Sci. Iran. Trans. A Civ. Eng. 21 (1) (2014) 101.

[36] L. Le Pen, W. Powrie, Contribution of base, crib, and shoulder ballast to the lateral sliding resistance of railway track: a geotechnical perspective, Proc. Inst. Mech. Eng. Part F: J. Rail Rapid Transit 225 (2) (2010) 113–128.

[37] M. Dipilato, et al., Ballast and Subgrade Requirements Study: Railroad Track Substructure-Design and Performance Evaluation Practices, Federal Railroad Administration, Office of Research and Development, United States, 1983.

[38] J. Irazábal, F. Salazar, E. Oñate, Numerical modelling of granular materials with spherical discrete particles and the bounded rolling friction model. Application to railway ballast, Comput. Geotech. 85 (2017) 220–229.

[39] G. Jing, et al., Ballast flying mechanism and sensitivity factors analysis, Int. J. Smart Sens. Intell. Syst. 5 (4) (2012) 928–939.

[40] Z. Zeng, et al., Ballast bed resistance characteristics based on discrete-element modeling, Adv. Mech. Eng. 10 (6) (2018). 1687814018781461.

[41] G. Jing, P. Aela, H. Fu, The contribution of ballast layer components to the lateral resistance of ladder sleeper track, Constr. Build. Mater. 202 (2019) 796–805.

[42] G. Jing, et al., Numerical and experimental analysis of single tie push tests on different shapes of concrete sleepers in ballasted tracks, Proc. Inst. Mech. Eng. Part F: J. Rail Rapid Transit (2018). 0954409718805274.

[43] G. Jing, H. Fu, P. Aela, Lateral displacement of different types of steel sleepers on ballasted track, Constr. Build. Mater. 186 (2018) 1268–1275.

[44] L.-M. Cléon, v360: High-Speed Laboratory. www.railway-research.org.

[45] S. Fischer, et al., Railway Construction, Széchenyi István University, 2015.

[46] P. Dogneton, The experimental determination of the axial and lateral track-ballast resistance, in: A.D. Kerr (Ed.), Railroad Track Mechanics and Technology, Pergamon, 1978, pp. 171–196.

[47] M. Esmaeili, S. Majidi-Parast, A. Hosseini, Comparison of dynamic lateral resistance of railway concrete, wooden and steel sleepers subjected to impact loading, Road Mater. Pavement Des. 20 (8) (2018) 1779–1806.

[48] C. Calla, Two Layered Ballast System for Improved Performance of Railway Track, Coventry University, 2003.

[49] R. Bogacz, W. Czyczuła, R. Konowrocki, Influence of sleepers shape and configuration on track-train dynamics, Shock. Vib. 2014 (2014).

[50] J.A. Zakeri, Lateral resistance of railway track, in: I.X. Perpinya (Ed.), Reliability and Safety in Railway, InTech, 2012, pp. 357–374.

[51] J.-A. Zakeri, B. Mirfattahi, M. Fakhari, Lateral resistance of railway track with frictional sleepers, in: Proceedings of the Institution of Civil Engineers-Transport, Thomas Telford, 2012.

[52] A. Pieringer, Innovative sleeper design with under sleeper pads as an efficient method to reduce railway induced vibration propagation. Research results of RIVAS, in: *Urbana-Champaign*, Crosstie & Fastening Symposium, Illinois, UIUC, 2014.

[53] Y. Koike, et al., Numerical method for evaluating the lateral resistance of sleepers in ballasted tracks, Soils Found. 54 (3) (2014) 502–514.

[54] K. Hayano, et al., Effects of sleeper shape on lateral resistance of railway ballasted tracks, in advances in soil dynamics and foundation, Engineering (2014) 491–499.
[55] M. Guerrieri, D. Ticali, M. Denaro, High performance bi-block sleeper for improvement the performances of ballasted railway track, AASRI Procedia 3 (2012) 457–462.
[56] D. McNaughton, S. Lowe, M. Brotzman, Railway Technical Handbook. http://www.railway-technical.com.
[57] G. Jing, et al., Numerical and experimental analysis of lateral resistance of bi-block sleeper on ballasted tracks, Int. J. Geomech. 20 (6) (2020) 04020051 (Accepted).
[58] N.K. Moses, R.A. Mcclung, Beams for Railroad Track Structure, 1967. Google Patents.
[59] M. Ma, et al., An experimental study of vibration reduction of a ballasted ladder track, Proc. Inst. Mech. Eng. Part F: J. Rail Rapid Transit 231 (9) (2017) 1035–1047.
[60] K. Hill, S. Relph, Steel Railroad Sleepers, 2001. Google Patents.
[61] A.M. Zarembski, Survey of Techniques and Approaches for Increasing the Lateral Resistance of Wood Tie Track, University of Delaware, Department of Civil and Environmental Engineering, Delaware, 2016.
[62] G. Jing, D. Ding, X. Liu, High-speed railway ballast flight mechanism analysis and risk management—a literature review, Constr. Build. Mater. 223 (2019) 629–642.
[63] J.A. Zakeri, et al., A numerical investigation on the lateral resistance of frictional sleepers in ballasted railway tracks, Proc. Inst. Mech. Eng. Part F: J. Rail Rapid Transit 230 (2016) 440–449.
[64] Y. Guo, et al., Effect of sleeper bottom texture on lateral resistance with discrete element modelling, Constr. Build. Mater. 250 (2020) 118770.
[65] C. Ciobanu, Behaviour of the Track in Hot Weather-Rail Thermal Forces for Jointed and CWR Track, Permanent Way Institution, 2017. www.slideshare.net.
[66] M. Esmaeili, et al., Investigating the effect of nailed sleepers on increasing the lateral resistance of ballasted track, Comput. Geotech. 71 (2016) 1–11.
[67] Y. Guo, et al., Experimental and numerical study on lateral and longitudinal resistance of ballasted track with nailed sleeper, Int. J. Rail Transp. 10 (1) (2021) 114–132.
[68] W. Ferdous, et al., Composite railway sleepers: New developments and opportunities, in: Proceedings of the 11th International Heavy Haul Association Conference: Operational Excellence (IHHA 2015), Perth, Australia, 2015.
[69] AREMA, "American Railway Engineering and Maintenance of Way Association", Engineered composite ties, part 5, in: Manual for Railway Engineering, vol. 1, AREMA, 2012 (Chapter 30).
[70] A. Manalo, et al., A review of alternative materials for replacing existing timber sleepers, Compos. Struct. 92 (3) (2010) 603–611.
[71] W. Ferdous, A. Manalo, G. Van Erp, Structural optimisation of composite railway sleeper, in: Composites Australia and CRC–ACS Conference, Composites Australia, 2014.
[72] Jinming, 2013. China Patent No. CN104164076A. Retrieved from: https://patents.google.com/patent/CN104164076A/en.
[73] S. Freudenstein, Investigation on FFU Synthetic Wood Sleeper, 2016. https://www.sekisui-rail.com/en/technical-research-on-ffu.html.
[74] KEBOS, Fiber Reinforced Foamed Urethane Sleeper, 2018. http://www.kebos.cn/.
[75] E. Kabo, A numerical study of the lateral ballast resistance in railway tracks, Proc. Inst. Mech. Eng. Part F: J. Rail Rapid Transit 220 (4) (2006) 425–433.
[76] B. Indraratna, S.K.K. Hussaini, J. Vinod, The lateral displacement response of geogrid-reinforced ballast under cyclic loading, Geotext. Geomembr. 39 (2013) 20–29.
[77] B. Indraratna, N.T. Ngo, C. Rujikiatkamjorn, Deformation of coal fouled ballast stabilized with geogrid under cyclic load, J. Geotech. Geoenviron. 139 (8) (2012) 1275–1289.

[78] S. Laryea, et al., Comparison of performance of concrete and steel sleepers using experimental and discrete element methods, Transp. Geotech. 1 (4) (2014) 225–240.
[79] W. Lim, G. McDowell, Discrete element modelling of railway ballast, Granul. Matter 7 (1) (2005) 19–29.
[80] G. McDowell, P. Stickley, Performance of geogrid-reinforced ballast, Ground Eng. 39 (1) (2006) 26–30.
[81] B. Indraratna, et al., Behavior of fresh and fouled railway ballast subjected to direct shear testing: discrete element simulation, Int. J. Geomech. 14 (1) (2012) 34–44.
[82] Y. Xu, et al., Discrete element method analysis of lateral resistance of fouled ballast bed, J. Cent. South Univ. 23 (9) (2016) 2373–2381.
[83] C. Ngamkhanong, S. Kaewunruen, C. Baniotopoulos, A review on modelling and monitoring of railway ballast, Struct. Monit. Maint. 4 (3) (2017) 195–220.

Diagnostics and management methods for concrete sleepers

Dan Li[a,b], Sakdirat Kaewunruen[a,b], Alex Remennikov[c], and Ruilin You[d]
[a]Department of Civil Engineering, School of Engineering, University of Birmingham, Birmingham, United Kingdom, [b]Birmingham Centre for Railway Research and Education, School of Engineering, University of Birmingham, Birmingham, United Kingdom, [c]School of Civil, Mining and Environmental Engineering, University of Wollongong, North Wollongong, NSW, Australia, [d]Railway Engineering Institute, China Academy of Railway Sciences, Beijing, China

It is commonly believed that railway is the world's safest transportation system for either passengers or goods. Track structures guide the safe, economical, and comfort ride of trains. Nowadays, the railway transportation is becoming increasingly important and railway engineer needs to improve the performance of railway infrastructure. The key components of a typical ballasted railway track are shown in Fig. 13.1 [1]. Conventional railway infrastructure can be divided into superstructure and substructure. The superstructure consists of sleepers, rail pads, fastening systems, and rails. The substructure includes ballast, sub-ballast and formation. Rail sleepers are a major component of railway structure as they play a major role in track performance, track stiffness, and operational safety.

The main functions of sleepers are: (1) to transfer and distribute loads from superstructure to substructure, (2) to hold the rails at the proper gauge through the rail fastening system, (3) to maintain rail inclination, and (4) to restrain longitudinal, lateral, and vertical movements of the rails [2]. The performance of concrete sleepers is a safety-concerned question of operation and maintenance departments within the railway organizations. A sleeper failure in critical locations such as switches and crossings, transom bridges, bridge ends, and rail joints could lead progressively to significant incidents in railway operations (e.g., train derailments, operational downtime, broken signaling equipment, etc.). The most common problems related to concrete sleepers around the world are surveyed and ranked in Table 13.1 [3,4].

Prestressed concrete is very suitable for railway sleeper because of its many superior advantages in design, construction, short- and long-term maintenance, sustainability, and life cycle cost. Nowadays, approximately 500 million railway sleepers in the world's railway networks are made from prestressed concrete and, every year, the demand for them constitutes more than 50% of total demand [5,6]. However, concrete sleepers are vulnerable to damage under different loading conditions, cracking, rail seat deterioration, and aggressive environments.

Improved knowledge has raised concerns about design techniques for prestressed concrete sleepers. Most current design codes for these rely on allowable stresses and

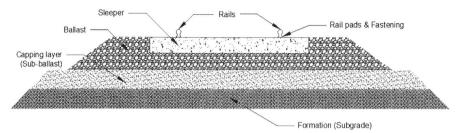

Fig. 13.1 Conventional railway structure.

material strength reductions. This chapter presents the potential failures of concrete sleepers and looks for sustainable solutions. It highlights the reliability approach and rationales associated with the development of short- and long-term maintenance and presents guidelines pertaining to maintenance management methods.

13.1 Sleeper design process

The design process of railway sleeper can be concluded as key steps and some significant variables. The vehicle applies its load to track through axles and wheels. Therefore, the axle load and wheel load are very important for railway sleeper design. At this stage impact effects have to be considered with significant variables: speed, vehicle type, and the levels of sprung and unsprung mass, the track structure; straight or curving; and maintenance. The ballast pressure should be obtained in order to consider rail stiffness, sleeper spacing, ballast packing and maintenance and the shape of the sleeper footprint. After these stages, the moments of sleeper can be estimated from the loads and ballast pressure. The unloaded resonant stresses have to be considered because of dynamic effects. Finally, a sleeper may be designed with given material strengths but with the remaining considerable freedom of varying the depth by profiling the top face to give optimum stresses throughout [7].

Table 13.1 Common damages of prestressed concrete sleepers (ranked from 1 to 8, with 8 being most critical).

Main causes	Damages	Rank
Lateral load	Abrasion of concrete material on rail seat	3.15
	Shoulder/fastening system wear or fatigue	5.5
Vertical dynamic load	Cracking from dynamic loads	5.21
	Derailment damage	4.57
	Cracking from center binding	5.36
Manufacturing and maintenance defects	Tamping damage (or impact forces)	6.14
	Other (e.g. manufactured defect)	4.09
Environmental considerations	Cracking from environmental or chemical degradation	4.67

13.2 Prestressed concrete railway sleeper subject to dynamic load

Premature cracking of PC sleepers has been found in railway tracks. The major cause of cracking is the infrequent but high-magnitude wheel loads produced by the small percentage of irregular wheels or rail-head surface defects. The allowable or permissible-stress design method makes use of an empirical function taking into account the static wheel load (P_0) with a dynamic impact factor ($\varnothing$) to account for dynamic vehicle/track interactions.

$$P_D = \varnothing P_0$$

where P_D is the design wheel load, P_0 the quasistatic wheel load, and $\varnothing$ the dynamic impact factor (>1.0).

Significant research effort has been devoted to the forces arising from vertical interaction of train and track because these dynamic transient forces are the main cause of railway track problems, when trains are operated at high speed and with heavy axle loads. It is important to note that wheel–rail interactions induce much higher frequency and much higher magnitude forces than simple quasi-static loads. These forces are referred to as "dynamic wheel–rail" or "impact" forces. The transient dynamic loading is the dynamic loading that occurred over a very short period of time. The causes of transient loading are usually from collisions, sudden loading, blast, shock, etc. Particularly in modern railway tracks, the transient loading is usually caused by abnormalities in wheel/rail interaction. The magnitude of the forces is very high within the very short impulse duration, for instance, 2–10 ms [8]. The effects of impact forces are very significant in the design and utilization of concrete sleepers as parts of the railway track structures [9].

13.2.1 Probabilistic dynamic loads on tracks

A maximum allowed impact force of 230 kN to be applied to the rail head by passing train wheels is commonly accepted through the Defined Interstate Network Code of Practice (Volume 5, Part 2, Section 8, 2002) [10]. Such impact loads may be caused by a variety of effects, including flats worn on the wheel tread, out-of-round wheels, defects in the wheel tread or in the rail head, or a derailment. The most severe impact forces are most likely from wheel flats [11], because such flats strike the rail head every revolution of the wheel, and severe flats have the potential to cause damage to track over many kilometers before detection. Despite the Code of Practice requirement, there are little published data showing the actual range and peak values of impact in the normal operation of trains, and certainly none were found for the defined interstate network. The value of 230 kN is therefore a desired upper limit rather than a measure of real maximum forces encountered on track.

A comprehensive investigation of actual impact forces was undertaken as a part of the Rail CRC project at Queensland University of Technology (QUT) [11]. Over

1 year, data were gathered from two Teknis Wheel Condition Monitoring stations located on different lines. The loading data from a total of nearly 6 million passing wheels were measured, primarily from unit trains with 26–28-ton axle loads, in both the full and empty states. The analysis of collected data is shown as a histogram in Fig. 13.2 [12]. The vertical axis shows the number of axles on a log scale, while the horizontal axis the measured impact force from the Teknis station. This impact force is the dynamic increment above the static wheel force (140 kN) exerted by the mass of the wagon on a wheel. Over 96% of the wheels created impact forces less than 50 kN. The bulk of the graph in Fig. 13.2 therefore, is derived from only the remaining 4% of wheels. However, that small percentage is still equivalent to more than 100,000 wheels throughout the year of the study, and they caused impact forces as high as 310 kN. The sloping dashed line in the plot represents a line of best fit to the data for these 100,000 wheel events.

In Fig. 13.2, the vertical line represents the Code of Practice maximum wheel impact force of 230 kN. The frequency of high-impact wheel forces in the histogram columns of figure lies along the sloping, dashed straight line, which means that the distribution would appear as a logarithmic curve on a graph with a linear scale on the vertical axis. In this case, the vertical axis in figure is the number of impacting wheels per year, so if the rate of occurrence of such impacts over the year of the study is a representative of impacts over a longer period, then extrapolation of that sloping dashed line will provide an estimate of the frequency of occurrence of impact forces greater than the largest measured value of 310 kN. On that basis, it could be predicted that an impact force of 380 kN would occur at the rate of 0.1 axles/year, or once in every 10 years; an impact of 450 kN would occur on an average once in every 100 years. This process naturally leads to the concept of a return period for impact force, which Murray and Leong [13] developed to produce equation:

$$\text{impact force}(kN) = 53(5.8 + \log R)$$

where R is the return period in years of a given level of impact. It should be emphasized that these impact forces are applied by the wheel to the rail head. To determine the impact force applied to other components in the track structure, such as the sleeper or ballast, appropriate measures should be applied which allow for force sharing among support elements and also allow for the not-insignificant dynamic behavior of the railway track. The impulse duration plays a key role in amplifying the dynamic responses of the concrete sleeper at the corresponding resonance. The strain rate tends to increase dynamic stiffness of materials to some extent. However, the dynamic magnification factors can be up to double of the static response and cause the sleeper to crack [14].

13.2.2 Capacities of prestressed concrete sleepers

To evaluate the performance of PC sleepers under impact loads, an experimental programme was conducted. The prestressed concrete sleepers supplied by Australian manufacturers Rocla and Austrak were used for tests. The sleepers were broad gauge

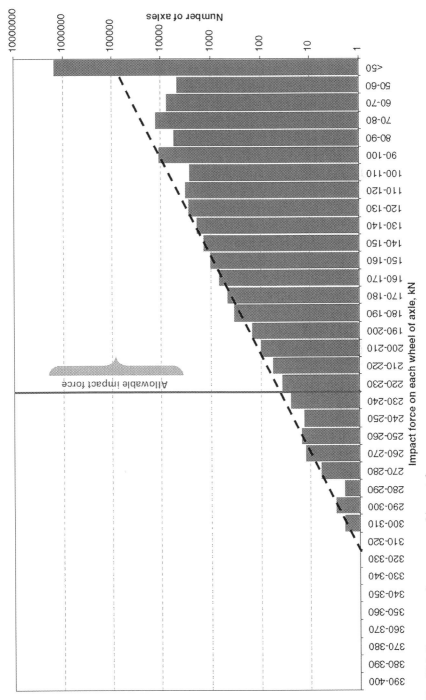

Fig. 13.2 Frequency of occurrence of impact forces per year.

(1.600 m) and standard gauge (1.435 m), both commonly used in heavy-haul coal lines [15,16]. A positive four-point bending moment test was conducted based on the assumption that the sleepers would behave similarly to those in situ [17]. It should be noted that the initial strain of prestressing wires is about 6.70 mm/m, and each prestressing wire has a specified minimum proof stress of 1860 MPa. The average compressive strength of cored concrete was 88 MPa. Fig. 13.3 illustrates the setup for a static testing. A load cell was used to measure the applied load, while a linear variable differential transformer was installed at the mid-span to obtain the corresponding deflections. Strain gauges were fixed to the top and bottom surfaces of the test sleeper and on both sides. The transducers were connected to a computer to record the experimental data.

The high-capacity drop-weight impact testing machine is shown in Fig. 13.4. Thick rubber mats were used to replicate the ballast support (static track spring rates: about 60 kN/mm for soft track and 135 kN/mm for hard track). It was found that the test setup could accurately represent the support conditions for PC sleepers found in typical track systems [14,17]. To apply impact loads, a drop hammer with a falling mass of 600 kg was used. The rail, with its fastening system for transferring the load to the specimens, was installed at the railseat. The drop hammer was hoisted mechanically to the required height and released. Impact load was recorded by the dynamic load cell.

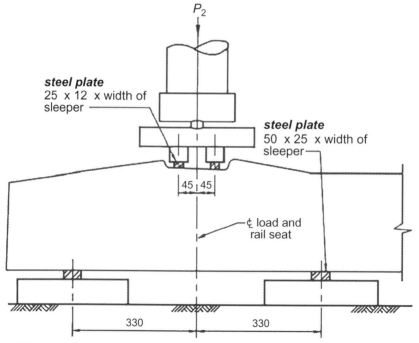

Fig. 13.3 Static test setup.

Diagnostics and management methods for concrete sleepers

Fig. 13.4 High-capacity drop-weight impact testing machine.

Fig. 13.5 shows a typical dynamic moment–deflection relationship at the rail seat for PC sleepers. The crack initiation load was observed visually during each test as well as determined by the use of the load–deflection relationships. Fig. 13.6 illustrates the crack propagation in a PC sleeper under static, monotonically increasing loading.

On the basis of the statistical data for the frequency of occurrence of impact loads and their magnitudes, separate impact tests on PC sleepers were designed to simulate wheel–rail interface forces by varying the height of drop and the contact stiffness to achieve the desired magnitudes and durations of the load pulses. Fig. 13.7 presents typical impact force–time histories measured by the dynamic load cell. Very small flexural cracks were initially detected starting from a drop height of 600 mm. Small shear cracks were also observed after several impacts from a drop height of 800 mm. However, no major failure could be observed in these single-impact load experiments [18].

Later, the prestressed concrete sleepers were also subjected to gradually increasing impact loads until they failed. The progressive impact behavior of a PC sleeper was conducted in the soft track environment. The crack widths at each stage were measured using a magnifier telescope. The crack widths were about 0.01–0.02 mm for impact loads between 150 and 600 kN. When subjected to impact loads with magnitudes between 700 and 1000 kN, the crack widths increased from 0.02 to 0.08 mm. At

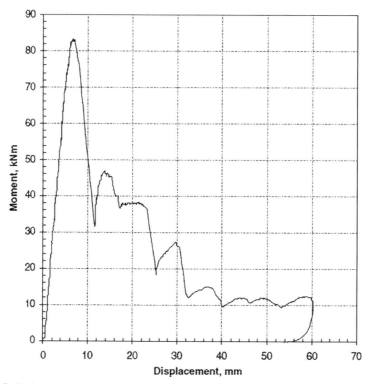

Fig. 13.5 Static moment–displacement relationship.

this stage, spalling of the concrete at the top of rail seat section could be detected, and the crack widths increased up to 0.5 mm when the impact forces were implemented up to 1500 kN. The ultimate impact load-carrying capacity was reached at about 1600 kN when the sleeper rail seat section disintegrated. The failure mode was associated with both flexural and longitudinal splitting actions. The splitting fractures were aligned along the prestressing tendons. The probabilistic analysis of dynamic loading suggests that the magnitude of the ultimate impact load that caused failure of the prestressed concrete sleeper would be equivalent to that with a return period of several million years.

13.3 Fatigue assessment for prestressed concrete sleeper

Fatigue failure can be defined as a failure that occurs below the stress limit of a material when it has been exposed to repeated loadings [19]. The effects of fatigue are based on the following considerations [20]:

- The magnitude of the stress range.
- The type and quality of the structural details.
- The number of applications (or cycles) of this stress range.

Diagnostics and management methods for concrete sleepers

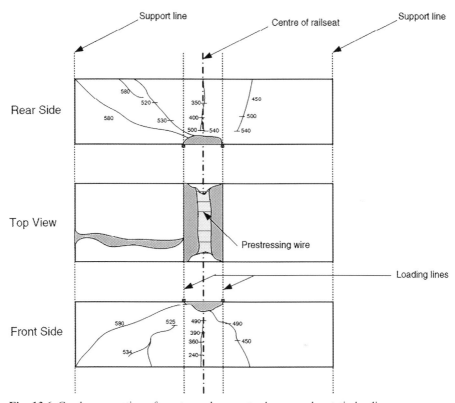

Fig. 13.6 Crack propagation of prestressed concrete sleeper under static loading.

The fatigue phenomenon was first observed in steel constructions. In 1830, Albert performed fatigue tests on welded mine hoist chains. The research work concerning fatigue of concrete and concrete structures started at about the turn of the 19th century. After this, much work concerning fatigue in concrete structures has been performed in many countries [21].

13.3.1 Structural performance over life cycle

The fatigue failures of member or structure consider being the process of accumulated damage due to the repeated loads over a long period of time. Therefore, the fatigue life of concrete sleeper can be determined by the service time to support the repeated train loads [22]. The fatigue life of concrete sleeper is controlled by the stress generated from repeat loads and resistance of it. Since the loads generated from wheel-rail interaction are random, and the dimension and strength of sleepers are random as well, the fatigue damage of concrete sleeper is random affairs. In addition, the service life of concrete sleeper is usually longer than 50 years [23], the fatigue loads generated from trains always increase because of the development of economy, but the resistance of sleeper always decrease due to aging and deterioration. Considering these reasons, the

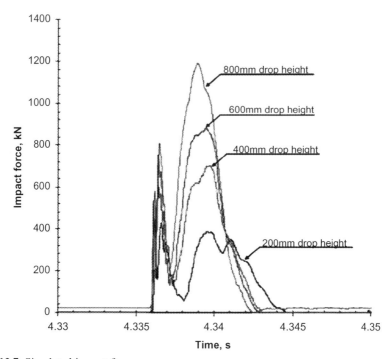

Fig. 13.7 Simulated impact forces.

life-cycle performance of concrete sleeper can be considered as shown in Fig. 13.8, in which uncertainties are associated with initial performance indicator, deterioration rate, fatigue loads and maintenance/repair, etc.

In Fig. 13.8, $R(t)$ is the probability distribution function of resistance, and $S(t)$ is the probability distribution function of loads effect. In general, the measure of risk associated with the specific event of $R(t) < S(t)$ can be considered as the probability of failure of concrete sleeper. As shown in Fig. 13.8, this interaction increases over time causing growth of the probability of failure. Structural models and their idealization, deterioration mechanisms, material resistances, geometries, and loads are uncertain. Therefore, probabilistic approach has been researched to quantify the reliability of concrete sleeper [24].

13.3.2 Fatigue properties of materials

13.3.2.1 Concrete

Fatigue of concrete is a progressive process of microcrack initiation and propagation toward the point at which failure occurs. The mechanical properties of concrete will change under.

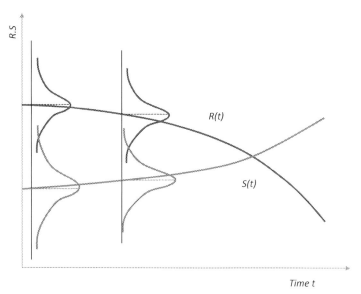

Fig. 13.8 Life-cycle performance of concrete sleeper.

Repeated cyclic loading, such as permanently increasing strain on the concrete which causes the stiffness to decrease. Cyclic loading may also cause a concentration of stress at the surface of the prestressed wires, which can lead to sudden fracture [25,26].

In the 1970s, a lot of research regarding concrete fatigue analysis was done for the construction of offshore structures in Scandinavia. The fatigue performance of concrete is commonly represented on graphs, with stress ranges plotted against the failure cycles number [27]. To assess fatigue performance, the Goodman diagram has been developed shown in Fig. 13.9 [27,28]. f_c' is the allowable compressive stress of concrete, N_f is the number of allowable load cycles, σ_t is the maximum stress caused by a dead load, and σ_h is the maximum stress caused by both live and dead loads.

The Fatigue Reference Compressive Strength $f_{ck,\,fat}$ of concrete (do Model Code 2010) is calculated based on the characteristic compressive strength f_{ck} as follows:

$$f_{ck,\,fat} = \beta_{cc}(t)\beta_{c,\,sus}(t,t_0)f_{ck}\left(1 - \frac{f_{ck}}{400}\right)$$

$$\beta_{cc}(t) = \left\{s\left[1 - \left(\frac{28}{t}\right)^{0.5}\right]\right\}$$

where $\beta_{c,\,sus}(t,t_0)$ is a coefficient which depends on the time under high sustained loads $t - t_0$ (days). The expression of $\beta_{cc}(t)$ describes the strength of the concrete (fib Model Code 2010), which is developing with time, in which s depends on the cement strength

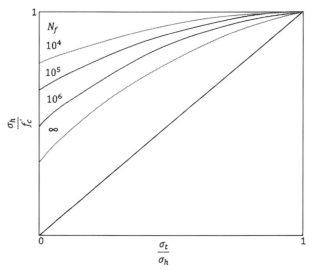

Fig. 13.9 The Goodman diagram.

class and t equals to the age of concrete age (in days) [29]. The typical S–N curves of concrete under pure compression are shown in Fig. 13.10. N is the number of cycles causing fatigue failure in plain concrete.

$$S_{c,\max} = |\sigma_{c,\max}|/f_{ck,\text{fat}}$$

$$S_{c,\min} = |\sigma_{c,\min}|/f_{ck,\text{fat}}$$

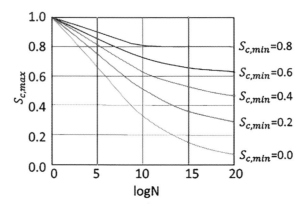

Fig. 13.10 S–N curves for concrete under pure compression.

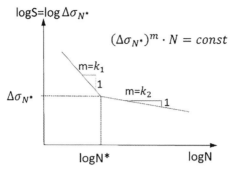

Fig. 13.11 S–N curves for steel.

13.3.2.2 Prestressed steel

In the fib Model Code 2010, the characteristic fatigue strength function for prestressed steel consists of two different slope segments (Fig. 13.11). The parameters of the prestressed steels S–N curve are shown in Table 13.2. The failure cycle of the prestressing steel under a constant amplitude cyclic loading can be estimated by:

If $\Delta\sigma_N > \Delta\sigma_{N^*}$ $\log N_f = \log N^* - k_1[\log(\Delta\sigma_N) - \log(\Delta\sigma_{N^*})]$
If $\Delta\sigma_N \leq \Delta\sigma_{N^*}$ $\log N_f = \log N^* + k_1[\log(\Delta\sigma_N) - \log(\Delta\sigma_{N^*})]$

where $\Delta\sigma_N$ is the stress range the prestressing steel, $\Delta\sigma_{N^*}$ is the stress range at N^* cycles which is given in Table 13.2.

13.3.3 Fatigue life assessment method

The damage accumulation method extended from Miner's rule has been widely used in Eurocode [30,31]. For multiple cycles with variable amplitudes, the damage will be added based on Damage Accumulation Method, and the cumulative damage index $\sum Di$ is given by:

$$\sum Di = \sum_i \frac{n(\Delta\sigma_i)}{N(\Delta\sigma_i)}$$

Table 13.2 Parameters of prestressed steel's S–N curves (fib Model Code 2010).

S–N curve of prestressed steel used for	Stress exponent			$\Delta\sigma_{N^*}$(MPa)
	N^*	k_1	k_2	at N^* cycles
Pre-tensioning	10^6	5	9	185
Post-tensioning				
– Single strands in plastic ducts	10^6	5	9	185
– Straight tendons or curved tendons in plastic ducts	10^6	5	10	150
– Curved tendons in steel ducts	10^6	5	7	120
– Splicing devices	10^6	5	5	80

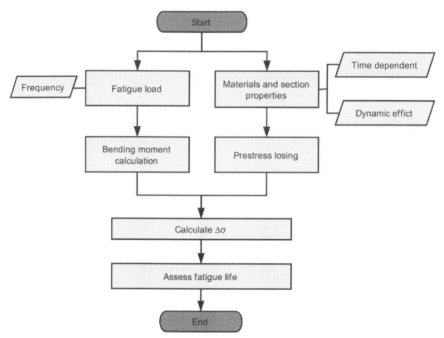

Fig. 13.12 Fatigue life assessment flow chart for concrete sleeper.

where $n(\Delta\sigma_i)$ is the applied number of cycles for a stress range $\Delta\sigma_i$, $N(\Delta\sigma_i)$ is the resisting number of cycles for a stress range $\Delta\sigma_i$.

Based on the fatigue loading conditions and the rulers mentioned above, the fatigue life of concrete sleeper can be calculated. The flow chart (Fig. 13.12) presents the analysis steps for predicting fatigue life of a prestressed concrete sleeper. The analysis of fatigue life is based on fatigue loads and material properties.

13.4 Rail seat abrasion

The performance deterioration of concrete sleepers is a safety-concerned question of operation and maintenance departments within the railway organizations. A sleeper failure in critical locations such as switches and crossings, transom bridges, bridge ends, rail joints, and so on can lead progressively to significant incidents in railway operations (e.g., train derailments, operational downtime, broken signaling equipment, etc.). Because it is rather different to predict a particular location of long continuous tracks where the sleeper will fail, the performance of concrete sleepers is generally defined by structural reliability obtained from the stress generated from repeat loads (or action) and sleepers' resistance (or capacity) [14]. In addition, the performance deterioration of concrete sleepers can be influenced by the lateral and vertical dynamic loads transferred from the rails, the manufacturing quality and maintenance defects, and the exposure to environmental conditions, etc.

Rail seat abrasion of a prestressed concrete sleeper can be related to the climatic and traffic conditions and the location of the concrete sleepers in the track. In particular, axle load, traffic volume, curvature and grade of the rail line, the presence of abrasive fines (e.g., locomotive sand or metal shavings), the behavior of the fastening system, and climate are the key factors that contribute to rail seat abrasion [32,33]. Based on North American heavy railway network experiences and concrete sleeper tests results, heavy axle loads, abrasive fines, moisture, and rail movement appear to be the most important factors [34,35].

Rail seat abrasion results in many problems of railway tracks such as loss of fastening toe load, gauge variation, improper rail cant, and eventually loss of rail fastening. There are many researches about rail seat abrasion; from the literal review above, most of previous works focused on the mechanisms of rail seat abrasion. However, there were very few studies that quantitatively examined the effects of rail seat abrasion on the loading capacity of the prestressed concrete sleepers. This implies that the maintenance of sleepers cannot be properly scheduled or planned in advance. In fact, the sleepers are generally embedded in ballast, it is almost impossible to inspect structural damage such as cracks, which are the warning sign toward structural failure. One sleeper failure can definitely lead to another failure. Therefore, it is necessary to develop the engineering guideline to determine the structural integrity and capacity of the aging and worn railway sleepers with one sleeper failure can definitely lead to another failure. Therefore, it is necessary to develop the engineering guideline to determine the structural integrity and capacity of the aging and worn railway sleepers with rail seat abrasion. Without the insight into the structural capacity, the structural reliability of sleepers cannot be determined and. Without the insight into the structural capacity, the structural reliability of sleepers cannot be determined and their safety margin cannot be quantified. This implies that railway operations would be based purely on radical assumptions and crude estimates. This study is thus the world's first to address this key challenge toward truly realistic condition-based predictive track maintenance. In this paper, a numerical study is rigorously executed to comprehensively evaluate structural capacity of railway prestressed concrete sleepers exposed to rail seat abrasion. Rigorous validations by both experimental tests and theoretical analyses (i.e., cross sectional analysis method) are conducted.

13.5 Time-dependent behavior of prestressed concrete sleepers

Durability and long-term behaviors of railway prestressed concrete sleepers depend largely on their creep and shrinkage response. Investigators have proposed various material models to predict creep and shrinkage but those were mostly based on the general reinforced concrete concept. The popular uses of prestressed concrete in long-span bridges, stadiums, silos, and confined nuclear power plants have led to the concern of practitioners about whether those predictive models could be realistically applied to prestressed concrete [36,37]. Due to high initial elastic shortening in prestressed concrete, the creep and shrinkage effects should be critically re-evaluated in flexural members. This study will investigate methods to evaluate creep and shrinkage effects in prestressed concrete railway sleepers. Comparison between design codes of EUROCODE2, ACI, and AS2009-3600 will provide insight into the durability of concrete sleepers. The outcome of the project will help rail track engineers to better design and maintain railway infrastructure, improving asset management efficacy.

13.5.1 Creep prediction on concrete railway sleeper

Neville [38] stated the strain increases with time when concrete loaded are due to creep. Therefore, creep can be defined as the increase in strain under sustained stress and it can be several times as large as the initial strain. Creep is a considerable factor in the concrete structure. Bhatt [39] stated creep is defined as the increase of strain with time when the stress is held constant. The displacement of concrete gradually increases with time when the load is left in place. The displacement reaches a value as large as three to four times of immediate elastic deformation. The inelastic deformation with constant load is known as creep deformation. Fig. 13.13 shows the behavior of the creep strain.

Creep develops in cement paste which is made of colloidal sheets formed by calcium silicate hydrates and evaporable water. Creep can be influenced by several factors such as concrete mix, environmental, and loading conditions. Creep is commonly inversely proportional to concrete strength. In addition, aggregate content and size; water/cement ratio can also affect creep strain. Environmental factors like relative humidity, temperature, and exposed condition are also significant to creep. Duration and magnitude of stress, and first loading age are external loading parameters in creep prediction [40]. When creep is taken into account, its design effects are always evaluated under quasi-permanent combination of actions irrespective of the design situation considered, i.e., persistent, transient, or accidental.

13.5.1.1 Eurocode 2

The total creep strain $\varepsilon_{cc}(t,t_0)$ of concrete due to the constant compressive stress of σ_c applied at the concrete age of t_0 is given by:

$$\varepsilon_{cc}(t,t_0) = \varphi(t,t_0) \times \frac{\sigma_c}{E_c}$$

where $\varphi(t,t_0)$ is the final creep coefficient. E_c is the tangent modulus.

$$\varphi(t,t_0) = \varphi_{RH} \times \frac{16.8}{\sqrt{f_{cm}}} \times \frac{1}{(0.1+t_0^{0.20})}$$

$$\varphi_{RH} = 1 + \frac{1-0.01 \times RH}{0.1+h_0^{0.333}}, f_{cm} \leq 35 MPa$$

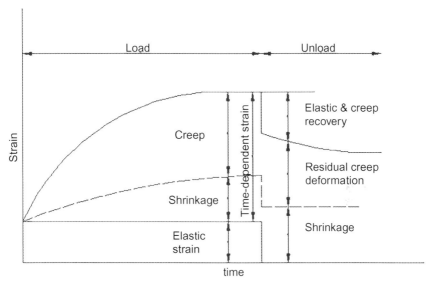

Fig. 13.13 Behavior of the concrete strain.

$$\varphi_{RH} = \left(1 + \frac{1 - 0.01 \times RH}{0.1 + h_0^{0.333}} \alpha_1 \right) \alpha_2, f_{cm} > 35 MPa$$

$$\alpha_1 = \left(\frac{35}{f_{cm}}\right)^{0.7}, \alpha_2 = \left(\frac{35}{f_{cm}}\right)^{0.2} f_{cm} = f_{ck} + 8 MPa$$

$$t_0 = t_{0,T} \left(\frac{9}{2 + t_{0,T}^{1.2}}\right)^{\alpha} \geq 0.5, \alpha = \{-1(S), 0(N), 1(R)\}$$

where RH is relative humidity in percentage; h_0 is the ratio of cross-sectional area and perimeter of the member in contact with the atmosphere, $h_0 = 2A_c/u$; S, R, and N refer to different classes of cement.

13.5.1.2 ACI code

According to ACI 209-92, the predicted parameter is creep coefficient $\varphi(t, t_0)$ and the equation is given by:

$$\varphi(t, t_0) = \frac{(t - t_0)^{\psi}}{d + (t - t_0)^{\psi}} \varphi_u$$

where $\varphi(t, t_0)$ is creep coefficient at any time t when a load applied at age t_0; d (in days) and ψ are considered constants for a given member shape and size that define the time-ratio part. ACI-209R-92 recommends an average value of 10 and 0.60 for d and ψ, respectively; φ_u is the ultimate creep coefficient. For the ultimate coefficient φ_u, the average value is suggested using 2.35. However, creep coefficient still needs to be modified by correct factors:

$$\varphi_u = 2.35\gamma_c$$

$$\gamma_c = \gamma_{c,t0}\gamma_{c,RH}\gamma_{c,vs}\gamma_{c,s}\gamma_{c,\psi}\gamma_{sh,\alpha}$$

where $\gamma_{c,\,t0}$ is loading age factor; $\gamma_{c,\,RH}$ is ambient relative humidity; $\gamma_{c,\,vs}$ is the volume to surface ratio of the concrete section; $\gamma_{c,\,s}$ is the slump; $\gamma_{c,\,\psi}$ is fine aggregate amount; $\gamma_{sh,\,\alpha}$ is the air content.

13.5.1.3 Australian standard 3600-2009

The creep coefficient at any time φ_{cc} can be determined by:

$$\varphi_{cc} = k_2 k_3 k_4 k_5 \varphi_{cc.b}$$

where k_2 is the development of creep with time; k_3 is the factor which depends on the age at first loading τ (in days); k_4 is the factor which accounts for the environment; k_5 is the factor which accounts for the reduced influence of both relative and humidity and specimen size; φ_{cc} is the basic creep coefficient.

The development of creep with time k_2 can be calculated by:

$$k_2 = \frac{\alpha_2 (t-\tau)^{0.8}}{(t-\tau)^{0.8} + 0.15 t_h}$$

$$\alpha_2 = 1.0 + 1.12 e^{-0.008 t_h}$$

$$t_h = 2 A_g / u_e$$

where t is any time in days; t_h is the hypothetical thickness; A_g is the cross-sectional area of the member; u_e is the portion of the section perimeter exposed to the atmosphere plus half the total perimeter of any voids contained within the section.

Factor k_3 which depends on the age at first loading τ can be shown as:

$$k_3 = \frac{2.7}{1 + \log(\tau)} \quad (\text{for } \tau > 1 \, day)$$

Factor k_4 which accounts for the environment:

$$k_4 = 0.7 \, for\, an\, arid\, environment$$

$$k_4 = 0.65 \, for\, an\, interior\, enviroment$$

Table 13.3 Basic creep coefficient.

f'_c (MPa)	20	25	32	40	50	65	80	100
$\varphi_{cc,\,b}$	5.2	4.2	3.4	2.8	2.4	2.0	1.7	1.5

$k_4 = 0.60$ for a temperate enviroment

$k_4 = 0.5$ for a tropical or near − coastal enviroment

Factor k_5 is given by:

$k_5 = 1.0, f'_c \leq 50 MPa$

$k_5 = (2.0 - \alpha_3) - 0.02(1.0 - \alpha_3)f'_c, 50 \leq f'_c \leq 100 MPa$

where $\alpha_3 = 0.7/(k_4 \alpha_2)$.

The basic creep coefficient φ_{cc} is shown in Table 13.3.

13.5.2 Shrinkage prediction on concrete railway sleeper

Concrete shrinkage is the time-dependent strain in an unloaded and unrestrained specimen at constant temperature. Shrinkage is not an entirely reversible process in comparison with creep. Both creep and shrinkage are influenced by similar factors such as relative humidity, concrete strength, and cement. Shrinkage can be divided into plastic shrinkage, chemical shrinkage, thermal shrinkage, and drying shrinkage. The shrinkage strain is usually considered to be the sum of the drying shrinkage and endogenous shrinkage. Drying shrinkage is the reduction in volume due to loss of water during the drying process. Endogenous shrinkage is sometimes used to refer to that part of the shrinkage of the hardened concrete that is not associated with drying (the sum of autogenous and thermal shrinkage) [39,41].

13.5.2.1 Eurocode 2

The total shrinkage strain ε_{cs} can be given by:

$$\varepsilon_{cs} = \varepsilon_{ds} + \varepsilon_{as}$$

where ε_{ds} is drying shrinkage strain; ε_{as} is autogenous shrinkage strain.

The drying shrinkage strain ε_{ds} can be estimated by:

$$\varepsilon_{ds} = \beta_{ds}(t, t_0) \times \varepsilon_{cd0} \times k_h$$

$$\varepsilon_{cd0} = 0.85[(220 + 110\alpha_{ds1}) \times \exp(-\alpha_{sd2} \times 0.1 f_{cm})] \times 1.55\left[1 - (0.01RH)^3\right]10^6$$

Table 13.4 Cement type and coefficient.

Cement type	α_{ds1}	α_{ds2}
S	3	0.13
N	4	0.12
R	6	0.11

$$\beta_{ds}(t,t_0) = \frac{(t-t_s)}{(t-t_s) + 0.04\sqrt{h_0^3}}$$

where k_h is a coefficient which depends on the national size h_0; RH is relative humidity in percentage; $h_0 = 2Ac/u$ (mm), Ac is cross-sectional area, u is the perimeter of the member in contact with the atmosphere. The values of parameter α_{ds1} and α_{ds2} as a function of the type of cement are shown in Table 13.4.

The autogenous shrinkage strain ε_{as} can be calculated from:

$$\varepsilon_{as} = \beta_{as}(t) \times \varepsilon_{ca}(\infty)$$

$$\varepsilon_{ca}(\infty) = 2.5 \times (f_{ck} - 10) \times 10^{-6}$$

$$\beta_{as}(t) = 1 - \exp(-0.2t^{0.5})$$

13.5.2.2 ACI code

The shrinkage strain $\varepsilon_{sh}(t)$ at age of concrete t (days), predicted from the start of drying at t_c can be calculated by:

$$\varepsilon_{sh}(t) = \frac{(t-t_c)^\alpha}{f + (t-t_c)^\alpha} \varepsilon_{shu}$$

$$\varepsilon_{shu} = 780 \times 10^{-6} \, mm/mm \, (in/in)$$

where f (in days) and α are considered constants for a given member shape and size that define the time-ratio factor, the average value for f is recommended 35 days of moist curing and α is suggested to use 1.0; ε_{shu} is ultimate shrinkage strain; $(t-t_c)$ is the time between end of curing and any time after curing.

For the ultimate shrinkage strain ε_{shu} the average value is suggested using 780×10^{-6} mm/mm. However, shrinkage strain still needs to be modified by correct factors:

$$\varepsilon_{shu} = 780\gamma_{sh} \times 10^{-6} \, mm/mm \, (in/in)$$

$$\gamma_{sh} = \gamma_{sh,tc}\gamma_{sh,RH}\gamma_{sh,vs}\gamma_{sh,s}\gamma_{sh,\psi}\gamma_{sh,c}\gamma_{sh,\alpha}$$

where $\gamma_{sh,\,tc}$ is initial moist curing coefficient; $\gamma_{sh,\,RH}$ is ambient relative humidity coefficient; $\gamma_{sh,\,vs}$ is the volume to surface ratio of the concrete section coefficient; $\gamma_{sh,\,s}$ is slump coefficient; $\gamma_{sh,\,\psi}$ is fine aggregate coefficient; $\gamma_{sh,\,c}$ is the cement content factor coefficient; and $\gamma_{sh,\,\alpha}$ is air content coefficient.

13.5.2.3 Australian standard 3600–2009

The total shrinkage strain ε_{cs} is shown below:

$$\varepsilon_{cs} = \varepsilon_{cse} + \varepsilon_{csd}$$

where ε_{cse} is autogenous shrinkage strain; ε_{csd} is drying shrinkage.

The autogenous shrinkage can be calculated by:

$$\varepsilon_{cse} = \varepsilon'_{cse}(1.0 - \exp\{-0.1t\})$$

$$\varepsilon'_{cse} = (0.6 f'_c - 1.0) \times 50 \times 10^{-6} \ (f'_c \ in\ MPa)$$

$$\varepsilon_{csd.b} = (1.0 - 0.008 f'_c) \times \varepsilon'_{csd.b}$$

where $\varepsilon'_{csd.b}$ depends on the quality of the local aggregates and may be taken as 800×10^{-6} for concrete supplied in Sydney and Brisbane, 900×10^{-6} in Melbourne and 1000×10^{-6} in elsewhere.

The drying shrinkage strain ε_{csd} after the beginning of drying $(t - \tau_d)$ can be estimated as:

$$\varepsilon_{csd} = k_1 k_4 \varepsilon_{csd.b}$$

where k_1 is the factor which describes the development of drying shrinkage with time; k_4 is the factor which accounts for the environment.

The factor k_1 can be given by:

$$k_1 = \frac{\alpha_1 (t - \tau_d)^{0.8}}{(t - \tau_d)^{0.8} + 0.15 t_h}$$

$$\alpha_1 = 0.8 + 1.2 \exp\{-0.005 t_h\}$$

The factor k_4 which accounts for the environment:

$k_4 = 0.7\ for\ an\ arid\ environment$

$k_4 = 0.65\ for\ an\ interior\ enviroment$

$k_4 = 0.60\ for\ a\ temperate\ enviroment$

$k_4 = 0.5\ for\ a\ tropical\ or\ near-coastal\ enviroment$

13.6 Summary

This chapter proposes the diagnostics and management methods associated with the practical issues. This chapter introduces the main duties of sleepers and railway sleeper design processes. The common damages of prestressed concrete sleepers: dynamic load, fatigue, rail seat abrasion, and time-dependent behavior are analyzed for potential risks. Typical dynamic loading (impact loading) refers to the shape, magnitude, and duration of impact loads. The wheel impact loads at different return periods should be evaluated. The ultimate impact resistance of prestressed concrete sleepers should also be studied required by the limit states design approach. Fatigue loads and material properties are the most important factors to analyze the fatigue life of concrete sleeper. The fatigue life of concrete sleeper can be calculated by using damage accumulation method. The fatigue failure of concrete sleeper at rail seat is decided by the positive bending moment and the fatigue of concrete sleeper at center is decided by the negative bending moment. Rail seat abrasion is one of the most common damages of prestressed concrete sleepers, especially in freight operations. Rail seat abrasion is commonly related to axle load, traffic volume, curvature and grade of the rail line, the presence of abrasive fines, the behavior of the fastening system, and climate, etc. Rail seat abrasion results in many problems of railway tracks including the deteriorated quality of the track geometry and the reduced loading carry capacity of aging, worn concrete sleepers. Time-dependent behavior can also influence the durability and long-term behaviors in the field. In real life, railway infrastructure experiences harsh environments and aggressive loading conditions from increased traffics and load demands, which means creep and shrinkage strains could have a more significant influence on the deformation of track components. When shortening and deflection occur in prestressed concrete sleepers, the track gauge could change with shortening and deflections. There is a danger that train derails might happen because of track gauge change. The outcome of this chapter contributes significantly to better insight into the diagnostics and management methods of prestressed concrete railway sleepers. The insight will improve track maintenance and inspection criteria.

References

[1] D. Li, S. Kaewunruen, Effect of extreme climate on topology of railway prestressed concrete sleepers, Climate 7 (1) (2019) 17.
[2] C. Esveld, Modern Railway Track, second ed., MRT Productions, Zaltbommel, The Netherlands, 2001.
[3] W. Ferdous, A. Manalo, Failures of mainline railway sleepers and suggested remedies—review of current practice, Eng. Fail. Anal. 44 (2014) 17–35.
[4] C. Stuart, International Concrete Crosstie and Fastening System Survey, Research Results, 2013.
[5] International Federation for Structural Concrete, fib bulletin 37: precast concrete railway track systems, state-of-art report, 2006.
[6] A. Palomo, A.F. Jiménez, C.L. Hombrados, J.L. Lleyda, Railway sleepers made of alkali activated fly ash concrete, Rev. Ing. Constr. 22 (2007) 75–80.

[7] H. Taylor, The railway sleeper: 50 years of pretensioned, prestressed concrete, Struct. Eng. 71 (1993) 281–295.
[8] M.L. Lee, W.K. Chiu, L.L. Koss, A numerical study into the reconstruction of impact forces on railway track-like structures, Struct. Health Monit. 4 (1) (2005) 19–45.
[9] G. Kumaran, D. Menon, N.K. Krishnan, Evaluation of dynamic load on railtrack sleepers based on vehicle-track modelling and analysis, Int. J. Struct. Stab. Dyn. 2 (3) (2002) 355–374.
[10] Australasian Railway Association, Common requirements, section 2: commissioning and recommissioning, 2201: Performance acceptance requirements, version 1, in: Code of Practice for the Defined Interstate Rail Network, vol. 5, Australasian Railway Association, Kingston, ACT, Australia, 2002. Rollingstock (Draft).
[11] S. Kaewunruen, A.M. Remennikov, Use of Reliability-Based Approach in the Conversion of the Existing Code AS1085.14 to Limit States Design Discussion Paper, CRC for Railway Engineering and Technologies, 2006. December.
[12] A.M. Remennikov, M.H. Murray, S. Kaewunruen, Reliability-based conversion of a structural design code for railway prestressed concrete sleepers, Proc. Inst. Mech. Eng. F J. Rail Rapid Transit 226 (2) (2012) 155–173.
[13] M. Murray, J. Leong, Discussion paper on the development of limit state factors: State 2, Research Report, CRC for Railway Engineering and Technologies, 2006. September.
[14] S. Kaewunruen, Experimental and Numerical Studies for Evaluating Dynamic Behaviour of Prestressed Concrete Sleepers Subject to Severe Impact Loading, University of Wollongong, School of Civil Mining and Environmental Engineering, Australia, 2007. PhD thesis.
[15] S. Kaewunruen, A.M. Remennikov, Relationship between interface impact force and rail seat moment of railway prestressed concrete sleepers, in: Proceedings of the SEM Annual Conference Exposition 2007, Society for Experimental Mechanics, Bethel, CT, 2007. (Springfield Massachusetts, USA) June 3–6.
[16] S. Kaewunruen, A.M. Remennikov, Probabilistic impact fractures of railway prestressed concrete sleepers, Adv. Mater. Res. 41–42 (2008) 259–264.
[17] Standards Australia, Railway Track Materials, Part 14: Railway Prestressed Concrete Sleepers, Australian Standard AS1085.14, Standards Australia, Australia, 2003.
[18] A.M. Remennikov, S. Kaewunruen, Simulating shock loads in railway track environments: experimental studies, in: Proceedings of the 14th International Congress on Sound and Vibration, 2007. (Cairns, Australia) July 9–12.
[19] H. Thun, Assessment of Fatigue Resistance and Strength in Existing Concrete Structures, Luleå Tekniska Universitet, Luleå, 2006.
[20] M.A. Grubb, J.A. Corven, K.E. Wilson, J.W. Bouscher, L.E. Volle, Load and Resistance Factor Design (LRFD) for Highway Bridge Superstructures—Design Manual, Federal Highway Administration, Washington, DC, 2007.
[21] K. Gylltoft, Fracture Mechanics Models for Fatigue in Concrete Structures, Luleå Tekniska Universitet, Luleå, 1983.
[22] S. Kaewunruen, A.M. Remennikov, M.H. Murray, Introducing a new limit states design concept to railway concrete sleepers: an Australian experience, Front. Mater. 1 (2014) 8.
[23] S. Kaewunruen, R. You, M. Ishida, Composites for timber-replacement bearers in railway switches and crossings, Infrastructures 2 (2017) 13.
[24] S. Kaewunruen, A. Remennikov, M.H. Murray, Limit states design of railway concrete sleepers, Proc. Inst. Civ. Eng. Transp. 165 (2012) 81–85.
[25] Mallet, G.P. n.d. Fatigue of Reinforced Concrete, State of the Art Review/2, Transport and Road Research Laboratory, HMSO. London, 8.

[26] A. Parvez, S.J. Foster, Fatigue of steel-fibre-reinforced concrete prestressed railway sleepers, Eng. Struct. 141 (2017) 241–250, https://doi.org/10.1016/j.engstruct.2017.03.025.
[27] A.M. Keyna, Structural Analysis of Reinforced Concrete, Isfahan University of Technology Press, Iran, 1998.
[28] J. Sadeghi, M. Fathali, Deterioration analysis of concrete bridges under inadmissible loads from the fatigue point of view, Sci. Iran. 14 (2007) 185–192.
[29] E.O. Lantsoght, C. van der Veen, A. de Boer, Proposal for the fatigue strength of concrete undercycles compression, Constr. Build. Mater. 107 (2016) 138–156, https://doi.org/10.1016/j.conbuildmat.2016.01.007.
[30] EN1992-2, Eurocode 2—Design of Concrete Structures. Part 2: Concrete Bridges—Design and Detailing Rules, European Committee for Standardization, Brussels, 2005.
[31] EN1993-1-9, Design of Steel Structures. Part 1-9: Fatigue, European Committee for Standardization, Brussels, 2009.
[32] J.C. Zeman, Hydraulic Mechanisms of Concrete-Tie Rail Seat Deterioration, University of Illinois at Urbana-Champaign, Illinois, America, 2010.
[33] C. Ngamkhanong, D. Li, S. Kaewunruen, Impact capacity reduction in railway prestressed concrete sleepers with surface abrasions, IOP Conf. Ser. Mater. Sci. Eng. 245 (2017) 032048. IOP Publishing.
[34] T. Bakharev, L. Struble, Microstructural feature of rail seat deterioration in concrete ties, J. Mater. Civ. Eng. 9 (1997) 146–153.
[35] J.C. Zeman, J.R. Edwards, D.A. Lange, C.P. Barkan, Investigation of Potential Concrete Tie Rail Seat Deterioration Mechanisms: Cavitation Erosion and Hydraulic Pressure Cracking, Proceedings of the Transportation Research Board 89th Annual Meeting, Transportation Research Board, Washington, DC, 2010.
[36] M. Ercolino, A. Farhidzadeh, S. Salamone, G. Magliulo, Detection of onset of failure in prestressed strands by cluster analysis of acoustic emissions, Struct. Monit. Maint. 2 (4) (2015) 339–355.
[37] Z. Li, S. Li, J. Lv, H. Li, Condition assessment for high-speed railway bridges based on train-induced strain response, Struct. Eng. Mech. 52 (2) (2015) 199–219.
[38] A.M. Neville, Properties of Concrete, fourth ed., Longman, Longman House, Burnt Mill, Harlow Essex, England, 1995.
[39] P. Bhatt, Prestressed concrete design to Eurocodes, first ed., Spon Press, Abington, UK, 2011.
[40] S. Vittorio, Time-Dependent Behaviour of Reinforced Concrete Slabs, University of Bologna, 2011.
[41] R.I. Gilbert, N.C. Mickleborough, G. Ranzi, Design of Prestressed Concrete to AS3600–2009, second ed., CRC Press, Boca Raton, USA, 2016.

Railway ballast

14

Yunlong Guo[a], Valeri Marikine[a], and Guoqing Jing[b]
[a]Delft University of Technology, Delft, The Netherlands, [b]Beijing Jiaotong University, Beijing, China

14.1 Introduction

14.1.1 Ballast definition

Railways play an important role in current transportation systems. The ballast bed (granular layer) is placed between the sleeper and the subballast as a shock absorber to reduce the stress from sleeper to an acceptable level (Fig. 14.1). Ballast bed is made of crushed rocks of certain size (in 20–60 mm range). Among the track components, ballast is the biggest part of the ballast track taking the largest volume.

As shown in Fig. 14.2, the ballast particles are placed between sleepers, under sleepers and on both sides of sleepers with certain profiles. The profiles include the ballast thickness (250–350 mm, from sleeper bottom), crib ballast (around 600 mm, between two adjacent sleepers) and shoulder ballast (300–500 mm).

14.1.2 Ballast function

The main purpose of ballast bed is to perform the following functions [1]:

- Providing an even load-bearing platform and supporting sleepers stably. Stable support and platform are necessary for safe train operation, furthermore, the track irregularity is mainly caused by unacceptable ballast bed deformation.
- Dissipating intense loads and reducing the stress magnitude at the subgrade surface. To avoid the stress concentration, the train loads from sleeper to the subgrade are minimized, dissipated, and uniformly distributed by the ballast. Nevertheless, ballast pockets are still developing due to high stresses on the ballast-soil interface.
- Keeping sufficient track stability by providing the sleeper resistance in vertical, longitudinal, and lateral directions. It needs to note that the lateral resistance is very important for continuous welded rail (CWR) track to reduce the buckling possibility. Ballast shear strength, influenced by ballast compaction and particle morphology, is the main characteristic affecting track stability.
- Providing necessary track elasticity and resiliency against dynamic loads. Losing the resiliency can lead to large differential settlements, and the proper elasticity can reduce damage to track components.
- Resisting sufficiently against biochemical contamination, mechanical contamination, and environment. The ballast disposes of not only the mechanical deterioration from the track structures (sleeper, ballast, and soil) and from the freight (coal, sands), but also the biochemical contamination (mostly the human excrement). Additionally, ballast needs to resist the weathering degradation (e.g., acid rain).

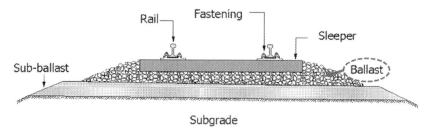

Fig. 14.1 Conventional ballast track.

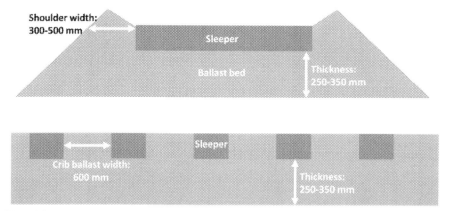

Fig. 14.2 Ballast bed profile size.

- Possessing adequate permeability for drainage. Newly built ballast bed can provide good drainage, and proper particle size distribution can increase the permeability. However, ballast fouling can reduce the permeability through jamming voids in ballast bed, and the fouling is inevitable.
- Absorbing noises. Compared with the slab track, the ballast layer can absorb noise and vibrations.
- Providing necessary electric insulation. The signaling needs the ballast layer to have enough electric insulation. Due to this, whether the steel slag can be used as railway ballast still remains a question.

14.1.3 Ballast research on current problems

With the development of high-speed railway and heavy haul railway, the main aspects in railway research are related to the development of new numerical methods, track evaluation standards, design philosophy, and maintenance strategies. For example, in the past two centuries, the design of ballast track almost remains the same, although the railway freight loads and speeds keep increasing [2,3].

Under cyclic loading due to passing trains, the ballast particles deteriorate due to breakage and abrasion, resulting in permanent plastic deformations of ballast bed. Accumulating the plastic deformation, the ballast bed cannot provide adequate performance.

The degradation becomes more severe due to the increasing axle loads (freight line) and train speed (passenger transport), which leads to frequent maintenance (e.g., tamping). More importantly, when the demand for higher speed and heavier haul is increasing, the unacceptable ballast bed performance can cause issues of passenger comfort and safety.

The current studies (research) on the ballast performed recently can be categorized into the four following issues [1,4].

14.1.3.1 Performance assessment

The performance characteristics of the ballast bed mainly contain durability, stability, shear strength, stiffness, and resilience [5]. In earlier studies, the factors influencing the performance (e.g., particle shape and size) were analyzed with the laboratory tests (e.g., direct shear test) or field tests (e.g., single sleeper push test, sleeper supporting stiffness measurement). Numerical models of the corresponding tests (e.g., direct shear test) were also applied. The studies for performance assessment of ballast bed have been relatively mature at both the basic knowledge and methodologies.

14.1.3.2 Ballast bed degradation mechanism

The mechanism of ballast bed degradation and the associated plastic deformations have not been revealed clearly, especially in some special railway structures, e.g., turnouts, transition zones. The problem becomes more complicated, due to the increasing train speed and heavier haul [6,7]. The main challenge for studying ballast bed degradation mechanism is that the factors are too much. For example, ballast particle degradation affects ballast bed degradation [1,4]. However, only a few factors are considered in one study in most cases, which causes the conclusions to be different.

14.1.3.3 Degradation mitigation and performance improvement of ballast bed

Using other materials in the ballast bed is an effective means for ballast degradation mitigation and performance improvement, e.g., using the under sleeper pads, geogrid, geocell, polyurethane [8], etc.

However, applying other materials changes the ballast bed properties, which possibly causes other issues of the ballast track. Therefore, the challenges are how to correctly use the new materials, reducing as much as possible their negative influence on ballast track.

14.1.3.4 Maintenance

Frequent maintenances to ballast lead to high costs. Earlier studies have shown that tamping (the most common maintenance) causes ballast particle degradation (breakage and abrasion) due to the impact from the insertion of the tamping tines into the ballast and the high squeezing force. Therefore, more studies should be performed toward more precise and correct maintenance.

14.1.4 Solutions to ballast problems

The solutions to the stressed ballast problems are summarized into three aspects, i.e., ballast degradation mitigation, ballast inspection improvement and better ballast condition assessment.

14.1.4.1 Ballast degradation

Ballast degradation usually describes the deterioration of ballast bed. Also, it has been used to describe the ballast particle degradation in recent studies. Ballast particle degradation mainly includes two mechanisms, namely breakage and abrasion [1]. Ballast particle degradation leads to ballast bed degradation ultimately influencing the performance of ballast bed and overall track performance. Therefore, it is necessary to study ballast particle degradation in order to better understanding the degradation mechanism and to improve the ballast bed performance.

Ballast particle degradation mitigation can be achieved by applying new ballast materials, for example, steel slag [9] and Neoballast [10], as well as using new other materials, such as, rubber chips [11], under sleeper pads [12], geogrid [13] and polyurethane [14]. These applications are explained more in Section 14.2.3.2.

14.1.4.2 Ballast inspection

Ballast inspection is to check the ballast bed condition (e.g., ballast bed profile, geometry and hanging sleeper) in order to allow the vehicle safely rides on the track. It is closely related to ballast performance assessment and maintenance.

More focuses should be paid on ballast inspection. In the past, ballast inspection was not focused enough as much as other track components (inspected by gauging rules). However, most of the dangerous situations (e.g., buckling and mud-pumping) is resulted from ballast bed failure (low resistance to sleeper and ballast particle degradation). In addition, the track irregularity in most cases is contributed by the deformation of ballast bed.

More technical means have been developed and used for ballast inspection, such as, GPR, inspection train with camera, drones, SmartRock, and satellite. How these technical means have been applied for ballast inspection is explained in Section 14.3.

14.1.4.3 Ballast condition assessment

Ballast condition assessment is performed usually based on the data from ballast inspection or using numerical simulations. The assessment can be categorized into three aspects, data/signal processing, and numerical simulation.

For the data/signal process, it is to process the data that are mostly measured by some traditional instruments, for example, ballast acceleration (accelerometer), ballast bed settlement (displacement meter), ballast bed stress (pressure sensor), and ballast degradation (sieving).

For the numerical simulations, more and more assessment has been performed by building numerical models, among which the discrete element method (DEM) has clear advantages of simulating ballast particles. For example, using the DEM, some detailed parametric studies that are often not feasible in laboratory tests, can be performed, e.g., interparticle friction and distribution of contact forces (contact force chain).

14.2 Ballast degradation

In this section, ballast degradation mechanism is firstly introduced and discussed, with the subsections of

- degradation mechanism: introduction of ballast degradation reasons, types, and consequences,
- degradation factors: factors influencing ballast degradation, loading type, parent rock material, compaction, particle size, and shape,
- and degradation mitigation: solutions to mitigate ballast degradation (new materials and sleeper innovations).

14.2.1 Ballast degradation mechanism

14.2.1.1 Ballast degradation reasons

Several reasons cause ballast degradation, such as, train loading (cyclic loading, impact loading), maintenance (tamping, stabilization), weather (extreme weather, acid rain), geology (earthquake, desert) and biology (plants). Among these reasons, the train loading and maintenance are two main reasons. Nevertheless, other reasons are also critical when the ballast beds are in the extreme conditions, such as, special structure (transition zone), air pollution (acid rain), and cold region (freeze-thaw).

For the train loading, the degradation mechanism has been already demonstrated in depth. For example, in [1,15,16], many studies show that cyclic loadings cause ballast degradation (breakage) using laboratory tests. The laboratory tests are mainly the triaxial test, ballast box test, and improved ballast box test (e.g., movable side walls [17]). In addition, the impact loading is admitted widely as the reason for accelerating ballast degradation, which for example has been studied in the research on transition zone and switch and crossing [18,19].

For the maintenance, the degradation mechanism during the tamping have been studied, and more studies are still strongly needed to deepen the understanding. For example, in [20–22], laboratory tests were performed to study ballast degradation caused by tamping. in [23,24], ballast degradation in the field tests caused by tamping were examined. These studies demonstrated that the tamping operation causes serious ballast degradation, which was firstly proposed in the book [25].

For the other reasons, limited studies have been performed, which means more focuses should be paid to fill in the research gaps. Some advices on performing the related studies are given as follows.

- Weather: the cold region can be simulated using a refrigeration house. For example, the study in [26,27] tested the ballast performance. Further study can be performed on the ballast degradation in the low temperature.
- Geology: using ballast box test is possible to study the ballast degradation of the mixture of ballast particles and sands (desert area), e.g., [28]. For the earthquake, the shake table can be used to simulate earthquake loadings.
- Biology: the plant damages to the ballast is very small, but it is possible to be serious to the high-speed vehicle. Another possible reason is the plants improve the stiffness of ballast bed, which reduces the resilience (or elasticity) of ballast bed. This can cause more ballast degradation.

14.2.1.2 Ballast degradation types

Ballast particle degradation is generally classified includes two main types, breakage, and abrasion. However, it may not be sufficient to classify ballast degradation by two types, because ballast breakage can have various types, such as, corner breakage, splitting in the middle, and breaking into several parts. Particularly, until now, few methods were reported for ballast abrasion evaluation.

Most importantly, the current evaluation methods for the breakage and abrasion, which are still insufficient and need improvement, cannot present ballast degradation types.

For instance, all the breakage evaluation methods are based on sieving, analyzing the change of the PSD or the percentage of particles passing some certain sieve size, when performing laboratory tests, e.g., the Los Angeles Abrasion test, the triaxial test, and the prismoidal triaxial test [1,29]. The breakage index B_g (proposed in [30]) calculates particle sizes between the initial and final particle size distributions. To be more specific, it is the sum of the difference in percentage retained on sieves, having the same sign. However, it may not be sufficient to evaluate ballast breakage only by calculating the PSD, since the final PSD results are obtained based on various types of ballast breakage.

Most of the current methods that can study ballast abrasion are related to image analysis. For example, in [31], the abrasion is evaluated by the changes of ballast particle morphology. The University of Illinois aggregate image analyzer (UIAIA) and a second-generation aggregate imaging system (AIMS) are utilized to capture changes of individual particles before and after the micro-Deval test [31].

Consequently, among the previous methods, image analysis is the most potential and effective one, which can be a significant method to study the ballast degradation types. More studies based on that should be performed for better understanding of the ballast degradation mechanism and further its effects on the performance and deformation of ballast bed [1,32].

14.2.1.3 Ballast degradation consequences

The ballast degradation causes several consequences, including deformation (differential settlement), low capacity (lateral and longitudinal resistances), ballast bed harden and drainage failure (mud-pumping).

Deformation
Ballast degradation has great influences on the shear strength leading to big deformation of ballast bed, which was measured in triaxial tests [33] or other laboratory tests (e.g., direct shear test [34]). In addition, the ballast fouling, produced by ballast degradation, in most cases reduces the shear strength.

In addition, particle breakage significantly influences the performance (e.g., shear strength) and the deformation of any kinds of ballast material [1]. Particle size would be changed after crushing and generally cause the densification and the contaminations clogging the voids, which may further increase the shear strength [35]. The ballast abrasion is demonstrated in [36] that permanent settlement is related to the ballast abrasion.

Hanging sleeper is a mostly-seen issue due to differential settlement, which is closely related to rapid ballast degradation (due to impact loading).

Low capacity
The low capacity of ballast bed means that ballast degradation reduces the lateral resistance and longitudinal resistance to the sleeper, which causes unstable ballast track. The buckling is caused by the insufficient lateral resistance of ballast bed to the sleeper. The longitudinal resistance insufficiency is an unsolved issue of tracks that are built on the long steep slope (mountainous areas).

Ballast bed hardening
Ballast bed harden is caused by the fouling, which means the ballast bed becomes like a cemented concrete with the fouling as binder. The fouling, except some special line (e.g., Australian freight line for coal transportation), results from ballast degradation. Ballast bed harden means the ballast bed lost the elasticity and resilience with high stiffness. The geometry of the ballast track (with hardened ballast bed) is normally irregular, which leads to rapid track component degradation.

Drainage failure
The drainage failure would also induce dramatic ballast settlement. As reported in [37], saturation increased settlement by about 40% of that of dry ballast. The fouling jams the voids of ballast bed, blocking water flow. The mixture of fouling and water is a lubricant causing low capacity of ballast bed. In addition, the drainage failure also causes the mud-pumping, which is also an important reason of subgrade failure.

14.2.2 Ballast degradation factors
14.2.2.1 Loading type
The loading type includes cyclic loading from train and some impact loading at some special railway structures, such as, the transition zone and switch and crossing. In addition, the cyclic loading can be divided based on transportation types, such as freight line, passenger line, and mixed line. Due to the different vehicle types, the loadings are quite different, such as, the loading frequency and amplitude. For

example, in some studies, the loading applied in the laboratory test simulates the normal speed train (e.g., 100 km/h) [38]. For the high-speed train, the loading frequency and amplitude are quite different, whose effects to the ballast was studied in [39,40].

14.2.2.2 Parent rock material

Generally, the material of the parent rock is analyzed using the petrographic methods. Ballast is typically made of crushed (from quarry) rock particles, e.g., limestone, volcanic, granite, quartzite, and sandstone. The parent rock types are different in each country up to the quality and availability.

For the parent rock material, limited studies have been published, because the ballast materials in different countries are quite different, which means there is not a universal degradation growth/prediction for each ballast material. However, the parent rock material is in most cases tested before using the material to make the ballast bed. For example, in China, the China Academy of Railway Sciences tests if the parent rock materials are suitable to make ballast bed following the Chinese standard [41], for example, the magnesium sulfate value and water absorption as a screening test for freeze-thaw resistance. In the British standard, the parent rock material is also required to meet certain assessment [42]. Other examples on parent rock material can be found in [29,43].

14.2.2.3 Compaction (bulk density)

The bulk density characterizes the compaction state of ballast bed, similar to porosity of ballast bed. The bulk density is calculated as the ballast mass divided by the total volume (ballast particles and voids), while the porosity is calculated as the ratio of void volume to the total volume.

The compaction state of ballast bed has a significant influence on the performance (e.g., track stability, shear strength, and stiffness), thus, the bulk density (or porosity) is a key indicator for ballast bed quality during the ballast track construction, as well as when performing field and laboratory tests, and numerical simulations. The bulk density (porosity) is easy to obtain in the numerical simulations (e.g., using the discrete element method (DEM)).

Particularly, the compaction of ballast bed is a key factor influencing ballast degradation. Because low compaction leads to low confining stress to ballast, which causes more ballast degradation as explained in [44]. When the ballast bed is well-compacted, which means the ballast particles have enough contacts with the adjacent particles (named as coordination), then a stable degradation process is shown. More studies on the compaction (or confining stress/pressure) can be found in [20,45,46].

14.2.2.4 Particle size and shape

The particle sizes of ballast beds are presented by the particle size distribution (PSD), also known as gradation, which is obtained by sieving. The PSD measures the percentage (by weight) of the particles in a certain size range. For example, in the British standard, the fractions of the Gradation A are 22.4, 31.5, 40, 50, and 63 mm.

The shape is normally evaluated roughly with the dimension ratio of the particles (elongation and flakiness) [47]. The flaky and elongated particles are calculated with the lengths of the three representative axes: the longest axis with the length L, the medium axis with the length I and the shortest axis with the length S. The ballast particles with S/I smaller than 0.6 or with L/I above 1.8 as the flaky or elongated ballast particles, respectively [48].

The particle shape is closely related to ballast degradation. For example, it was found that ballast specimens with flaky or elongated particles can cause lower resilience [35]. However, a limited percentage of flaky or elongated particles leads to higher shear strength and thus a lower rate of settlement accumulation [49]. Nevertheless, it was also reported that adding flaky or elongated particles results in more severe degradation and higher deformation of ballast bed [50].

Besides the elongation and flakiness, the angularity and surface texture are the other two main shape characteristics. However, limited studies on ballast degradation related to these two characteristics were found. Because the assessment of angularity and texture requires more accurate measurement tools than simply sieving, for example, laser scanning.

14.2.3 Ballast degradation mitigation

In this section, methods for ballast degradation mitigation are summarized from three aspects, ballast material, new geo-inclusion materials, and sleeper innovations.

- Ballast material: in these methods, solutions from the ballast bed itself are considered, for example, using high-density material and using more cubic ballast particles.
- New geo-inclusion materials: solutions from adding new materials in the ballast bed are considered, for example, geogrid, geocell, polyurethane, and ballast mat.
- Sleeper innovations: solutions from changing the sleeper material and shapes are considered, for example, ladder sleeper and plastic sleeper.

14.2.3.1 Ballast material

The ballast material is introduced in the Section 14.2.2, which has great influence on the ballast degradation, mainly the particle size and shape, as well as parent rock material. Therefore, optimizations on ballast material have been performed in many studies.

For example, in [51], different ballast materials are used for different sections of the bridge approach (transition zone) to balance the stiffness difference of the transition zone. The results show that using a smooth transition with different ballast materials at each section is effective to reduce or mitigate differential settlement, which can reflect the ballast degradation is mitigated.

Another example is using the steel slag as ballast particles. The densities of different steel slag are not always the same, but mostly higher than ballast material. Most importantly, it has higher resistance to degradation than rock, which was proved in [52]. More studies on the steel slag can be found in [53–56].

By bonding rubber chips to the ballast particles, the Neoballast is created. The degradation of the Neoballast particles was proved to be less than normal ballast in [57]. The shear strength and dynamic performance of Neoballast particles were studied in [10,58].

14.2.3.2 New geo-inclusion materials

The new geo-inclusions materials are able to reduce ballast degradation.

Under sleeper pads is a plastic pad attached between ballast particles and sleeper for softening their contacts and finally reducing ballast degradation [59]. It is a popular solution, which has attracted lots of studies, especially for the special structures, such as transition zone [60,61], switch and crossing [62,63], and joints [64].

Geogrid has been successfully applied to build railway tracks at weak subgrades, improve ballast layer stability, and reduce track settlement. However, to what extent it can reduce the ballast degradation has not been reported in any publications.

Geocell has been applied to improve the subgrade stability and reduce subgrade settlement [65]. In recent years, using it to enhance the performance of subballast was studied [66,67]. Some studies attempted to use the geocell in ballast layer to improve the ballast layer performance, such as [68,69]. However, until now the ballast degradation after using geocell has not been confirmed yet.

Polyurethane is also named as ballast glue, and in the review paper [14], studies on polyurethane-reinforced ballast bed are explained in details. Two types of polyurethane for now have been used, foam and glue. Polyurethane protects ballast particles and reduce the ballast degradation by reducing the relative motions between the contacted particles.

Ballast mat is rubber mat that is placed usually under the ballast layer (above subgrade/bridge). Ballast mat can absorb some energies from the vehicle, and dissipate the loadings more uniformly to the subgrade. It reduces the ballast degradation by storing some energies from the vehicle loading, then slowly giving part of the energies back to ballast. More studies about ballast mat can be found in [70–73].

14.2.3.3 Sleeper innovations

Sleeper innovations are new sleepers, which have been proposed to improve the CWR (continuous welded rail) track stability, such as, winged-shape sleeper [74,75], ladder sleeper [76], nailed sleeper [77], Y-shape sleeper [78], bi-block sleeper, sleeper anchor [79], and steel sleeper [80], as shown in Fig. 14.3. The innovative sleepers focus on improving the sleeper materials and shapes, and according to the results [74,75,77,78,81], they can provide larger lateral resistance. They can possibly also be used for ballast degradation mitigation. More studies can be performed in this direction.

For example, the winged-shape sleeper (Fig. 14.3E) was designed as a mono-block sleeper with wings on the bottom, end side, and middle side [82]. The sleeper was designed as "H-shape." The ladder sleeper is designed the shape as a "ladder." These two sleepers can reduce ballast degradation by increasing the contact area between ballast and sleeper, which can more uniformly dissipate the energies from cyclic loadings.

Railway ballast 305

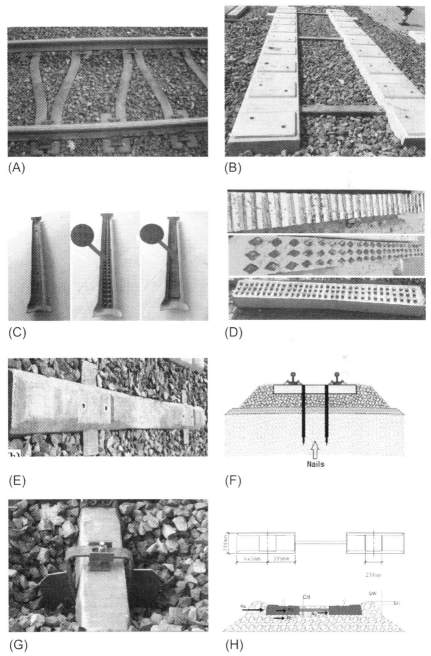

Fig. 14.3 Different types of innovated sleepers. (A) Y-shape sleeper (reproduced from G. Jing, H. Fu, P. Aela, Lateral displacement of different types of steel sleepers on ballasted track, Construct. Build. Mater. 186 (2018) 1268–1275). (B) Ladder sleeper (reproduced from G. Jing, P. Aela, H. Fu, The contribution of ballast layer components to the lateral resistance of ladder

(continued)

14.3 Ballast inspection and assessment

14.3.1 Ballast inspection

Ballast inspection provides guidance for ballast tamping. Because tamping breaks ballast particles, therefore, condition-based and predictive tamping are needed to reduce unnecessary tamping.

Ballast inspection with high-tech is the development trend toward achieving smart railway maintenance. Currently, many early-stage studies have been performed on smart monitoring, such as, structure health monitoring, smart sleeper, and smart rock [83,84]. More and more inspection methods are also developed, such as, using drones, inspection train, and satellite [85], as shown in Fig. 14.4.

Fig. 14.4A presents the SmartRock, which is made of a 3D printed plastic cover (ballast particle shape) and an accelerometer. The plastic cover has rock-similar characteristics, such as density and surface roughness. The accelerometer can measure the angular acceleration and axial acceleration (three orthogonal directions). The SmartRock can be used during the tamping to show the ballast accelerations, and comparing the acceleration of SmartRock with that of ballast particles in the DEM models. In addition, the acceleration of ballast particles can also show the ballast bed condition, such as the fouling, stiffness, which can be used as an indicator of maintenance. This is possible when the data of accelerations are well correlated with ballast bed conditions.

In addition, the drones with a camera to video the track has been studied. The track images are analyzed to assess the ballast bed condition and also the track geometry. This method has the advantage of less interruption to the train operation and safer to railway staffs. Because inspection train should be scheduled during the maintenance period (early in the morning usually), but the drones can operate and take videos at any time. In addition, some measurement of track geometry is operated by railway workers, which is a heavy work and also dangerous when trains are passing. Using the high quality (resolution) images can analyze not only the track geometry, but also

---

Fig. 14.3, cont'd sleeper track, Construct. Build. Mater. 202 (2019) 796–805). (C) Steel sleeper (reproduced from G. Jing, H. Fu, P. Aela, Lateral displacement of different types of steel sleepers on ballasted track, Construct. Build. Mater. 186 (2018) 1268–1275). (D) Frictional sleeper (reproduced from Y. Guo, H. Fu, Y. Qian, V. Markine, G. Jing, Effect of sleeper bottom texture on lateral resistance with discrete element modelling, Construct. Build. Mater. 250 (2020)). (E) Winged-shape sleeper (reproduced from T. Ichikawa, K. Hayano, T. Nakamura, Y. Momoya, Lateral resistance of ballasted tracks for various shapes of sleepers based on limit equilibrium methods, Jpn. Geotech. Soc. Spec. Publ. 2(46) (2016) 1632–1635). (F) Nailed sleeper (reproduced from M. Esmaeili, A. Khodaverdian, H.K. Neyestanaki, S. Nazari, Investigating the effect of nailed sleepers on increasing the lateral resistance of ballasted track, Comput. Geotech. 71 (2016) 1–11). (G) Sleeper anchor (reproduced from A. Zarembskis, *Survey of Techniques and Approaches for Increasing the Lateral Resistance of Wood Tie Track*, Department of Civil and Environmental Engineering, University of Delaware, Newark, DE). (H) Bi-block sleeper (reproduced from G. Jing, P. Aela, H. Fu, M. Esmaeili, Numerical and experimental analysis of lateral resistance of biblock sleeper on ballasted tracks, Int. J. Geomech. 20(6) (2020) 04020051).

Railway ballast

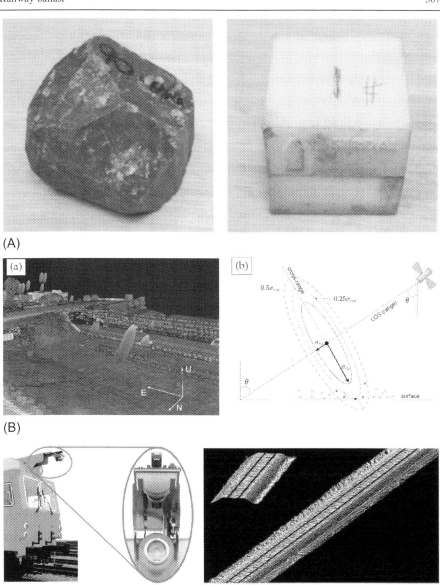

Fig. 14.4 Cutting-edge technique for smart monitoring. (A) SmartRock (reproduced from K. Zeng, T. Qiu, X. Bian, M. Xiao, H. Huang, Identification of ballast condition using SmartRock and pattern recognition, Construct. Build. Mater. 221 (2019) 50–59). (B) Satellite (reproduced from L. Chang, N.P. Sakpal, S.O. Elberink, H. Wang, Railway infrastructure classification and instability identification using sentinel-1 SAR and laser scanning data, Sensors (Basel) 20(24) (2020)). (C) Inspection train with laser scanner (reproduced from J. Sadeghi, M.E. Motieyan Najar, J.A. Zakeri, C. Kuttelwascher, Development of railway ballast geometry index using automated measurement system, Measurement 138 (2019) 132–142).

the rail, fastener and particularly ballast particles (particle size distribution, particle roughness, etc.).

Fig. 14.4B presents that using the satellite analyses track geometry, for now, the settlement is mainly focused. Using the satellite can involve a lot of factors that are very important to the maintenance, for example, the geology (water, desert, and mountain) and weather (snow, rain). In addition, it can record the track geometry change (revolution) of the whole railway lines, which is very helpful to make the maintenance plans.

Fig. 14.4C is the inspection train with a camera to video track geometry. It is similar to using the drones. The advantage is that it can measure the track geometry when the track is loaded by the inspection train. A promising idea is to combine the dynamic responses (of the train-track) with the rail, sleeper and ballast performances. For example, the ballast bed profile has some defects (e.g., hanging sleeper), which possibly causes some acceleration abrupt change. This defect can also be reflected by the big displacements of ballast particles (seen through the video). The abrupt change means this part needs maintenance to repair the ballast bed profile (correcting the geometry).

The inspection data of the track have been collected for over 100 years. Using deep learning and machine learning to analyze the data is still on the early stage, which can be developed further for the condition-based and predictive ballast bed maintenance. In addition, this method can also be used for track quality prediction.

14.3.2 Ballast condition assessment

14.3.2.1 Data/signal processing

Data process

Through the ballast inspection, large amounts of data are obtained, for which rapid and accurate means for data process and analysis are required. The accuracy and efficiency of the data process is dependent on the quality of the labeled useful data. Specifically, the data have been collected from sensors (e.g., acceleration), manually the data are labeled in most cases. Therefore, using the data-driven algorithms to label the data automatically, e.g., unsupervised learning models, is able to contribute to data process accuracy and efficiency. Moreover, the data from inspection is highly imbalanced, which can be alleviated by using the data-driven algorithms by identifying more faulty samples.

Deep learning

Deep learning algorithms are the new trend of machine learning, which abstracts neural networks with more and more layers. Using the algorithms, it is not necessary to perform the data preprocessing, due to they can learn the representation directly. The algorithms have been applied to analyze many complicated data from all kinds of measurement, such as image, audio, and video [2]. The algorithms have been used for analyzing the data (railway track) measured during the inspection, which is also for railway ballast. The deep learning algorithms that have been applied in railway track include convolutional neural networks (CNNs), recurrent neural networks (RNNs), and long-short-term memory (LSTM) models. More explanations can be found in [86].

The convolutional neural network (CNN) is the most widely-used algorithms. The CNN has been used for the detection of track defects with computer-based vision. The CNN models can reach human-level ability (often used as a proxy for the Bayes error rate) in image recognition tasks [87].

Unsupervised learning models
Unsupervised learning aims to find patterns automatically from unlabeled data. Clustering methods and dimensionality reduction techniques are the most widely used unsupervised methods in railway track engineering [86].

Autonomous maintenance projects
Operational Technology, Canadian National Railway Company (CN) is moving out of the lab and into the field with new real-time technology platforms that increase the safety, execution and efficiency of our operations.

CN's Autonomous Track Inspection Program (ATIP) is a fully automated rail car that employs wireless communications to test and monitor real-time geometric track parameters without interrupting normal railroad operations. Powered by solar panels and a generator and traveling at revenue service track speed, our Autonomous Track Inspection Program uses the latest sensor and AI technology to deploy fully automated track inspections 24/7/365.

14.3.2.2 Numerical simulations—Discrete element method

The discrete element method (DEM) has been used in plenty of ballast-related studies and proved to be an effective numerical method [51,88–90]. The DEM is a numerical model or computer simulation approach that can simulate granular materials. It describes the mechanical behavior of assemblies of spheres (disks in 2D) or polyhedrons (polygons in 2D) and considers the individual particles in granular materials and their interactions (e.g., contacts, motions) [91,92]. Nowadays, it has become a powerful and efficient tool to reproduce the performance and deformation of granular materials [93]. Particularly, the DEM is widely applied in the ballast-related studies due to the advantage that an identical sample can be performed with various test conditions (e.g., loading). Moreover, using the DEM can perform some detailed parametric studies that are often not feasible in laboratory tests, e.g., interparticle friction and distribution of contact forces. More importantly, it can record the complete particle information (e.g., displacement, acceleration) during the numerical simulations, consider the characteristics of ballast particles (e.g., size, density), and understand the effects of ballast particle degradation (i.e., breakage and abrasion) on the performance and deformation of the ballast assemblies.

Regarding the above-mentioned research problems (Section 14.1.3), the DEM has been effectively applied to study them. For instance, the performance evaluation of ballast assemblies under various conditions (e.g., particle size distribution, fouling/

contamination) can be performed with the models of direct shear tests [94–101], ballast box test [36,90,102–104], or the triaxial tests [13,17,32,105–110]. Alternatively, the performance evaluation can be analyzed with the model of field tests, e.g., the single sleeper push test model [76,89,93,111–113] and the in-situ ballast track model [88,97,114–124].

When considering the particle degradation in the DEM models, setting the breakage and abrasion criterions is the first step [32,109,115,120,125–128]. With the criteria, the corresponding plastic deformation or fouled ballast bed performance can be presented, e.g., [36,116,122,129]. Particularly, the DEM models have also been applied in the dynamic performance and degradation study of ballast bed at the transition zone [51,130].

As for the ballast degradation mitigation and performance improvement, the under sleeper pads [90], the geogrid [13,104,130–137], the geocell [65,67], and the polyurethane [8,14] are the widely used geomaterials. Plenty of studies with DEM models have been performed to demonstrate their effectiveness and propose application advices.

Regarding the track maintenance, tamping is the most common means operated on ballast layer to restore the track elastic and geometry. Using the DEM models, the studies mainly concern the tamping frequency, compaction, and performance after tamping, etc. [113,138–144].

References

[1] B. Indraratna, W. Salim, C. Rujikiatkamjorn, Advanced Rail Geotechnology: Ballasted Track, CRC Press, London, 2011.
[2] Y. Guo, V. Markine, X. Zhang, W. Qiang, G. Jing, Image analysis for morphology, rheology and degradation study of railway ballast: a review, Transp. Geotech. 18 (2019) 173–211.
[3] Y. Guo, C. Zhao, V. Markine, G. Jing, W. Zhai, Calibration for discrete element modelling of railway ballast: a review, Transp. Geotech. 23 (2020), 100341.
[4] D. Li, J. Hyslip, T. Sussmann, S. Chrismer, Railway Geotechnics, CRC Press, 2002.
[5] W. Zhai, Vehicle–Track Coupled Dynamics Theory and Applications, Springer, Singapore, 2020.
[6] A. Danesh, M. Palassi, A.A. Mirghasemi, Evaluating the influence of ballast degradation on its shear behaviour, Int. J. Rail Transp. 6 (3) (2018) 145–162.
[7] W. Zhai, K. Wang, J. Lin, Modelling and experiment of railway ballast vibrations, J. Sound Vib. 270 (4) (2004) 673–683.
[8] D. Gundavaram, S.K.K. Hussaini, Polyurethane-based stabilization of railroad ballast—a critical review, Int. J. Rail Transp. 7 (3) (2019) 219–240.
[9] G. Jing, J. Wang, H. Wang, M. Siahkouhi, Numerical investigation of the behavior of stone ballast mixed by steel slag in ballasted railway track, Constr. Build. Mater. 262 (2020) 120015.
[10] Y. Guo, Y. Ji, Q. Zhou, V. Markine, G. Jing, Discrete element modelling of rubber-protected ballast performance subjected to direct shear test and cyclic loading, Sustainability 12 (7) (2020) 2836.
[11] Y. Guo, V. Markine, W. Qiang, H. Zhang, G. Jing, Effects of crumb rubber size and percentage on degradation reduction of railway ballast, Constr. Build. Mater. 212 (2019) 210–224.

[12] Y. Guo, J. Wang, V. Markine, G. Jing, Ballast mechanical performance with and without under sleeper pads, KSCE J. Civ. Eng. 24 (11) (2020) 3202–3217.
[13] Y. Qian, D. Mishra, E. Tutumluer, H.A. Kazmee, Characterization of geogrid reinforced ballast behavior at different levels of degradation through triaxial shear strength test and discrete element modeling, Geotext. Geomembr. 43 (5) (2015) 393–402.
[14] G. Jing, L. Qie, V. Markine, W. Jia, Polyurethane reinforced ballasted track: review, innovation and challenge, Constr. Build. Mater. 208 (2019) 734–748.
[15] B. Indraratna, D. Ionescu, H. Christie, Shear behavior of railway ballast based on large-scale triaxial tests, J. Geotech. Geoenviron. 124 (5) (1998) 439–449.
[16] B. Indraratna, T. Ngo, Ballast Railroad Design: SMART-UOW Approach, CRC Press, 2018.
[17] C. Chen, B. Indraratna, G. McDowell, C. Rujikiatkamjorn, Discrete element modelling of lateral displacement of a granular assembly under cyclic loading, Comput. Geotech. 69 (2015) 474–484.
[18] H. Wang, V. Markine, X. Liu, Experimental analysis of railway track settlement in transition zones, Proc. Inst. Mech. Eng. F J. Rail Rapid Transit 232 (6) (2018) 1774–1789.
[19] X. Liu, V.L. Markine, H. Wang, I.Y. Shevtsov, Experimental tools for railway crossing condition monitoring (crossing condition monitoring tools), Measurement 129 (2018) 424–435.
[20] B. Aursudkij, A Laboratory Study of Railway Ballast Behaviour Under Traffic Loading and Tamping Maintenance, University of Nottingham, 2007.
[21] S.C. Douglas, Ballast quality and breakdown during tamping, in: Proceedings of the 2013 American Railway Engineering and Maintenance-of-Way Association Conference, Inidanpolis, IN, 2013.
[22] C. Paderno, Improving ballast tamping process, in: Proc., World Congress on Railway Research, 2011, pp. 1–6.
[23] Y. Descantes, R. Perales, G. Saussine, N. Milesi, On the damaging effects of the ballast tamping operation, AIPCR, 2011. p. sp.
[24] O. Barbir, D. Adam, F. Kopf, J. Pistrol, F. Auer, B. Antony, Development of condition-based tamping process in railway engineering, ce/papers 2 (2–3) (2018) 969–974.
[25] E.T. Selig, J.M. Waters, Track Geotechnology and Substructure Management, Thomas Telford, 1994.
[26] J. Liu, P. Wang, G. Liu, J. Dai, J. Xiao, H. Liu, Study of the characteristics of ballast bed resistance for different temperature and humidity conditions, Constr. Build. Mater. 266 (2021) 121115.
[27] J. Liu, P. Wang, G. Liu, M. Zhang, J. Xiao, H. Liu, Uniaxial compression characteristics of railway ballast combined with ice, Constr. Build. Mater. 263 (2020) 120554.
[28] M. Esmaeili, P. Aela, A. Hosseini, Experimental assessment of cyclic behavior of sand-fouled ballast mixed with tire derived aggregates, Soil Dyn. Earthq. Eng. 98 (2017) 1–11.
[29] R. Nålsund, Railway Ballast Characteristics, Selection Criterion and Performance, Department of Civil and Transport Engineering, Norwegian University of Science and Technology, Trondheim, 2014.
[30] R.J. Marsal, Mechanical Properties of Rockfill, Wiley (John) and Sons, Incorporated, 1973.
[31] M. Moaveni, E. Mahmoud, E. Ortiz, E. Tutumluer, S. Beshears, Use of advanced aggregate imaging systems to evaluate aggregate resistance to breakage, abrasion, and polishing, Transp. Res. Rec. 2401 (2014) 1–10.
[32] G.R. McDowell, H. Li, Discrete element modelling of scaled railway ballast under triaxial conditions, Granul. Matter 18 (3) (2016) 1–10.

[33] B. Indraratna, T. Ngo, C. Rujikiatkamjorn, Performance of ballast influenced by deformation and degradation: laboratory testing and numerical modeling, Int. J. Geomech. 20 (1) (2019) 04019138.

[34] W. Jia, V. Markine, Y. Guo, G. Jing, Experimental and numerical investigations on the shear behaviour of recycled railway ballast, Constr. Build. Mater. 217 (2019) 310–320.

[35] D. Ionescu, Evaluation of the Engineering Behaviour of Railway Ballast, University of Wollongong, Faculty of Engineering, 2004. https://ro.uow.edu.au/theses/421. (Doctor of Philosophy thesis).

[36] M. Lu, G. McDowell, Discrete element modelling of ballast abrasion, Géotechnique 56 (9) (2006) 651–655.

[37] B. Indraratna, D. Ionescu, D. Christie, R. Chowdhury, Compression and degradation of railway ballast under one-dimensional loading, Aust. Geomech. J. (1997) 48–61 (December).

[38] Y.L. Guo, C.F. Zhao, V. Markine, C. Shi, G.Q. Jing, W.M. Zhai, Discrete element modelling of railway ballast performance considering particle shape and rolling resistance, Rail. Eng. Sci. 28 (4) (2020) 382–407.

[39] X. Zhang, C. Zhao, W. Zhai, Importance of load frequency in applying cyclic loads to investigate ballast deformation under high-speed train loads, Soil Dyn. Earthq. Eng. 120 (2019) 28–38.

[40] Q.D. Sun, B. Indraratna, S. Nimbalkar, Deformation and degradation mechanisms of railway ballast under high frequency cyclic loading, J. Geotech. Geoenviron. 142 (1) (2016) 04015056.

[41] T.P.M.o. Railways, Railway Ballast, TB/T2140-2008, China Railway Publishing House, Beijing, 2008.

[42] B.s.p.B.E. British Standards Institution, Aggregates for Railway Ballast, British Standards Institution, London, 2013.

[43] A. Yitayew Alemu, Survey of Railway Ballast Selection and Aspects of Modelling Techniques, Trita-VBT, 2011. Available from: http://urn.kb.se/resolve?urn=urn:nbn:se:kth:diva-87466. (Internet; Dissertation).

[44] B. Indraratna, J. Lackenby, D. Christie, Effect of confining pressure on the degradation of ballast under cyclic loading, Geotechnique 55 (4) (2005) 325–328.

[45] J. Lackenby, B. Indraratna, G. McDowell, D. Christie, Effect of confining pressure on ballast degradation and deformation under cyclic triaxial loading, Géotechnique 57 (6) (2007) 527–536.

[46] Z. Yu, D.P. Connolly, P.K. Woodward, O. Laghrouche, E. Tutumluer, Railway ballast anisotropy testing via true triaxial apparatus, Transp. Geotech. 23 (2020) 100355.

[47] L. Gates, E. Masad, R. Pyle, D. Bushee, Aggregate imaging measurement system 2 (AIMS-II): final report, Highway for Life Technology Partnership Program, Federal Highway Administration, US Department of Transportation, 2011.

[48] Y. Guo, V. Markine, J. Song, G. Jing, Ballast degradation: effect of particle size and shape using Los Angeles Abrasion test and image analysis, Constr. Build. Mater. 169 (2018) 414–424.

[49] C. Dunn, P. Bora, Shear strength of untreated road base aggregates measured by variable lateral pressure triaxial cell, J. Mater. 7 (2) (1972).

[50] Y. Gur, E. Shklarsky, M. Livneh, Effect of coarse-fraction flakiness on the strength of graded materials, in: Asian Conf Soil Mech & Fdn E Proc/Is/, 1967.

[51] Y. Qian, E. Tutumluer, Y.M.A. Hashash, J. Ghaboussi, D.D. Davis, Ballast settlement ramp to mitigate differential settlement in a bridge transition zone, Transp. Res. Rec. 2476 (2476) (2015) 45–52.

[52] A. Paixão, E. Fortunato, Abrasion evolution of steel furnace slag aggregate for railway ballast: 3D morphology analysis of scanned particles by close-range photogrammetry, Constr. Build. Mater. 267 (2021) 121225.

[53] S. Mehran Khoshoei, H. Mortazavi Bak, S. Mahdi Abtahi, S. Mahdi Hejazi, B. Shahbodagh, Experimental investigation of the cyclic behavior of steel-slag ballast mixed with tire-derived aggregate, J. Mater. Civ. Eng. 33 (2) (2021) 04020468.

[54] W. Jia, V.L. Markine, G. Jing, Analysis of furnace slag in railway sub-ballast based on experimental tests and DEM simulations, Constr. Build. Mater. 288 (2021) 123114.

[55] B. Guimarães Delgado, A. Viana da Fonseca, E. Fortunato, P. Maia, Mechanical behavior of inert steel slag ballast for heavy haul rail track: laboratory evaluation, Transp. Geotech. 20 (2019), 100243.

[56] Y. Qi, B. Indraratna, A. Heitor, J.S. Vinod, Effect of rubber crumbs on the cyclic behavior of steel furnace slag and coal wash mixtures, J. Geotech. Geoenviron. 144 (2) (2018) 04017107.

[57] V. Fontserè, A.L. Pita, N. Manzo, A. Ausilio, NEOBALLAST: new high-performance and long-lasting ballast for sustainable railway infrastructures, Transp. Res. Proc. 14 (2016) 1847–1854.

[58] M. Sol-Sánchez, F. Moreno-Navarro, M. Rubio-Gámez, N. Manzo, V. Fontseré, Full-scale study of Neoballast section for its application in railway tracks: optimization of track design, Mater. Struct. 51 (2) (2018) 43.

[59] T. Ngo, B. Indraratna, Mitigating ballast degradation with under-sleeper rubber pads: experimental and numerical perspectives, Comput. Geotech. 122 (2020), 103540.

[60] A. Paixão, C. Alves Ribeiro, N. Pinto, E. Fortunato, R. Calçada, On the use of under sleeper pads in transition zones at railway underpasses: experimental field testing, Struct. Infrastruct. Eng. 11 (2) (2014) 112–128.

[61] A. Paixão, J.N. Varandas, E. Fortunato, R. Calçada, Numerical simulations to improve the use of under sleeper pads at transition zones to railway bridges, Eng. Struct. 164 (2018) 169–182.

[62] L. Le Pen, G. Watson, A. Hudson, W. Powrie, Behaviour of under sleeper pads at switches and crossings – field measurements, Proc. Inst. Mech. Eng. F J. Rail Rapid Transit 232 (4) (2018) 1049–1063.

[63] H. Loy, Under sleeper pads in turnouts, Railw. Tech. Rev. 2 (2009) 35.

[64] S. Kaewunruen, A. Aikawa, A.M. Remennikov, Vibration attenuation at rail joints through under sleeper pads, Procedia Eng. 189 (2017) 193–198.

[65] Y. Liu, A. Deng, M. Jaksa, Three-dimensional modeling of geocell-reinforced straight and curved ballast embankments, Comput. Geotech. 102 (2018) 53–65.

[66] B. Indraratna, M.M. Biabani, S. Nimbalkar, Behavior of geocell-reinforced subballast subjected to cyclic loading in plane-strain condition, J. Geotech. Geoenviron. 141 (1) (2014) 04014081.

[67] N.T. Ngo, B. Indraratna, C. Rujikiatkamjorn, M. Mahdi Biabani, Experimental and discrete element modeling of geocell-stabilized subballast subjected to cyclic loading, J. Geotech. Geoenviron. 142 (4) (2016) 04015100.

[68] Y. Liu, A. Deng, M. Jaksa, Three-dimensional discrete-element modeling of geocell-reinforced ballast considering breakage, Int. J. Geomech. 20 (4) (2020) 04020032.

[69] S.K. Dash, A.S. Shivadas, Performance improvement of railway ballast using geocells, Indian Geotech. J. 42 (3) (2012) 186–193.

[70] G. Wettschureck, M. Heim, M. Tecklenburg, Long-term properties of Sylomer® ballast mats installed in the rapid transit railway tunnel near the Philharmonic Hall of Munich, Germany, Rail Eng. Int. 31 (4) (2002) 6–11.

[71] S. Mohammadzadeh, A. Miri, M. Nouri, Enhancing the structural performance of masonry arch bridges with ballast mats, J. Perform. Constr. Facil. 31 (5) (2017) 04017089.

[72] S.K. Navaratnarajah, B. Indraratna, Use of rubber mats to improve the deformation and degradation behavior of rail ballast under cyclic loading, J. Geotech. Geoenviron. 143 (6) (2017) 04017015.

[73] A.D.O. Lima, M.S. Dersch, Y. Qian, E. Tutumluer, J.R. Edwards, Laboratory fatigue performance of under-ballast mats under varying loads and support conditions, Proc. Inst. Mech. Eng. F J. Rail Rapid Transit 233 (6) (2019) 606–613. 0954409718795920.

[74] Y. Koike, T. Nakamura, K. Hayano, Y. Momoya, Numerical method for evaluating the lateral resistance of sleepers in ballasted tracks, Soils Found. 54 (3) (2014) 502–514.

[75] T. Ichikawa, K. Hayano, T. Nakamura, Y. Momoya, Lateral resistance of ballasted tracks for various shapes of sleepers based on limit equilibrium methods, Jpn. Geotech. Soc. Spec. Publ. 2 (46) (2016) 1632–1635.

[76] G. Jing, P. Aela, H. Fu, The contribution of ballast layer components to the lateral resistance of ladder sleeper track, Constr. Build. Mater. 202 (2019) 796–805.

[77] M. Esmaeili, A. Khodaverdian, H.K. Neyestanaki, S. Nazari, Investigating the effect of nailed sleepers on increasing the lateral resistance of ballasted track, Comput. Geotech. 71 (2016) 1–11.

[78] V. Ungureanu, A. Mariş, Y shape railway sleepers from fiber reinforced foamed urethane, Bull. Transilv. Univ. Bras. CIBv 6 (55, Special Issue No. 1) (2013).

[79] G. Jing, Y. Ji, P. Aela, Experimental and numerical analysis of anchor-reinforced sleepers lateral resistance on ballasted track, Constr. Build. Mater. 264 (2020) 120197.

[80] J.A. Zakeri, R. Talebi, F. Rahmani, Field investigation on the lateral resistance of ballasted tracks with strengthened steel sleepers using the multi sleeper push test, Proc. Inst. Mech. Eng. F J. Rail Rapid Transit 234 (9) (2020) 969–975. 095440971987777.

[81] J. Ali Zakeri, M. Esmaeili, A. Kasraei, A. Bakhtiary, A numerical investigation on the lateral resistance of frictional sleepers in ballasted railway tracks, Proc. Inst. Mech. Eng. F J. Rail Rapid Transit 230 (2) (2014) 440–449.

[82] L.M. Domingo, J.I.R. Herraiz, C. Zamorano, T.R. Herraiz, Design of a new high lateral resistance sleeper and performance comparison with conventional sleepers in a curved railway track by means of finite element models, Lat. Am. J. Solids Struct. 11 (7) (2014) 1238–1250.

[83] K. Zeng, T. Qiu, X. Bian, M. Xiao, H. Huang, Identification of ballast condition using SmartRock and pattern recognition, Constr. Build. Mater. 221 (2019) 50–59.

[84] G. Jing, M. Siahkouhi, J. Riley Edwards, M.S. Dersch, N.A. Hoult, Smart railway sleepers – a review of recent developments, challenges, and future prospects, Constr. Build. Mater. 271 (2021) 121533.

[85] L. Chang, R.P. Dollevoet, R.F. Hanssen, Nationwide railway monitoring using satellite SAR interferometry, IEEE J. Sel. Top. Appl. Earth Obs. and Remote Sens. 10 (2) (2017) 596–604.

[86] J. Xie, J. Huang, C. Zeng, S.-H. Jiang, N. Podlich, Systematic literature review on data-driven models for predictive maintenance of railway track: implications in geotechnical engineering, Geosciences 10 (11) (2020) 425.

[87] C. Szegedy, A. Toshev, D. Erhan, Deep neural networks for object detection, Adv. Neural Inf. Process. Syst. 26 (2013).

[88] X. Zhang, C. Zhao, W. Zhai, Dynamic behavior analysis of high-speed railway ballast under moving vehicle loads using discrete element method, Int. J. Geomech. 17 (7) (2016) 04016157.

[89] G. Jing, H. Fu, P. Aela, Lateral displacement of different types of steel sleepers on ballasted track, Constr. Build. Mater. 186 (2018) 1268–1275.
[90] H. Li, G.R. McDowell, Discrete element modelling of under sleeper pads using a box test, Granul. Matter 20 (2) (2018) 1–12.
[91] P.A. Cundall, O.D. Strack, A discrete numerical model for granular assemblies, Geotechnique 29 (1) (1979) 47–65.
[92] C. O'Sullivan, Particulate Discrete Element Modelling: A Geomechanics Perspective, CRC Press, 2014.
[93] J.I. González, Numerical Analysis of Railway Ballast Behaviour Using the Discrete Element Method, Universitat Politècnica de Catalunya, 2017.
[94] B. Suhr, S. Marschnig, K. Six, Comparison of two different types of railway ballast in compression and direct shear tests: experimental results and DEM model validation, Granul. Matter 20 (4) (2018) 70.
[95] D. Mishra, S.N. Mahmud, Effect of particle size and shape characteristics on ballast shear strength: a numerical study using the direct shear test, in: 2017 Joint Rail Conference, American Society of Mechanical Engineers, 2017. V001T01A014.
[96] Z. Wang, G. Jing, Q. Yu, H. Yin, Analysis of ballast direct shear tests by discrete element method under different normal stress, Measurement 63 (2015) 17–24.
[97] H. Huang, E. Tutumluer, Discrete Element Modeling for fouled railroad ballast, Constr. Build. Mater. 25 (8) (2011) 3306–3312.
[98] B. Suhr, K. Six, Parametrisation of a DEM model for railway ballast under different load cases, Granul. Matter 19 (4) (2017) 64.
[99] B. Indraratna, S.S. Nimbalkar, N.T. Ngo, T. Neville, Performance improvement of rail track substructure using artificial inclusions – experimental and numerical studies, Transp. Geotech. 8 (2016) 69–85.
[100] B. Indraratna, N.T. Ngo, C. Rujikiatkamjorn, J. Vinod, Behavior of fresh and fouled railway ballast subjected to direct shear testing: discrete element simulation, Int. J. Geomech. 14 (1) (2012) 34–44.
[101] X. Bian, W. Li, Y. Qian, E. Tutumluer, Micromechanical particle interactions in railway ballast through DEM simulations of direct shear tests, Int. J. Geomech. 19 (5) (2019) 04019031.
[102] S. Laryea, M. Safari Baghsorkhi, J.F. Ferellec, G.R. McDowell, C. Chen, Comparison of performance of concrete and steel sleepers using experimental and discrete element methods, Transp. Geotech. 1 (4) (2014) 225–240.
[103] M. Lu, G.R. McDowell, The importance of modelling ballast particle shape in the discrete element method, Granul. Matter 9 (1–2) (2006) 69–80.
[104] C. Chen, G.R. McDowell, N.H. Thom, Discrete element modelling of cyclic loads of geogrid-reinforced ballast under confined and unconfined conditions, Geotext. Geomembr. 35 (2012) 76–86.
[105] J. Harkness, A. Zervos, L. Le Pen, S. Aingaran, W. Powrie, Discrete element simulation of railway ballast: modelling cell pressure effects in triaxial tests, Granul. Matter 18 (3) (2016) 1–13.
[106] M. Lu, G.R. McDowell, Discrete element modelling of railway ballast under monotonic and cyclic triaxial loading, Géotechnique 60 (6) (2010) 459–467.
[107] J. Xiao, D. Zhang, K. Wei, Z. Luo, Shakedown behaviors of railway ballast under cyclic loading, Constr. Build. Mater. 155 (2017) 1206–1214.
[108] S. Ahmed, J. Harkness, L. Le Pen, W. Powrie, A. Zervos, Numerical modelling of railway ballast at the particle scale, Int. J. Numer. Anal. Methods Geomech. 40 (5) (2016) 713–737.

[109] B. Indraratna, P.K. Thakur, J.S. Vinod, Experimental and numerical study of railway ballast behavior under cyclic loading, Int. J. Geomech. 10 (4) (2009) 136–144.
[110] S. Liu, T. Qiu, Y. Qian, H. Huang, E. Tutumluer, S. Shen, Simulations of large-scale triaxial shear tests on ballast aggregates using sensing mechanism and real-time (SMART) computing, Comput. Geotech. 110 (2019) 184–198.
[111] F. Khatibi, M. Esmaeili, S. Mohammadzadeh, DEM analysis of railway track lateral resistance, Soils Found. 57 (4) (2017) 587–602.
[112] Z. Zeng, S. Song, W. Wang, H. Yan, G. Wang, B. Xiao, Ballast bed resistance characteristics based on discrete-element modeling, Adv. Mech. Eng. 10 (6) (2018). 1687814018781461.
[113] E. Tutumluer, H. Huang, Y. Hashash, J. Ghaboussi, Aggregate shape effects on ballast tamping and railroad track lateral stability, in: AREMA Annual Conference, Loisville, KY, September 2006, 2006, pp. 17–20.
[114] D. Nishiura, H. Sakai, A. Aikawa, S. Tsuzuki, H. Sakaguchi, Novel discrete element modeling coupled with finite element method for investigating ballasted railway track dynamics, Comput. Geotech. 96 (2018) 40–54.
[115] B. Dahal, S.N. Mahmud, D. Mishra, Simulating ballast breakage under repeated loading using the discrete element method, in: 2018 Joint Rail Conference, American Society of Mechanical Engineers, 2018. V001T01A003.
[116] E. Tutumluer, Y. Qian, Y.M.A. Hashash, J. Ghaboussi, D.D. Davis, Discrete element modelling of ballasted track deformation behaviour, Int. J. Rail Transp. 1 (1–2) (2013) 57–73.
[117] S. Ji, S. Sun, Y. Yan, Discrete element modeling of dynamic behaviors of railway ballast under cyclic loading with dilated polyhedra, Int. J. Numer. Anal. Methods Geomech. 41 (2) (2017) 180–197.
[118] N.T. Ngo, B. Indraratna, C. Rujikiatkamjorn, Simulation ballasted track behavior: numerical treatment and field application, Int. J. Geomech. 17 (6) (2016) 04016130.
[119] E. Mahmoud, A.T. Papagiannakis, D. Renteria, Discrete element analysis of railway ballast under cycling loading, Procedia Eng. 143 (2016) 1068–1076.
[120] S. Lobo-Guerrero, L.E. Vallejo, Discrete element method analysis of railtrack ballast degradation during cyclic loading, Granul. Matter 8 (3–4) (2006) 195–204.
[121] S. Liu, H. Huang, T. Qiu, L. Gao, Comparison of laboratory testing using SmartRock and discrete element modeling of ballast particle movement, J. Mater. Civ. Eng. 29 (3) (2017) D6016001.
[122] I. Deiros, C. Voivret, G. Combe, F. Emeriault, Quantifying degradation of railway ballast using numerical simulations of micro-deval test and in-situ conditions, Procedia Eng. 143 (2016) 1016–1023.
[123] H. Huang, S. Chrismer, Discrete element modeling of ballast settlement under trains moving at "Critical Speeds", Constr. Build. Mater. 38 (2013) 994–1000.
[124] W.T. Hou, B. Feng, W. Li, E. Tutumluer, Ballast support condition affecting crosstie performance investigated through discrete element method, in: Pr ASME Joint Rail C, 2018.
[125] B. Wang, U. Martin, S. Rapp, Discrete element modeling of the single-particle crushing test for ballast stones, Comput. Geotech. 88 (2017) 61–73.
[126] P. Thakur, J.S. Vinod, B. Indraratna, Effect of particle breakage on cyclic densification of ballast: a DEM approach, IOP Conf. Ser.: Mater. Sci. Eng. 10 (1) (2010) 012229. IOP Publishing.
[127] J. Qian, J. Gu, X. Gu, M. Huang, L. Mu, DEM analysis of railtrack ballast degradation under monotonic and cyclic loading, Procedia Eng. 143 (2016) 1285–1292.
[128] Z. Hossain, B. Indraratna, F. Darve, P. Thakur, DEM analysis of angular ballast breakage under cyclic loading, Geomech. Geoengin. 2 (3) (2007) 175–181.

[129] E. Tutumluer, H. Huang, Y. Hashash, J. Ghaboussi, Discrete element modeling of railroad ballast settlement, in: AREMA Conference, 2007.
[130] C. Chen, G.R. McDowell, An investigation of the dynamic behaviour of track transition zones using discrete element modelling, Proc. Inst. Mech. Eng. F J. Rail Rapid Transit 230 (1) (2014) 117–128.
[131] C. Chen, G.R. McDowell, N.H. Thom, A study of geogrid-reinforced ballast using laboratory pull-out tests and discrete element modelling, Geomech. Geoengin. 8 (4) (2013) 244–253.
[132] B. Indraratna, S. Nimbalkar, C. Rujikiatkamjorn, From theory to practice in track geomechanics – Australian perspective for synthetic inclusions, Transp. Geotech. 1 (4) (2014) 171–187.
[133] S.N. Mahmud, D. Mishra, D.O. Potyondy, Effect of geogrid inclusion on ballast resilient modulus: the concept of 'geogrid gain factor', in: 2018 Joint Rail Conference, American Society of Mechanical Engineers, 2018. V001T01A005.
[134] Y. Qian, E. Tutumluer, H. Huang, A validated discrete element modeling approach for studying geogrid-aggregate reinforcement mechanisms, in: Geo-Frontiers 2011: Advances in Geotechnical Engineering, 2011, pp. 4653–4662.
[135] N.T. Ngo, B. Indraratna, C. Rujikiatkamjorn, Modelling geogrid-reinforced railway ballast using the discrete element method, Transp. Geotech. 8 (2016) 86–102.
[136] N.T. Ngo, B. Indraratna, C. Rujikiatkamjorn, DEM simulation of the behaviour of geogrid stabilised ballast fouled with coal, Comput. Geotech. 55 (2014) 224–231.
[137] C.-x. Miao, J.-j. Zheng, R.-j. Zhang, L. Cui, DEM modeling of pullout behavior of geogrid reinforced ballast: the effect of particle shape, Comput. Geotech. 81 (2017) 249–261.
[138] J.-F. Ferellec, R. Perales, V.-H. Nhu, M. Wone, G. Saussine, Analysis of compaction of railway ballast by different maintenance methods using DEM, in: EPJ Web of Conferences, EDP Sciences, 2017, p. 15032.
[139] G. Saussine, E. Azéma, P. Gautier, R. Peyroux, F. Radjai, Numerical modeling of the tamping operation by Discrete Element Approach, in: World Congress Rail Research, 2008, pp. 1–9.
[140] T. Zhou, B. Hu, J. Sun, Z. Liu, Discrete element method simulation of railway ballast compactness during tamping process, Open Electr. Electr. Eng. J. 7 (2013) 103–109.
[141] R. Perales, G. Saussine, N. Milesi, F. Radjai, Numerical investigation of the tamping process, in: 9th World Congress on Railway Research, 2011.
[142] X.J. Wang, Y.L. Chi, W. Li, T.Y. Zhou, X.L. Geng, The research of the numerical simulation on the granular ballast bed tamping, Adv. Mater. Res. (2012) 1395–1398. Trans Tech Publ.
[143] T.M.P. Hoang, P. Alart, D. Dureisseix, G. Saussine, A domain decomposition method for granular dynamics using discrete elements and application to railway ballast, Ann. Solid Struct. Mech. 2 (2–4) (2011) 87–98.
[144] D.S. Kim, S.H. Hwang, A. Kono, T. Matsushima, Evaluation of ballast compactness during the tamping process by using an image-based 3D discrete element method, Proc. Inst. Mech. Eng. F J. Rail Rapid Transit 232 (7) (2018). 0954409718754927.

Railway turnouts and inspection technologies

Mehmet Z. Hamarat[a], Mika Silvast[b], and Sakdirat Kaewunruen[a]
[a]Department of Civil Engineering, School of Engineering, University of Birmingham, Birmingham, United Kingdom, [b]Loram Finland Oy, Tampere, Finland

15.1 Introduction

Turnouts, or so-called switches and crossings, are an essential infrastructure to provide flexibility in railway management. Their function is to divert traffic from one route to another route by altering the position of specially designed rails. Despite their relatively simple technology, turnouts are still one of the great challenges in railways. They are critically dependent on maintenance and inspection to operate, which imposes constraints on railway management such as lower operational speeds.

Including different turnout configurations, a turnout consists of three sections, namely panels, such as switch, closure, and crossing panels (Fig. 15.1). Each panel is named after its roles in terms of kinematic behavior. The switch panel is the section where one of the wheels is diverted through specially designed rails, namely switchblades that have a varying profile in the longitudinal direction. Diversion is provided with actuators automatically or manually by positioning switchblades until the thin tip of the switchblade, which allows the wheel to follow the new path under flange guidance, sticks to the stock rail. Simultaneously, the other switchblade is separated from the other stock rail to open a gap that enables the other wheel to pass. The term "stock rail" represents the rails on which no wheel diversion happens depending on the traveling route. Following the diversion or switching, the wheels reach the closure panel, the purpose of which is to connect switch and crossing panels. All rails of the closure panel have constant rail profiles and have no distinctive characteristics associated with their profile. It is noteworthy that the rails between two stock rails are called closure rails. Due to the closure rails layout, their extensions intersect at a point on the track, the so-called theoretical crossing point. That section where the theoretical crossing point exits is named as crossing panel. The transition through a theoretical crossing point is provided by two methods such as fixed crossings and movable crossings. Fixed crossings have a geometrical discontinuity, a gap, that allows the wheel flange to pass through the crossing point. During transition over a fixed crossing, high-frequency high impact forces are usually observed, which produce undesirable component damages. Therefore, fixed crossings are preferred under 200 km/h speed, commonly, in the case of normal and medium axle loads. Movable crossings have specially designed rails, namely points that bridge between wing rails (extension of closure rail) and stock rails of a new route by switching its position. Movable crossings

Rail Infrastructure Resilience. https://doi.org/10.1016/B978-0-12-821042-0.00016-2
Copyright © 2022 Elsevier Inc. All rights reserved.

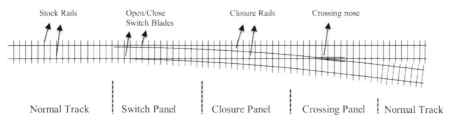

Fig. 15.1 A typical turnout layout.

work similar to switch rails. The advantage of movable crossing is to mitigate impact forces over the crossing nose by providing an uninterrupted running surface for the wheel.

Turnouts could be assumed as a pure mechanic structure due to its technology, where manual mechanical methods could be used to switch the rails. Therefore, they must be frequently inspected and well maintained to preserve structural integrity. On modern railway tracks, many turnouts are controlled by a signaling system that drives and locks the switch rails and moveable crossing noses. Furthermore, the signaling system could be employed to detect specific failures. However, signaling systems have a small share in total works that are conducted on turnouts. Consequently, this chapter is dedicated to only track works.

15.2 Components of turnouts

15.2.1 Rails

Turnouts are made of three different rail profiles such as switch, crossing, and standard flat-bottom rail profiles. Stock rails and closure rails are ordinary rails having a standard rail profile as in a normal track section and their role is also to support and guide wheels. The manufacturing process of a rail with a standard profile follows these steps: steel-making, forming rail profile by hot rolling of bulk material, straightening rails in vertical and lateral directions, finishing operation such as cutting and drilling, acceptance, and delivery.

Switch rail profiles start with a thin profile that sticks to stock rail and end with a normal rail profile (Fig. 15.2). As a consequence, switch rails are composed of three sections that could be named as switchblade, heel, and closure parts. The profile of switchblades varies in a longitudinal direction and is obtained by machining ordinary rails. The end of the switchblade has a full cross-section and is called the heel. Neglecting the occasional use of normal rails, in practice, switchblades are manufactured from different rail profiles rather than normal rail profiles. Hence, the heel should be suitable to be welded to the normal rail profile. To provide the transition from switchblade profile to normal rail profile, the heel is heated and forged. Following the forging operation, the switchblades are subjected to heat treatment, straightening, and fine machining with finish tolerances. Afterward, switchblades are welded by flash-butt welding, to the closure part having a standard rail profile. In the final step of

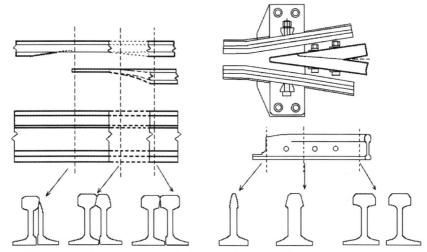

Fig. 15.2 Typical views of: a switch (left); a crossing nose (right).

manufacturing, grinding and shot blasting are applied to have the desired surface quality.

The crossing nose has a wedge-type profile due to the intersection of running surfaces (Fig. 15.2). In common crossings, the front of the wedge profile is lower in the vertical direction to provide smooth wheel transition whereas movable crossing has constant height along the running surface. The manufacturing of common crossings has two main processes such as fabrication and casting. Fabricated crossings are produced by welding two specially profiled normal rails. They are more suitable for low-speed operations due to poor performance under impact loads in terms of plastic deformation. Another method is the mono-block casting of the crossing nose, where an effective surface hardening such as explosion depth hardening could be used. On the other hand, fabrication is the only tool for movable crossings which have a similar structure with switches where rails with a special profile are assembled to be able to bend.

Pearlitic steel is a commonly used material in the production of rail steel. The manufacturing process is simple and there are no major alloying elements. The mechanical properties of the pearlite structure are relatively good and can be adjusted by controlling the formation of the lamellar structure to provide the desired strength and hardness [1]. Hence, pearlitic steel is also available for high-speed turnout applications. Recently, carbide-free bainitic steel has been increasing its popularity due to higher steel grades and fatigue behavior [2]. These steels are commonly used for stock, closure, and switch rails. However, they are not suitable for a common crossing part as well as pearlitic steel. It should be reminded that material technology on using free carbide bainitic steel is still developing and assumed promising [3]. Nevertheless, there is no consensus on the advantage of using bainitic steel, particularly for a crossing nose due to insufficient experience [1,3]. Austenitic manganese steel is widely used during manufacturing cast crossing. Despite raw material properties that have low strength and hardness, the ability of high work hardening provide significant hardness over time, and therefore, it gains a good resistance to wear and plastic deformation after each use [4]. Furthermore, hardening ability could be used with explosive depth hardening methods, which produce significant hardness on the surface. The main drawback of manganese crossing is to suffer from spalling. That is believed to be a result of thermal shrinkage during casting that produces inevitable voids under the subsurface that forms cracks [3,5]. Furthermore, those cracks cannot be detected by ultrasonic testing due to the large grain size in the microstructure. Lastly, there are researches to replace austenitic manganese and other steels with new materials such as maraging, chrome-vanadium, and manganese-molybdenum steels. Nonetheless, experience and data on using those materials are limited.

15.2.2 Bearers

Bearers (Fig. 15.3) are the beams laid under rails to support a turnout system and transfer wheel loads toward to ballast layer. They are anchor points of rails and provide isolation between two rails, which is important in terms of signaling. Four types of materials such as timber, concrete, steel, and composite could be found in the field as turnout bearers. Timber is a historically common bearer material for a turnout system. It is a simple technology and suitable for any tracking system due to its excellent ability to mitigate vibrations imposed by dynamic train loads. Moreover,

Fig. 15.3 Different bearer types: timber (top left); concrete (top right); steel (bottom left); composite (FFU) (bottom right).
Courtesy: RailCorp NSW.

installing, repairing, and replacing a timber bearer is relatively effortless. However, they are prone to biological attacks from the environment, and therefore, environmentally hazardous chemicals are used to prevent and slow down biological degradation. Besides, their mechanical degradation is also faster than other material types. Particularly, long timber bearers above 4 m suffer from large deformations and localized weakness [6]. The application of timber turnout is gradually disappearing due to concerns on availability and environmental impact. The popular material replacing timber is mono-block prestressed concrete. It is noteworthy that prestressed concrete is not an alternative to replace a timber bearer in the case of spot replacement on an active track due to inconsistency in geometrical dimensions and properties. The main advantages of concrete bearers are cheap to manufacture, heavy to provide good track stability, and relatively durable to lower replacement and maintenance costs [7]. Particularly, track stability is crucially important for high-speed turnouts. Hence, it is the most common bearer type for high-speed rails. The disadvantage of prestressed concrete bearers is a considerably heavy structure that requires special equipment while replacing and maintaining concrete bearers. Furthermore, the performance of prestressed concrete bearers under impact loads is poor, resulting from the stiffness of concrete bearer and resulting in faster ballast degradation.

Steel bearers are lighter and easy to transport. Their resistance to impact loads seems to be promising in the vertical direction [8]. Likewise, they provide significant lateral track resistance, particularly Y-shape steel bearers [7]. However, their performance in terms of track stability is poor, and therefore, they are suitable for tracks applying the speed limit below 160 km/h [7]. Furthermore, they suffer from corrosion in the field and maintenance of steel bearers particularly during tamping is challenging. Another challenge in the application of steel bearers is to provide good rail insulation not to disturb the track circuit, which increases maintenance costs. In conclusion, the cost of steel bearers is distinctly higher than concrete bearers.

Nowadays, composite materials have been tested to be an alternative material for bearers, particularly, timbers. Expected mechanical properties from a composite bearer are sufficient strength, good flexibility, low maintenance cost, and have high durability [9]. Furthermore, they are lightweight, resistant to heat, chemical, and biological agents. One of the composite materials used in the field is fiber-reinforced foamed urethane (FFU). FFU has a good ability to distribute bending moments and it shows good performance at high-frequency high impact loading areas such as crossing nose [8]. Despite their superior properties, the cost of composite bearers is extremely high, and therefore, FFU bearers are only suitable to be applied at particular locations such as switches and crossings. Furthermore, composite bearers require extra care in terms of safety since they are quickly flammable and dangerous for human health while cutting, drilling due to the possibility of inhalation of fiberglass particles.

15.2.3 Track bed

Track bed supports rails and sleepers/bears dynamic track loads, particularly at turnouts where high dynamic force occurs. Both ballasted and ballastless track beds are available for turnouts. In general, there is no significant difference between normal track bed and turnout track bed in terms of ballasted track and ballastless track properties. Hence, they mimic similar behavior. Ballasted tracks are inexpensive structures with high damping, and fewer noise properties, whereas ballastless tracks require lower maintenance and provide higher stability and durability. Ballasted turnouts could be laid with any kind of bearer materials on the contrary to ballastless turnouts using embedded concrete bearers or slabs.

The exception of track bed behavior at turnouts is that ballast structure tends to deteriorate faster than normal track section due to high stiffness originating from long bearers and impact forces acting on a crossing nose [10]. In addition, the occurrence of foreign material between switchblades and stock rails is likely to be higher due to cluttering ballast, which fails the driving mechanism.

15.2.4 Fasteners

The role of fastenings is to mitigate vibrations, provide isolation for track circuits and maintain rail position in the longitudinal, lateral, and vertical direction within certain limits. A typical fastener includes a base plate, a resilient rail pad, gauge block, elastic clips, insulators, and bolts/nuts/spike screws to anchor on a bearer. It should be noted that there are numerous types of fastening systems associated with technologies of manufacturers in which they can avoid some components such as insulators, steel base plates, and so on. A trade-off is considered between cost, easy adjustment, easy maintenance, effective track interaction, good noise emission, target stiffness while designing and purchasing a fastening system. Modern designs hold the requirements for both conventional and high-speed train lines [11]. In practice, stiffness values of fastenings are relatively low for high-speed rails in comparison to conventional tracks that have high damping and low stiffness properties. A particular difference in fastenings of turnouts is that sliding plates are used while shifting switchblades and points. Two types of sliding plates are used in practice such as simple and roller supported sliding plates [12]. Roller-supported plates have small switching resistance and force, mostly lubricant-free and expensive in comparison to simple sliding plates (Fig. 15.4).

15.2.5 Driving mechanisms

To change the position of a switchblade and enable wheel diversion, turnouts are equipped with driving mechanisms that could be categorized into four groups based on types of actuators (i.e., manual, electric, pneumatic, and hydraulic). In modern tracks, electric, pneumatic, and hydraulic driving mechanisms are controlled remotely via signaling systems in most applications where the demand is intensive for heavy traffic and high safety standards. The use of manual driving mechanisms, so-called levers, could be observed on low-traffic low-risk tracks.

The components of a common driving mechanism are actuators/switch machines, a stretch bar, a drive rod, a locking mechanism, sensors, and heaters. The force produced by switch machines and transferred via drive rods and stretch bars opens and closes switchblades. Following the switching process, the locking mechanism ensures the switchblades are fully closed and opened in the correct position. There are two types

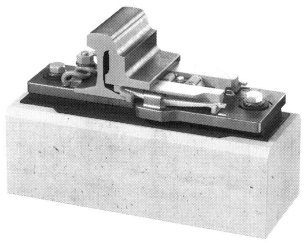

Fig. 15.4 A sliding chair with rollers.
Courtesy: Voestalpine VAE.

of locking mechanisms. Internal locking mechanisms are part of actuators by which the forces on switches are mitigated. On the other hand, the external locking mechanism is a separate system that does not transfer any forces to actuators and therefore, is used in a high-risk area, particularly a requirement for high-speed rail due to high reliability [13]. Another way to ensure the switch blades are at the desired location is the application of sensors. By analyzing the parameters such as distance, temperature, current, voltage, etc., it is possible to detect switch blades position as well as its condition. In extreme weather, based on temperature readings, electric heaters are used to dissolve the ice and snow, not to damage components and cause disfunction switchblades.

In some applications, the driving mechanism is embedded in a hollow bearer, aiming at efficient and effective maintenance and protection of equipment. Apart from that, in practice, there are several configurations of turnouts in terms of driving methods. Driving methods could be "single drive from single point," "single drive from multiple points," and "multiple drive from multiple points." Multiple driving mechanism is beneficial to distribute driving force along with switchblade. However, it requires frequent maintenance to keep in good condition to provide simultaneous movement of the driving mechanism.

15.3 Inspection

Inspection of turnouts shows variations in application among the countries around the world [14,15]. Inspection and maintenance procedures are mostly based on the individual experience of the infrastructure managers. Therefore, it is impractical to summarize or categorize all activities that are conducted in terms of inspection. Consequently, an effort was made to present the general application in this regard, and in some cases, particular examples are used to provide general ideas. A summary of activities conducted over turnouts is illustrated in Fig. 15.5.

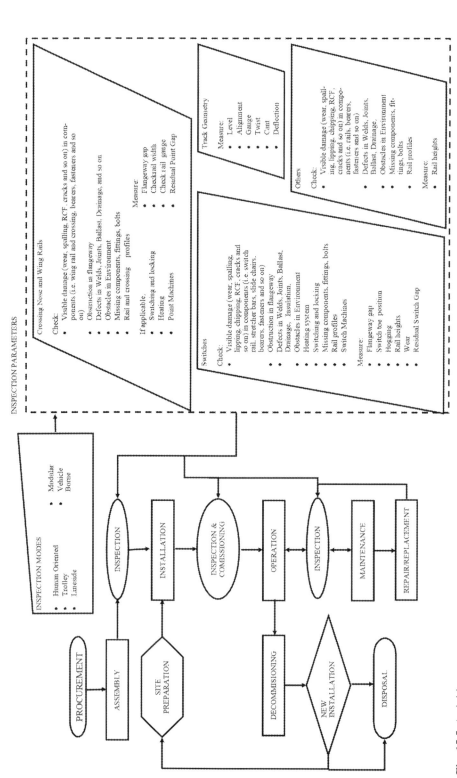

Fig. 15.5 Activities over turnouts.

15.3.1 Inspection period

Neglecting the limited applications of continuous observations for turnouts, the common practice is old-school routine inspections owing to their reliability and feasibility. The inspection periods are influenced by inspection types, track category, available sources as well as the experience of the infrastructure managers. Table 15.1 shows the inspection periods for the United Kingdom. Track categories in the table are decided based on permissible track speed and equivalent million gross tons per annum. For instance, track category 1a is applicable for the tracks having a permissible track speed over 120 mph. Thus, it is vital to inspect those tracks more frequently in comparison to the level 6 track having a speed limit below 20 mph.

15.3.2 Inspection modes

Today, the expectation from a railway track in terms of safety and riding comfort is incredibly high, which results in considerable demand for inspection. Over the years, several inspection modes have been offered to inspect a railway track, which is also applicable for turnouts. Infrastructure managers have been using a combination of inspection modes in practice and trying to figure out the optimal inspection process by seeking a trade-off between modes.

Human-oriented inspection is the core of the inspection process. Here, the so-called human-oriented inspection means that an individual or a team conduct inspection activity manually on site. The inspection process whether nondestructive testing or data-driven measurement is based on human skills, performance, and experience. Thus, human-oriented inspection offers significant advantages such as inspection of inaccessible areas, being able to clean the inspection area, determining damage level and surface quality, inspection repeatability of ambiguous defects, detection of uncatalogued defects, low-cost application, and so on. The inspection tools are hand gauges, chords, tapes, and hand-based measurement gadgets such as MiniProf [17] (Fig. 15.6 left).

Table 15.1 Track inspection frequency in the United Kingdom [16].

Inspection interval (weeks)	Track category						
Inspection types	1a	1	2	3	4	5	6
Visual	1	1	1	2	2	4	4
Ultrasonic	8	12	25	52	52	52	104
Geometry train	12	12	25	52	52	52	52
Supervisor cab ride	4	8	12	25	25	23	25
Supervisor visual	8	8	8	12	12	12	12
Engineer cab ride	8	12	24	52	52	52	52
Engineer visual	104	104	104	104	104	104	104

The main disadvantage of human-oriented inspection is the higher likelihood of failure to obtain correct measurements and conduct accurate testing due to human error. Moreover, performing measurements over a large structure like turnouts are a time-consuming activity. Therefore, inspection trolleys have been used to increase inspection speed and accuracy while measuring parameters (i.e., track gauge, flange way width, and depth, check rail gauge, cant, twist, cross-level, wear) or conducting an ultrasonic test. The trolley system could be manually [18] or automatically driven [19] (Fig. 15.6 right).

Modular inspection systems are developed to measure track and turnout geometry and model track environment [20] (Fig. 15.7 left). The advantage of such a system is that high-speed inspection by using the same inspection tool on different vehicles is possible. In other words, there is no need for an allocated track measurement car that requires track possession. Therefore, the cost of inspection drops significantly. The basic working principle of modular inspection is that the modular inspection tool is mounted on coupler/buffer scans rails by the laser-camera-GPS system. Then, a geometrical model of a track including turnouts and its surroundings is produced in a computer environment where parameters could be measured, and defects could be inspected on a desk at a remote location. The system also offers video recording for the validation purpose of the geometrical model.

Vehicle-borne inspection systems are widely used essential inspection systems (Fig. 15.7 right), available for both track and turnouts. Indeed, they are the only tool with high inspection speeds that makes it feasible to inspect large railway track networks. It should be noted that vehicle-borne systems generally are utilized for track measurements. However, they are also capable of inspecting turnouts. It should be emphasized that there are not many vehicle-borne inspection systems customized for turnouts owing to structural complexity and a limited number of parameters that can be measured by vehicle-borne systems. There are three types of vehicle-borne

Fig. 15.6 A gauge for human-oriented inspection (left) [17]; an automatic turnout inspection trolley—Felix (right) [19].
Courtesy: Left image, Greenwood Engineering A/S and right image, Loccioni.

Fig. 15.7 Modular inspection system (left) [20]; track inspection system (right) [21]. Courtesy: Left image, Fugro and right image, Ensco Rail, Inc.

systems such as self-propelled, towed car, and modified hi-rail vehicles [21]. Vehicles are equipped with numerous sensors such as lasers, accelerometers, GPS sensors, displacement sensors, gyroscopes, etc. The evaluation of data obtained from those sensors provides the information of track geometry properties, ride quality and rail corrugation as well as vehicle position on the track. Technological developments enable also to scan rail profiles and conduct a visual inspection. Vehicle-borne systems use both inertial measurement and chord-based measurement methods. Inertial systems use accelerometers and gyroscopes to measure the linear and angular accelerations, velocities, and displacements. Properties of track geometry could be solved by the interpretation of those measurements. Chord-based measurements are established on the versine method in which wheel positions are assumed as measurement points and the wheels in the middle provide the level of deterioration in the track [22]. In some cases, a mechanical chord could be applied to measure track properties, specifically, under extreme weather conditions [23]. The parameters (i.e., twist) that cannot be measured directly by the versine method are calculated by using information gathered from additional sensors such as the inclination sensor that detect the angular difference between two rails.

In vehicle-borne measurement, it is crucial to detect the location of the vehicle while inspection. Hence, vehicle-borne systems are also equipped with navigational systems that could be categorized as inertial navigation systems and global navigation satellite systems. Inertial navigation systems provide position information from the interpretation of the information obtained by accelerometers and gyroscopes as well as odometers with respect to well-known reference points [23]. On the other hand, global navigation provides location information from the interpretation of radio signals broadcasted by satellites. Finally but yet importantly, the measurements gathered by train-borne systems are classified as loaded track inspection whereas all other measurements are considered as unloaded measurements.

Remote inspection mode is commonly selected to observe turnout conditions continuously instead of measuring them in specific intervals. Turnouts are equipped with sensors (i.e., current, force, temperature, humidity, displacement, power, proximity, and accelerometers) and cameras to provide continuous data to a remote data center where the data is processed to detect failures and imminent failures. The method has been becoming popular as it offers flexibility in the inspection process, despite drawbacks such as high investment costs and data reliability. Particularly, remote inspection seems to be beneficial whilst monitoring the turnouts with heavy traffic.

15.3.3 Detection methods

Detection methods to identify any anomalies in a turnout system that might be harmful could be categorized into two groups such as nondestructive testing and data-driven detection. It is noteworthy that an interpretation of data obtained by any measurement method is considered under the topic of data-driven detection whereas any detection method that could also include measurements in theory but cannot quantify damage in numbers, is considered under nondestructive testing. For instance, a visual assessment of a geometry scanned by optical laser or ultrasonic testing belongs to the NDT section. By contrast, data-driven detection includes simple measurements, whether any measurements are in the range or not, as well as estimations by using sophisticated data analysis tools.

15.3.3.1 Nondestructive testing

Visual inspection is still the backbone of inspection activities in railways, relatively an inexpensive and reliable method to assess the condition of a turnout, and its components. The process of inspection is founded on the experience and visual acuity of an inspector who discerns consequential surface damages. Inherently, failure types detected by visual inspection of turnouts are limited to observable damages (excessive wear, lipping, chipped rails, RCF, and other rail defects; broken or deformed components, ballast failures, and so on), missing components (bolts, nuts, spikes), wrong installation/adjustment of components (clips, switch blades, check rails), obstacles (vegetation, foreign materials in drainage system), and nonfunctional components. Despite the prevalence, visual inspection is a heavy burden for infrastructure managers owing to constraints such as the number of inspected components per patrol, work-force capacity, track availability, inspection repeatability, and track safety as well as inherent drawbacks of visual inspection (i.e., damage classification, time allocation, environmental challenges, social and individual factors, etc.). Thus, infrastructure managers have been trying to apply alternative visual inspection concepts. Although "visual inspection" mostly reminds in-situ inspection activities, it also defines off-site activities. One of the alternative visual inspection methods is off-site visual inspection. The track and components are recorded in a video format by onboard or line-side digital cameras, commonly with charge-coupled device (CCD) technology that produces high-quality images. Later, an inspector watches recording and detects the failures. A more sophisticated way of analyzing the video records is to use image processing tools that can process and analyze the video frame by frame and detect the failures. Despite offering flexibility and repeatability, off-site visual inspection is still under development and cannot replace in-situ visual inspection completely. One of the main drawbacks of off-site visual inspection is the lack of ability to clean the inspection area since turnouts are laid in the open air and exposed to dirt. Furthermore, limited depth of field, resolution, and field of view prevent inspections at specific positions for specific failure types. Another off-site inspection concept is off-site geometrical modeling, where the turnout geometry and environment are scanned by scanning methods and reproduced virtually in a computer environment. The twin model is then assessed by an investigator. The level of details in modeling is directly related to the inspection target and available scanning methods for the target. For instance, turnout rails are scanned by an optical laser system to produce high-resolution high accuracy model on which an inspector can detect rail defects such as lipping, chipped, broken rails, large gaps. By contrast, the model of a large turnout environment is built by lidar scanners (Fig. 15.8 right), on which the inspector checks potential obstacles (i.e., vegetation). The disadvantage of geometric modeling is the limited level of details due to the necessity for high-capacity computer systems being capable of processing large data as well as other off-site parameters such as vibration, dirt, and so on.

Visual inspection is ineffective in terms of detecting small-scale surface and any subsurface cracks, and internal flaws of turnout rails. In that case, ultra-sonic and eddy current testing could be applied by infrastructure managers (Fig. 15.8 left). The basic definition of ultrasonic testing is an evaluation of propagation and reflection of sound waves produced and collected by probes. Cracks and similar defects in rails switch blades, fabricated crossings, welded areas, heel, and joints interrupt sound waves and reflect them back. By processing reflected waves, the damage could be detected and observed via an interface. The inspection process, particularly in manual testing, is heavily dependent on inspectors' skills and experience, and therefore, is subjected to a certification. In addition, it consumes a significant amount of time, which means low testing capacity. Recent attempts to increase ultrasonic testing capacity are to develop train-borne UT systems. Nonetheless, there is a tradeoff between higher testing speed and the accuracy of detection [25]. The main advantage of ultrasonic testing is to be capable of detecting in-depth damages in comparison to Eddy-current testing. Eddy current testing is based on deviations in surface and near-surface currents and is capable of detecting shallow damages such as fatigue cracks that cannot be detected by ultrasonic testing due to insufficient crack length.

In some cases, safety critical parts and components that might present imperceptible defects are inspected by visual inspection enhanced with methods such as liquid penetrant and magnetic particles. Particularly, austenitic manganese crossing, widely

Fig. 15.8 Ultrasonic testing trolley (left) [24]; lidar technology to scan the environment (right). Courtesy: Left image, Pandrol and right image, Loram Finland OY.

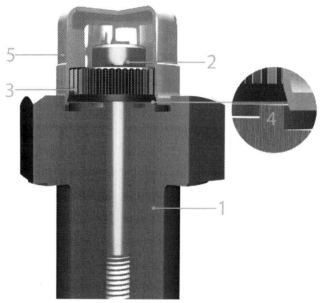

Fig. 15.9 Smart bolt with indicator cap [28].
Courtesy: James Walker

used around the globe, cannot be tested with eddy current testing and traditional ultrasonic testing due to nonmagnetic structure and coarse grain size in microstructure that attenuates sound waves. Despite recent reports of developing UT method for inspection of cast manganese crossing [26], visual inspection with liquid penetrant method seems to be the only method to inspect such a crossing. It should be emphasized that the liquid penetrant method is only able to detect surface cracks. Due to the absence of subsurface inspection, austenitic manganese crossings are one of the significant challenges in turnout inspection and maintenance.

Last but not least, new technologies focusing on smart components to facilitate visual inspection increase its popularity. For instance, smart bolts allow users to check bolt tension via a color indicator [27] or a rotating indicator cap [28], during visual inspection (Fig. 15.9). Furthermore, particular smart washers and bolts can measure force and transfer data to remote data centers [29,30] or ensure required tightening until it is damaged [31].

15.3.3.2 Data-driven detection

Data-driven detection is used to quantify the condition of turnouts and could be divided into two groups such as conventional measurement and condition monitoring. In conventional measurement, the measurements are collected manually or automatically by hand gauges or on-board sensors and evaluated based on infrastructure managers' manuals that enforce certain limits, including track geometry limits with reference to national and international standards, to maintain the quality of turnouts in terms of safety and ride comfort. Any measurements beyond the limits are reported and repaired to prevent any unfortunate accidents. The number of parameters to be measured shows variations in application among different countries. For instance, turnout parameters such as superelevation, twist, alignment are not measured in the United Kingdom the contrary to the Netherlands and Australia [15,32]. The reason for such a difference could be related to the fact that turnouts have stiff and plain geometries that are expected to be resilient against those defects. Similar processes to track geometry measurement is followed for turnout geometry measurements. For this purpose, chord-based measurements and inertial-based measurements are available. As well-known, a chord is laid between two points and the deviation from the midpoint determines the level of damage in the concept of chord-based measurements. Differently in inertial-based measurements, the measurements are collected from accelerometers and converted into deviation information. It is noteworthy that infrastructure managers might prefer a combination of both measurement methods [23]. Apart from turnout geometry measurements, there are other measurements peculiar to turnouts such as flangeway gap, check rail gauge, and so on. The number of parameters considered by infrastructure managers in that manner also show variations around the globe. Indeed, even the inspection approach for a particular parameter could be different such that the parameter could be assessed by visual inspection or neglected [15,32]. In some cases, they use specific hand gauges, having specific dimensions to indicate whether measurements are in the range or not. However, in that case, it might not be considered as a data-driven detection owing to the assessment based on visual perception. Finally, in conventional measurement, the data could be stored in the form of physical, virtual records, or both.

Condition monitoring is continuous measurements of turnout parameters, which aims at the detection of any failure in advance to increase reliability and safety and reduce costs and delays in traffic. Condition monitoring could be a supportive detection method as it cannot replace visual inspection and conventional measurements due to limitations in failure detection. Despite being capable of covering specific failures that can be detected by both visual inspection and conventional measurements, the detection method usually provides an estimation for failures. For instance, the deterioration of track bed could be detected by accelerometers installed at a specific location [33]. Nevertheless, it is a great challenge to extract information about types of the deterioration. Therefore, condition monitoring is more effective for failures such that any obstacles blocking switch rails or distance between switch rail and stock rails can be detected effortlessly. Similarly, it is efficient to activate a particular component on time by providing feedback. For instance, temperature measurements are used to switch on turnout heating systems as well as estimation of potential failures. The basic structure of condition monitoring consists of sensors, a gateway, an internet connection, a data center, and a visual interface for computers. Measurements such as force, current, acceleration, and so on are collected by sensors and transferred to a remote data center via an internet connection. A particular software or user evaluates the data and detects potential failure based on reference values, historical data, and experience. Due to its advantages such as remote and continuous measurement, the demand for condition monitoring is escalating, particularly the turnouts exposed to heavy traffic and extreme conditions.

15.4 Maintenance

Turnouts are high maintenance critical component that requires periodic maintenance and repair. The maintenance strategy for turnouts is mainly focused on the prevention of failure.

15.4.1 Removal of turnouts

In some cases, turnouts might become redundant. Following the decommissioning, redundant turnouts are removed swiftly as they require maintenance on the track and might pose a threat to operation in terms of reliability. In the case of the late removal process, moving parts of the turnout (i.e., switchblades) are secured to any movements opposed to the direction of the desired path [32].

15.4.2 Installation/replacement of turnouts

Three methods such as assembled in situ, preassembled close to the installation location, and assembled in the factory are used during the installation/replacement of turnouts. It should be reminded that terminology could be different in some countries. For instance, in China, the same definition is used for two methods, preassembled close to installation location and assembled in the factory, and called "ex-situ laying" [13]. According to DB, most of the turnouts are installed in Germany by using the "preassembled close to installation location" method [14]. Assembly in situ is mostly avoided due to challenges in providing required track quality, workforce, and track availability for conventional turnouts. However, in-situ installation is common for high-speed turnouts where the length of turnout is relatively long and the track structure is mostly slab (Fig. 15.10). "Assembled in factory" method, namely modular

Fig. 15.10 Installation of a turnout.
Courtesy: Turkish State Railways.

installation, is not common due to challenges in transportation. Nevertheless, modular installation is expected to be expanded since the drawbacks of the "preassembled closed to installation location" method such that the size of the site, access to the site, obstructions such as overhead wires, distance from installation point, ground problems such as not flat and level, methods of installation, other working difficulties makes the installation harder. It should be emphasized that the installation of high-speed turnouts could be also done segment by segment. The segmentation is due to difficulties in lifting heavy and long high-speed turnout structures. The segments could be constructed on a site close to the installation location or in the factory. The advantages of segment by segment are efficiency in working hours, less track occupation, the lower probability to damage the components, particularly long rails.

The installation process for conventional turnouts follows a general pattern although there are differences in the application of countries or the installation of different turnout types.

In Australian practice [32], the installation of a turnout starts with drawing a reference line. On this reference line, bearer positions are marked for every fifth bearer. Afterward, bearers are laid in the desired layout (fan shape or perpendicular to reference line) along with the reference point. In the following order, stock rails, switchblades, and remaining rails are laid and adjusted based on the layout drawing. Particular attention is given to switchblades that must be perpendicular to the reference line. Then, all the plates, fastenings, and check rails are installed and bolted. Before installation, geometric properties such as turnout dimensions, alignment, gauge, and so on are measured and checked whether the assembly provides the standard. Afterward, turnouts to be replaced are cut, removed, and dismantled and groundworks of installation location are completed. Assembled turnout is transported to the installation area and connected by welding. The welding process is also standardized to avoid potential harm. For instance, switchblades are buttwelded components and any welding close to the buttwelding area is prohibited. Following the connection, the ballast bed is laid and packed with respect to the standard of ballast laying and packing. In the final stage, all required track quality dimensions and limits are checked and visual inspection is conducted to detect the possible surface damage, distortion, buckling, and so on. Having passed the checks, signal systems and spring assists are installed and tested. Then, the track is certified.

The procedure of installation high-speed turnout varies based on the support structure. The ballasted high-speed turnouts are laid on the ground like conventional turnouts on ballasted track. Hence, the process in the previous paragraph is valid for high-speed turnouts. The installation of high-speed turnouts on slab track shows differences based on slab track types. In general, the slab track could be categorized under three sections such as wet-poured, precasted, and others [34]. In wet-poured slab systems, turnouts are assembled first similar to conventional turnouts and then the concrete slab is cast. On the contrary, slab structure is constructed first and then turnout is installed in situ in precasted turnout systems. The acceptance limits for high-speed turnouts are demanding in comparison to conventional systems, which affect construction costs significantly [13].

15.4.3 Repair activities in turnouts

15.4.3.1 Switch blades

The faults observed at switchblades are plastic deformations, cracks, loose fasteners, foreign materials between switchblades and stock rails, dimensions below acceptance level, dead driving mechanism, deformed switchblades (bent, crippled, or hogged), damaged components(heel blocks, baseplates, sliding chairs, insulations, bearers, stretch bars) and soft spots. The repair of damaged components and deformed switchblades is likely to be a low probability. Particularly, soft spots originating from incorrect heat treatment cannot be repaired. During the repair process, components are replaced with a new component most of the time. In some cases, baseplates with a stopper could be repaired. Similarly, a chipped or broken switchblade toe could be fixed by grinding. However, certain limits are defined for grinding. For instance, in Australian practice, the damaged section longer than 200 mm cannot be fixed by using grinding [32]. Grinding could be used also remove the general rail defects such as spalling, shelling, head checks, squads, wheel burn, corrugation, imprints. In some cases, welding of damaged sections could be necessary before grinding. Other defects such as loose fasteners, dimensions below acceptance limit, dead driving mechanism, foreign materials between rails, or incorrect switchblades height could be removed by tightening, adjusting, or cleaning of components, tamping, and packing ballast. In some cases, adjustment might require the replacement of components with a suitable one or the use of adjustment shims. Apart from those, it should be emphasized that all the components are mostly manufactured as "ready to install" and therefore do not need any modification by using methods such as drilling, cutting, or welding unless necessary and permitted by manuals. Avoiding such modifications is crucial in all maintenance activities in terms of component and track safety. For instance, the drilling operation should be done considering the distance between holes and height from the running surface. Inappropriate drilled holes could cause stress concentration and result in broken rails that could lead to a derailment.

15.4.3.2 Stock and closure rails

The defects observed on stock rails are similar to the normal track. Rail defects such as corrugation, wear, squats, and so on, could be fixed to a certain level by grinding. Damaged components such as baseplates, bolts, bearers are replaced by new ones. Apart from those, turnouts without an anticreep device might suffer from rail creep. In that case, the rail adjustment is conducted as stated in the manual.

15.4.3.3 Crossing

As aforementioned, a common crossing nose is subjected to a high magnitude impact force that causes defects such as plastic deformation, spalling, transverse crack more frequently in comparison to other turnout panels or normal rails. Those defects expose high risk due to the risk of inappropriate flange contact that may lead to a derailment. The repair process of the crossing nose is done by mostly welding and grinding

Fig. 15.11 Repair activities; welding (left); grinding (right).

processes on site (Fig. 15.11). The process follows as; remove the damage section by cutting, filled the removed area by wire-feed welding, grind the welded area until the desired profile is achieved. In some countries, the repair of cast manganese on site is not recommended [32]. The surface hardness of cast manganese in service, particularly explosive cast manganese, is higher than fabricated crossing noses. Hence, achieving a high-quality crossing nose by welding and grinding is relatively lower. Furthermore, welding and grinding might not extend the life cycle of crossing the nose, as it depends on the skills of the workforce, the position of the damaged area, degree of change in material, and geometry [35]. Consequently, infrastructure managers follow the procedure enriched with their experience. Apart from the damage on the crossing nose, common rail-track defects are observed at the crossing and they are repaired similarly at the crossing panel. Differently, crossing might suffer from casting defects when cast crossings are used. Furthermore, the crossing panel is laid over long bearers that makes the crossing section stiffer and causes a large stiffness variation along with turnout. To prevent that problem, tamping is done cautiously to provide even track stiffness along with the turnout by using special tamping machines [36]. Lastly, check rails are mostly used to restrict the wheel's lateral position to control and provide a safe transition between wing rails and crossing the nose. The flangeway of check rails must be within certain limits defined in the manual [32]. In the case of a narrow flangeway, excessive wear could be observed. The repair process for excessively damaged check rails is adjustment and replacement.

The technology used in the movable crossing nose is a combination of both switch and crossing technology, which means that movable crossings suffer from problems observed at both sections such as incorrect lateral attachment, insufficient flangeway, foreign material between wing rail and crossing nose, rail defects, plastic deformation (particularly lipping), and so on. However, the frequency and magnitude of damage are lower than common crossings due to lower dynamic forces over the crossing nose and smooth transition between wing rail and crossing nose. The defects observed at the movable crossing are removed similarly to the removal of the switch, rail, and crossing defects.

15.5 Lifecycle cost

The manufacturing process for turnouts is quite complex and costly due to dimensions, variable rail profiles, and the requirements for high precision and high-quality surface works. Furthermore, turnouts are maintenance-dependent components on the track. Hence, the operation of a turnout is a costly activity for infrastructure managers. Most of the time, cost estimations are given in total expenditure for turnouts and extracted from the spending accounts of infrastructure managers. In Sweden, it is reported that maintenance activities of turnouts use 13% of the maintenance budget [37]. In terms of the cost per turnout, a good way is to analyze turnout's lifecycle cost, which is an estimation method including costs of acquisition, operation, maintenance, and disposal. In a deterministic study [38], the acquisition cost of a turnout was estimated lower than the cost of maintenance and inspection. Furthermore, it was mentioned that material costs for maintenance and acquisition are similar level however there is a significant difference in work force cost. In Fig. 15.12, the breakdown of maintenance costs in a turnout lifecycle is presented. As can be seen from figure, the cost of two activities (tamping and general maintenance activities such as cleaning, tightening, and so on) are significant. These activities are frequently conducted and mostly work-force-based activities. It should be reminded that life-cycle cost analysis gives insight into a particular country. In some countries, the maintenance activities could be lower due to the lower cost of work force. Indeed, the life-cycle cost of turnout is quite a complex process where maintenance activities, cost of acquisition, culture, experience, and more parameters affect the cost. In conclusion, general experience shows that the maintenance of a turnout is a costly activity compared to acquisition costs [14].

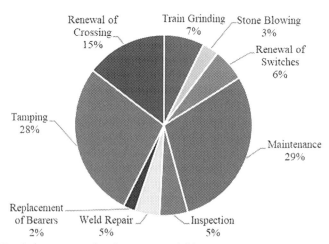

Fig. 15.12 Break-down costs of maintenance activities per turnout.
Reproduced from M. Hamarat, S. Kaewunruen, M. Papaelias, Life-cycle cost analysis of railway turnouts exposed to climate uncertainties, in Proceedings of the 3rd World Multidisciplinary Civil Engineering—Architecture—Urban Planning Symposium, Prague, The Czech Republic, 2018.

15.6 Conclusions

Despite their complex behavior under train loads and being critically dependent on inspection and maintenance, railway turnouts have been overlooked until recently. Parallel to technological advances in the last two decades, infrastructure managers, governments, and private companies have been investing to develop more efficient and effective processes to inspect, maintain and manufacture turnouts. Particularly, in several European Projects, turnouts have been a focal point to achieve improved performance from railways [3,5,39,40]. It seems that a new turnout design replacing current technology will not be available in near future. Consequently, more resources should be allocated to improve current inspection processes toward fully automated inspection processes.

Acknowledgments

The authors would like to express their sincere gratitude to employees and managers, Fehmi Ucan and Aydin Ozdemir in particular, of Turkish State Railways (TCDD), Ensco, Vostalpine, Rota Bolt, Pandrol, Loram OY, Greenwood Engineering and Fugro for the pictures' courtesy.

References

[1] M. Schilke, Degradation of Railway Rails From a Materials Point of View, Chalmers University of Technology, 2013.
[2] Limited, B.S., https://britishsteel.co.uk/what-we-do/rail/flat-bottomed-vignole-rail/bainitic-low-fatigue-blf/ [cited 2020 28.08.2020].
[3] Trafikverket, Innovative concepts and designs for resilient S&Cs, in: UIC (Ed.), Capacity4rail, 2017.
[4] R. Skrypnyk, et al., Prediction of plastic deformation and wear in railway crossings—comparing the performance of two rail steel grades, Wear 428 (2019) 302–314.
[5] Trafikverket, Operational failure modes of switches and crossings, in: UIC (Ed.), Capacity4rail, 2015.
[6] S. Kaewunruen, R. You, M. Ishida, Composites for timber-replacement bearers in railway switches and crossings, Infrastructures 2 (4) (2017) 13.
[7] A. Manalo, et al., A review of alternative materials for replacing existing timber sleepers, Compos. Struct. 92 (3) (2010) 603–611.
[8] M. Hamarat, et al., New insights from multibody dynamic analyses of a turnout system under impact loads, Appl. Sci. 9 (19) (2019) 4080.
[9] P. Sengsri, et al., Damage detection in fiber-reinforced foamed urethane composite railway bearers using acoustic emissions, Infrastructures 5 (6) (2020) 50.
[10] S. Kaewunruen, Monitoring structural deterioration of railway turnout systems via dynamic wheel/rail interaction, Case Stud. Nondestruct. Test. Evaluation 1 (2014) 19–24.
[11] Vossloh AG, https://www.vossloh.com/en/products-and-solutions/product-finder/product_11031.php [cited 2020 26.08.2020].
[12] Schwihag AG, 2012. https://www.schwihag.com/uploads/media/SCHWIHAG_Produkt broschuere_EN_2012.pdf. (cited 2020 26.08.2020).
[13] P. Wang, Design of High-Speed Railway Turnouts: Theory and Applications, Academic Press, 2015.

[14] Mainline, Maintenance, renewal and improvement of rail transport infrastructure to reduce economic and environmental impacts, European Project. 7th Framework Programme, 2014.
[15] M.F. Rusu, Automation of Railway Switch and Crossing Inspection, University of Birmingham, 2017.
[16] A. Cornish, Life-Time Monitoring of in Service Switches and Crossings Through Field Experimentation, Imperial College London, 2014.
[17] Greenwood Engineering A/S, https://greenwood.dk/railway/ [cited 2020 28.08.2020].
[18] Vossloh AG, https://www.vossloh.com/en/products-and-solutions/product-finder/product_11137.php [cited 2020 28.08.2020].
[19] Loccioni, https://www.loccioni.com/wp-content/uploads/2018/08/loccioni-felix-a2-brochure.pdf [cited 2020 28.08.2020].
[20] Fugro, https://www.fugro.com/our-services/asset-integrity/raildata/rila-track-rila-360 [cited 2020 28.08.2020].
[21] Ensco, https://www.ensco.com/rail/track-inspection-vehicles [cited 2020 28.08.2020].
[22] T. Hisa, et al., Rail and contact line inspection technology for safe and reliable railway traffic, Hitachi Rev. 61 (7) (2012) 325–330.
[23] J. Nielsen, et al., Overview of methods for measurement of track irregularities, RIVAS Railway Induced Vibration Abatement Solutions Collaborative Project, 2013.
[24] https://www.pandrol.com/product/ultrasonic-rail-testing/, 2020 [cited 2020 01.10.2020].
[25] Network Rail, https://www.networkrail.co.uk/wp-content/uploads/2019/06/Ultrasonic-Rail-Inspection-v1.pdf [cited 2020 28.088.2020].
[26] TWI, https://www.twi-global.com/media-and-events/insights/railsaft-ultrasonic-inspection-of-cast-austenitic-manganese-steel-railway-crossings [cited 2020 28.08.2020].
[27] SmartBolts, http://www.smartbolts.com/ [cited 2020 28.08.2020].
[28] JamesWalker, https://www.rotabolt.co.uk/ [cited 2020 28.08.2020].
[29] Nord-Lock, https://www.nord-lock.com/insights/knowledge/2019/superbolt-smart-load-sensing-tensioner/ [cited 2020 28.08.2020].
[30] Technologies, S.C., https://smartcomptech.com/smart-washer/ [cited 2020 28.08.2020].
[31] Group, W, https://www.wsgroupglobal.com/ [cited 2020 28.08.2020].
[32] RailCorp, TMC 251-Turnouts, in: Engineering Manual-Track, 2019. Australia.
[33] E. Kassa, A. Skavhaug, A.M. Kaynia. http://www.destinationrail.eu/ajax/DownloadHandler.php?file=2123. (cited 2020 28.08.2020).
[34] PWI, https://www.thepwi.org/technical_hub/presentations_for_tech_hub/150520_london_half_day_seminar_ballast_beyond/05_150520_london_half_day_ballast_beyond_john_porrill [cited 2020 28.08.2020].
[35] L. Xin, Long-Term Behaviour of Railway Crossings: Wheel-Rail Interaction and Rail Fatigue Life Prediction, Delft University of Technology, Netherlands, 2017.
[36] Plasser&Theurer GmbH, https://www.plassertheurer.com/en/machines-systems/unimat-09-324s-dynamic-e3.html [cited 2020 28.08.2020].
[37] A. Nissen, Development of Life Cycle Cost Model and Analyses for Railway Switches and Crossings, Luleå tekniska universitet, 2009.
[38] M. Hamarat, S. Kaewunruen, M. Papaelias, Life-cycle cost analysis of railway turnouts exposed to climate uncertainties, in: Proceedings of the The 3rd World Multidisciplinary Civil Engineering—Architecture—Urban Planning Symposium, Prague, The Czech Republic, 2018.
[39] A. Ekberg, B. Paulsson, INNOTRACK: Concluding Technical Report, International Union of Railways (UIC), 2010.
[40] http://www.s-code.info/ [cited 2020 28.08.2020].

Risk-based maintenance of turnout systems

16

Serdar Dindar[a] and Sakdirat Kaewunruen[b]
[a]Civil Engineering, Izmir Katip Celebi University, Izmir, Turkey, [b]School of Civil Engineering, University of Birmingham, Birmingham, United Kingdom

16.1 Introduction

Current risk analysis tools are often intended to concentrate on all hazard types (e.g., derailment hazards, collision hazards, fall hazards, fire hazards, slip/trip hazards, electrocution hazards, train strike hazards, and platform/train interface hazards) and the railway risks as a whole system, regardless of the distinguishing characteristics of the systems forming the railway and hazards, resulting in different consequences. Nevertheless, the ability to manage a specific hazard occurring within an individual system, such as derailments at turnouts, would facilitate more accurate failure estimation and would provide the opportunity to establish more effective maintenance approaches for diminishing the risks levels related to the specific hazard within the specific system [1].

Risk analysis reliability is based on the following factors:

- The level of reliability and credibility these models have for turnout operations in terms of the development of strategic risk priorities.
- The capacity to further develop these models, to help rail companies, to identify and justify their strategic risk priorities.

Risk analysis is performed satisfactorily through trustable sources and logical hazard scenarios. As regards the trustable sources, a comprehensive analysis of incidents and accidents experienced internationally has been conducted [2]. In cases where this analysis has revealed new incidents, they have been included in the modeling process to generate estimations of frequencies and outcomes in the United Kingdom and United States. As regards the logical hazard scenario, a particular causal factor of train derailments at turnouts can be defined as an infrequent occurrence. Rather than a holistic strategy (where all risks are combined in the analysis). A particular risk group alone should be investigated first, and after this investigation, the impact of this particular group on another one should be examined [3]. This could enable the risk analysis reliability to be improved.

This book chapter is to develop frameworks (up to a point) capable of modeling risk, risk monitoring, and risk management at railway turnouts using a range of engineering approaches for better understanding of existing risk as well as possibly increasing the risk in the future, and establishing an optimized maintenance strategy, compiling from own research by the authors. On the other hand, derailment cases as logical hazard scenarios are investigated to perform Risk-based Maintenance of Turnout systems.

16.2 Railway turnouts

Turnouts in the railway system alternatively called switches and crossings in British vernacular, are comprised of a mechanical rail system that consists of two or multiple movable rails whose function is to guide railway vehicles on their designated route (see Fig. 16.1). The considerably complex nature of railway switches has led to emerging operational risks for railway networks [4,5] This has been demonstrated by the increasing amount of derailed trains on or close to railway crossings and switches [6,7]. These accidents generate operational interruptions and financial damages, and occasionally even deaths. A suitable estimation of the type of risks inherent to systems of railway turnouts is necessary for those responsible for managing railway systems so that the whole railway network can function while interconnecting with different infrastructures of the railway [8]. As a result of the increasing amount of strain placed on railway networks, the management of risks associated with railway switch systems will necessitate more in-depth examination to provide a concrete rail operation.

Switches and crossings on railway systems are fundamental components of any interconnected railway network. The transition of a train from one rail to another is achieved through the use of S&C. In general, they comprise approximately 30% of the overall budget allocated to maintaining and constructing railway systems, which equates to around 0.3 km of 1 km regular plan track [9]. Within the European Union, it is estimated that countries have installed S&C at a rate of slightly more than one

Fig. 16.1 Fundamental components of a turnout.

switch per kilometer of rail [10]. Therefore, it is expressed that the rail industry should always target improvement of safety measurements as rail turnouts are pivotal infrastructures.

16.2.1 Component definition of a turnout

The different components of turnouts are illustrated in Fig. 16.1. A brief summary of each part in addition to its technical description is given below.

- Stock rail: the primary rail of the track on which the switch blades are tightly positioned.
- Points (alternatively called switches): steel blocks constructed by the mixture of a pair of stock and switch blades along with the essential linkages and components.
- Crossing nose: a composition of rails incorporated at the junction where a pair of rails intersect, which enable a vehicle's wheel flange to be transitioned between tracks.
- Switch motor: a system that runs on electric, hydraulic or pneumatic power that is utilized in the alignment of the switch with one of the potential tracks.
- Stretcher bar: a steel bar utilized to ensure that the switch rails are correctly positioned under a moving vehicle.
- Sleepers: they in general are situated perpendicularly to the rails and function by transferring loads from these rails to the track ballast and subgrade. Their additional advantages are that they maintain the upright position of the rails and ensure that the track gauge is appropriate.
- Closure rail: the element of a fixed rail that is situated between the points and the crossing of a turnout to facilitate the transition of a railway vehicle from the switch panel to the crossing panel.
- Heel block: a unit that delivers a splice with the adjacent closure rail and a location for the switch point rail.
- Check rail: a short piece of rail situated adjacent to the stock rail opposing the crossing to guarantee that the suitable flange way through the crossing is trailed by wheels.

16.3 Identification of a risk analysis method

16.3.1 Risk and safety

Railway systems have highly complicated geometries, as a significant amount of constituents are interconnected. Derailments largely occur as a result of knowledge deficiencies caused by this level of complexity [11,12]. One such system is the turnout, which is essentially a mechanical structure that facilitates the transition of flanged vehicles between tracks [13].

Due to the fact that turnout systems are characterized by their complexity, it is important to focus not only on the component failures that occur in the railway system but additional failures related to operations (i.e., communication system deficiencies), environmental aspects (e.g., inclement weather), and interaction issues (i.e., searches when identifying the probability of any incident in every rail turnouts) [13,14]. This is because every system has specific technical attributes. Hence, it could be claimed that each turnout has a variety of different kinds of possible causes that could lead to derailments irrespective of the quality of their construction, monitoring, and maintenance [15].

In such situations, the process of risk analysis, an important stage in the management of risk is an essential factor in the reduction or potentially the elimination of train derailments in specific instances. A variety of different risk analysis methods have been developed, where each could have benefits in comparison to the others in the context of the railway sector for various reasons. Analysts must select a technique that provides more pragmatic results, otherwise, unwanted outcomes, wasted time, and excessive costs will likely occur [16].

Therefore, to fully comprehend the prevailing risks at turnout systems, suitable kinds of risk analysis techniques must be determined within the railway sector to achieve this objective.

16.3.2 Risk analyses

A railway-related risk analysis is often scarce, incomplete or, sometimes even has missing data [17]. The weakness in building a satisfying database arises mainly from building new lines, the new materials used in railway tech, and climate and traffic density changes over the years [18]. As a result, many precise safety estimates for the ensuing years need to be carried out, as many of the changes mentioned above have already occurred or will.

This book chapter deals with derailment causes arising from a complex nature of rail infrastructure and cannot be matched by a simple multiple-criterion decision-making framework, without the insights into multilayer asset vulnerabilities derived from expert opinions. Some of them are observed to depend largely on statistical techniques to deal with variables for risk analyses. In addition, the limitations and expert opinions associated with the risk analysis methods are discussed throughout the chapter. To assess emerging risks in the railway sector, a large number of risk analysis methods, which might be used for a turnout in the railway industry, have been evaluated comprehensively [19]. Table 16.1 shows the findings by risk analysis methods including risk matrix (RM), failure mode and effect analysis (FMEA), a reliability block diagram (RBD), fault tree analysis (FTA), event tree analysis (ETA), Markov analysis (MA), hazard function (HA), Bayesian analysis (BA).

In summary, there is another issue for classical risk analysis methods, e.g., FTA and ETA, typically decomposing a system into subsystems and basic elements. Investigating risks for a turnout system with strong interdependencies in nature has to go beyond the convention cause-consequence analysis to concentrate on spill-over clusters of failures [20]. Indeed, the sum of the behavior of individual components in a turnout cannot be expected to describe implicitly the behavior of the whole system. This renders questionable the suitability of such risk analysis techniques. Moreover, predefined causal chains, e.g., defined by ETA, are likely to be inappropriate to identify hidden risks.

It is ultimately worth noting that each technique might provide different parameters or outputs that may be particularly useful regarding intended solutions to the problem. Therefore, a risk analysis method should be chosen, not only based on hazard, but also the consideration of the capabilities of each technique.

Table 16.1 Comparative evaluations of risk analysis methods in the railway industry.

Methods	Life cycle phases	Strengths	Weakness	Availability prediction	Common cause failures	Effect of uncertainty in data	Proactive use
RM	All phases	Quick preparation; suitable in the case of subjective data, e.g., expert opinion on only ties degradation of a turnout	Inadequate for complex systems; cannot identify dependencies such as signaling errors vs environmental effects	Yes	No	No	Yes
FMEA	After the design is finalized	Good for identifying single-point failures, e.g., electrification process of a switch mechanism	Human error cannot be addressed; unable to reflect system redundancies, interactions, and common cause failures	No	No	No	Yes
ETA	All phases	Excellent tool to model temporal escalation of events such as high speed-based derailment; ideally suited to model efficiency of safety-critical tasks and emergency	High dependencies on the correct capture of event escalation; need scarce data for such complex systems as aging any railway components through FTA	No	Yes	Yes	Yes
MA	After the design is finalized	Good for complex systems; a good tool for identifying process inefficiencies	Unable to reflect redundancies and common cause failures	Yes	No	Yes	No

Continued

Table 16.1 Continued

Methods	Life cycle phases	Strengths	Weakness	Availability prediction	Common cause failures	Effect of uncertainty in data	Proactive use
HF	Design of emergency preparedness plans and evaluation of safety-critical tasks	Evaluates existing safeguards and identifies ultimate consequences; may be a good tool to derive a safety-based maintenance model of a turnout	For a human-based failure, quantification may be misleading since such failures are quite difficult to model; due to its reliance on scarce date to model, gathering of data might be difficult	Yes	No	Yes	Yes
MC	To establish properly reliability of system, ideally during consolidated design, but could be used in all phases	Once the model is built, input distributions are quickly updated to yield new results; an intuitive process, helping users to add some qualitative data into a mathematical model, which describes the risk parameter; provides a range of consequences, enabling better estimation of risk	The creation of a mathematical model can be challenging; relies on computerized methods, e.g., spreadsheeting; satisfaction of the analysis highly depending on the complexity	Depends on model	Depends on model	Yes	Yes

| FTA | Throughout all stages of operation | It may be excellent for complex systems where interaction and combination of events and failure needs to be considered; use properly statistical data of component failures of a turnout to evaluate probability for unwanted top event; provides a visual model of a safety system; provides ranked lists of critical turnout components; an excellent tool based on a qualitative or quantitative application to model redundancies and fault tolerance (vulnerability) | Despite databases unsuitable for specific application, e.g., aging of rail track, failure information might be supported using FORM methods; unable to model temporal events of a turnout such as changing weather conditions. Dependencies on correct capture of faults and failure mechanisms and interaction to predict system behavior | Yes | Yes | Yes | Yes |

Continued

Table 16.1 Continued

Methods	Life cycle phases	Strengths	Weakness	Availability prediction	Common cause failures	Effect of uncertainty in data	Proactive use
BA	Especially, a well-known relationship of failures	Provides a solid decision theoretical framework, forming a prior distribution for future analysis; useful method for fuzzy integration; easily adoptable after new data; shows relationship of nodes	Requires time and skill to prepare a solid network	Yes	Yes	No	Depends on model

As a summary of outputs, desired outputs can be simple lists of individual failures (FMEA, RM), numerical estimates of system failure probabilities (MA, RBD, FTA), listings of event scenarios and their likelihoods (ETA), numerical system failure probabilities and sensitivities to input variables (BA, MC), or unique combinations resulting from a combination of these methods (e.g., use of both FTA and ETA).

16.4 Establishment of risk-based maintenance

A risk-based maintenance (RBM) approach helps in determining the most appropriate maintenance plans to minimize the risk resulting from railway turnout failures. Considerable work is being carried out on this type of maintenance approach, especially in the United States and the United Kingdom, albeit mostly concentrated on static equipment (pipes, vessels, etc.) in petrochemicals. For example, Network Rail plans to change completely its current maintenance strategies into RBM [21].

The primary objective of this RBM is to facilitate the railway industry's work to ensure that railway turnouts are operated within safety limits and, thus, to reduce risk to passengers, employees, and goods. RBM's secondary objective is to reduce the maintenance expenditure stemming from a variety of reasons, including planning and scheduling problems, operational problems, errors in cost estimations and arrangement. As this paper is shortened, the secondary objective is absent. However, the extended version of the paper includes the secondary objective, which entails a well-integrated comprehensive discussion on optimization strategies in its concluding remarks.

Fig. 16.2 shows a set of significant milestones to firmly establish the best fitting maintenance practice into a particular railway infrastructure, turnouts.

As seen, RSB begins with hazard identification. Of the many techniques, HAZOP is suitable for geometry restoration due to firstly being usable at varying times during a turnout's life cycle and, secondly, being easily modifiable (for new hazards identified assessment and monitoring). It is followed by consequence prediction, which might be derailment or collision. Then, the frequency of failure occurrence in a particular turnout system is calculated through a set of methods, such as fault tree analysis and expert opinion, while FTA determines the frequencies by stepping through a series of events logically, whose values are revealed as maintenance, failure or accident reports. Expert opinion is needed where there is a lack of solid knowledge to determine the frequencies through different approximation methods, e.g., Buckley's expert opinion based on the fuzzy logic method. In the fourth step of the maintenance chain, risk assessment can be conducted through a variety of techniques, including ETA, FMEA,

Fig. 16.2 Phases of a risk-based maintenance strategy for railway turnout system.

risk matrices, and Bayes network (BN). Each of these techniques has unique characteristics and the ability to identify and assess factors that may jeopardize the operation of railway turnouts. BN is claimed to be suitable for turnouts owing to the well-representing of causal relationships between turnout variables and uncertainty, handling missing observations, easy integration of new evidence, variety in input data, and its strength in structural and parameter learning [20]. The residual risks have to be approved by a specifically appointed team and, then, the execution of the mitigating actions has to be carefully carried out to monitor and identify the consequences as well as likelihood prediction to learn to what degree action should be taken against them [22]. This can be achieved through a dynamic-structured BN. Maintenance planning is one of the most significant steps in the chain and covers planning decides the what, how and time estimate for a specific maintenance event considering cost and operational shut-downs [23]. The last step covers "Plan B," which is to develop and implement contingency/recovery plans during times of severe disruption to operations.

16.5 Environmental impact consideration into a maintenance chain

Section 16.3.2 has identified a large number of appropriate risk analysis techniques for railway turnout systems. As the relationship of failures has already been determined [23–25], Bayesian analysis-based risk analysis is chosen to perform environmental impact-related failures at turnouts. This technique might be underlined to be a quantitative assessment of the probability of failure (PoF) and the consequence of failure (CoF) providing operators risk ranking, savings, equipment inspection requirements, and inspection guidelines. While PoF is found out through suggested a Bayesian-based framework, CoF is assigned to be derailment at turnouts.

There is a long record of weather-caused derailments at turnouts, which has enhanced the knowledge of what causes most give rise to derailment. However, we have no idea regarding the interaction of these causes or about what the probability distribution is going to be like in a situation in which one of these causes is impossible to happen, e.g., tornado in areas with a mild climate. Hence, there is a need for a generic BN-based weather-caused flow diagram to be developed.

For the implementation of weather-related derailment estimates at turnouts, a fuzzy Bayesian network (FBN) is developed as Fig. 16.3.

In this proposed approach, the following three steps are adopted:

- Step (1) Problem definition: Search available databases which refer to all kinds of weather-related derailments; judge data to identify all anticipated weather-based causes/factors to potential derailment accidents at turnouts; pay attention to causal relationships among those causes/factors.
- Step (2) BN module construction: Define both variables (nodes) having a finite set of mutually exclusive states as identified root nodes (RNs) or intermediate nodes (INs) to represent the identified hazards; develop failure logic through conditional probability distribution

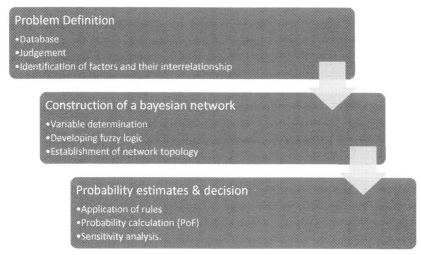

Fig. 16.3 The framework of Bayes network-based derailment prediction for railway turnouts.

(CPD); establish a network topology to describe conditional independence relationships of defined variables.
- Step (3) Probability estimates and decision: Specify states and assign input values for probability estimation of RNs; calculate probabilities based upon Buckley's alpha cut methods via Eq. (4.7); update the values of all nodes by calculating posterior probabilities; perform sensitivity analysis to reveal the performance of each variable's contribution to the occurrence of a derailment accident at turnouts.

FBN is quite a prominent technology with huge potential for various applications across many engineering domains. This study discusses FBN and its application in railway turnout systems. The proposed FBN approach uses the probabilities of environmental-related causes of accidents to perform Bayesian inference, which is established by causal relationships through accident reports. Therefore, the BN provides the model structure, fuzzy prior probability and likelihood calculation, and inference and interpretation. Aside from the BN, there have been many other techniques that are suggested to risk, occurrence, or consequence analysis of any type of accident across railway systems. In this case, the identified PoF through FBN is ready integrated to likelihood estimation, which is the third step in Fig. 16.3.

16.6 Concluding remarks

The overall aim of this chapter was to provide an informed answer to the question of how a RBM strategy for railway turnouts can be developed to enhance the operability of railway turnout systems. This chapter is limited, excluding the discussion on planning and scheduling problems, operational problems, errors in cost estimations, and arrangement.

Failures on turnouts and possible relationships between geometry-related failures and precursors have been assessed briefly. The fundamental steps of the maintenance are shown, explained, and discussed with a recent reference. In conclusion, the developed RBM model is considered to be likely helpful for the railway industry to reduce and mitigate problems turnouts, and the financial burden arising from such.

As a case study, environmental impacts are investigated as being one of the causal groups responsible for train derailments (DoF) at rail turnouts. Therefore, this chapter illustrates a Bayesian network to deal with the risk of derailment associated with environmental impacts. The chapter's proposal is established and is shown as to how to implement PoFs into the maintenance chain.

References

[1] S. Dindar, S. Kaewunruen, Investigation of risk-based maintenance strategies for turnout geometry restoration, in: The 1st Asian Conference on Railway Infrastructure and Transportation, Jeju, Korea, 2016.
[2] O.H.B. Mohd, S. Kaewunruen, S. Dindar, Disruption management of resource schedule in transportation sector: understanding the concept and strategy, Procedia Eng. 161 (2016) 1295–1299.
[3] S. Dindar, S. Kaewunruen, O.H. Mohd, Review on feasibility of using satellite imaging for risk management of derailment related turnout component failures, IOP Conf. Ser. Mater. Sci. Eng. 245 (4) (2017) 042025.
[4] Á.G. Barrera, S. Dindar, S. Kaewunruen, D. Ruikar, LOD BIM element specification for railway turnout systems risk mitigation using the information delivery manual, IOP Conf. Ser. Mater. Sci. Eng. 245 (4) (2017) 042022.
[5] C. Satish, Railway Engineering, first ed., Oxford University Press, London, 2007.
[6] FRA, Accident by State/Railroad, Federal Railroad Administration, 2016.
[7] S. Dindar, S. Kaewunruen, M. An, J.M. Sussman, Bayesian network-based probability analysis of train derailments caused by various extreme weather patterns on railway turnouts, Saf. Sci. 110 (2018) 20–30.
[8] S.L.B. Sazrul, S. Kaewunruen, D. Jaroszweski, S. Dindar, Operational risks of Malaysia-Singapore high speed rail infrastructure to extreme climate conditions, in: The 1st Asian Conference on Railway Infrastructure and Transportation, Jeju, Korea, 2016.
[9] S. Dindar, S. Kaewunruen, M. An, Rail accident analysis using large-scale investigations of train derailments on switches and crossings: comparing the performances of a novel stochastic mathematical prediction and various assumptions, Eng. Fail. Anal. 103 (2019) 203–216.
[10] M.F. Ishak, S. Dindar, S. Kaewunruen, Safety-based maintenance for geometry restoration of railway turnout systems in various operational environments, in: Proceedings of the 21st National Convention on Civil Engineering, 21st National Convention on Civil Engineering, Thailand, 2016, pp. 1–8.
[11] G.D. Chattopadhyay, D. Raman, M.R. Alam, A study of derailment in Australia: analysing risk gaps with remote data monitoring, in: Engineering Asset Management, Springer, London, 2014.
[12] S. Dindar, S. Kaewunruen, M. An, A hierarchical Bayesian-based model for hazard analysis of climate effect on failures of railway turnout components, Reliab. Eng. Syst. Saf. 218 (2022) 108130.

[13] S. Kaewunruen, A.M. Remennikov, S. Dindar, Influence of asymmetrical topology on structural behaviours of bearers and sleepers in turnout switches and crossings, in: Sustainable Civil Infrastructures: Innovative Infrastructure Geotechnology, Springer, 2017, pp. 51–60.
[14] S. Dindar, S. Kaewunruen, M. An, Bayesian network-based human error reliability assessment of derailments, Reliab. Eng. Syst. Saf. 197 (2020) 106825.
[15] S. Kaewunruen, A.M. Remennikov, S. Dindar, Influence of asymmetrical topology on structural behaviours of bearers and sleepers in turnout switches and crossings, in: J. Pombo, G. Jing (Eds.), International Congress and Exhibition "Sustainable Civil Infrastructures: Innovative Infrastructure Geotechnology", Springer, 2017, pp. 51–60.
[16] S. Dindar, S. Kaewunruen, M. An, A. Gigante-Barreara, Derailment-based fault tree analysis on risk management of railway turnout systems, IOP Conf. Ser. Mater. Sci. Eng. 245 (4) (2017) 042020.
[17] A. Mullai, Risk management system—a conceptual model, in: Supply Chain Risk, Springer, 2009, pp. 83–101.
[18] J. Rabatel, S. Bringay, P. Poncelet, SO_MAD: SensOr mining for anomaly detection in railway data Berlin, in: Industrial Conference on Data Mining, Springer, 2009, pp. 191–205.
[19] NetworkRail, Asset Management Strategy. Technical Report, Department for Transport, London, 2016.
[20] P. Rungskunroch, S. Dindar, S. Kaewunruen, Life cycle assessment of ground borne vibration mitigation strategies using subgrade stiffening, soft-filled barriers and open trenches, in: Proceedings of the 47th International Congress and Exposition on Noise Control Engineering, Chicago, IL, USA, 2018.
[21] S. Dindar, S. Kaewunruen, M. An, Identification of appropriate risk analysis techniques for railway turnout systems, J. Risk Res. 21 (8) (2018) 974–995.
[22] S. Bruni, D. Serdar, K. Sakdirat, Best practices on advanced condition monitoring of rail infrastructure systems, volume II, Front. Built Environ. 126 (2021) 748846.
[23] S. Dindar, S. Kaewunruen, M. An, Identification of appropriate risk analysis techniques for railway turnout systems, J. Risk Res. 21 (8) (2016) 974–995.
[24] NetworkRail, Asset Management Strategy. Technical Report, NR, London, 2014.
[25] S. Kaewunruen, S. Dindar, The effect of climate change on service life and cost investigation of rail turnouts with various mitigation methods, in: Inter-Noise and Noise-Con Congress and Conference Proceedings, Vol. 258, 2018, pp. 6091–6101. no. 1.

Railway bridge under increased traffic demands

17

Shervan Ataei* and Amin Miri
School of Railway Engineering, Iran University of Science and Technology, Tehran, Iran

17.1 Introduction

An important aspect in management of railway networks is to find suitable solutions for increasing the capacity of the network. Increasing the allowable axle load, or operational speeds are two of the methods used for this purpose. Doing so, however, requires assessment of infrastructures such as bridges, to make sure such structures are capable of withstanding increased forces. It is possible to determine the capacity of bridges and determine strengthening or retrofitting methods when bridges are not capable of performing under higher traffic loads.

Moreover, structural monitoring of bridges is of grave importance, which is highlighted by the collapse of Morandi bridge in Italy [1]. Structural monitoring allows engineers to have an in-depth understanding of the behavior of the structure to various loadings and acknowledge the safe operation of the bridge. This, however, requires an adequate understanding of the bridge structure and methods of analysis of that structure.

Among different types of bridges, masonry arch bridges are the oldest, and most complex of all to analyze. There are a number of issues associated with such bridges, the first being their remaining structural capacity. As the aim of the operators is to increase the axle load on the bridge, it is vital for the managers to have a clear understanding of the remaining structural capacity of the bridge. This can ensure safe and optimized operation of the bridges in a network.

There is also a lack of availability of plans of some of the bridges as most of them are relatively old, which makes the understanding of the accurate physical dimensions and material characteristics of the bridges a rather time and money-consuming task. There is the possibility of internal members (which in most cases are hidden inside the barrels) which with the lack of proper plans could be hard to detect with an external observation. These members are shown to have a significant effect on the load-carrying capacity of arch bridges.

A common approach for assessment of masonry arch bridges is dynamic load testing and using field measurements for calibration of numerical models [2]. To develop numerical models, masonry and geometrical characteristics are acquired from field experiments and measurement and used in finite element (FE) software. FE model

---

* Shervan Ataei: ORCID: 0000-0002-7436-5446.

is then calibrated using recorded response of bridge during field experiments. It is then possible to use simulations to certify whether the bridge can accommodate higher axle loads and operational speeds.

Iranian railway network has a large portion of masonry arch bridges which were built over 70 years ago. Most of these bridges are in good shape and have been maintained adequately over the years. However, new challenges have arisen over the years regarding the assessment of these bridges which are mainly due to a lack of proper knowledge of the structural characteristics of such structures. Moreover, the network operators are pushing due to higher demands for the rail sector. Railway network managers have therefore launched an extensive monitoring program in Iran with the main aim of improving the knowledge of masonry arch bridges.

17.1.1 Aims and scope of study

The aim of the research is to provide the technical details required for the assessment of masonry arch bridges in Iran. The outcomes of this research will help managers with decisions regarding future maintenance and upgrading of masonry arch bridges in the network. The main aims of the research program are:

- Selection of up to eleven masonry arch bridges form the network.
- Reviewing the available documents of selected bridges (plans and material testing).
- Visual field inspection of selected structures.
- Developing 3D finite element models based on the available documents and field investigation results.
- Selecting proper sensors and instrumentation strategy for all of the selected bridges.
- Field instrumentation and dynamic load testing of the selected bridges.
- Filtering the raw data and preliminary analysis of the results.
- Calibration and verification of developed FE models based on the results of field tests.
- Structural capacity assessment of the bridges incorporating the calibrated FE models.

Due to the large variety of masonry arch bridges in Iranian railway network and budget constraints for the research program, it was decided that up to 12 bridges had to be chosen as representative of the bridges in the network. It was, therefore, necessary to select the bridges such that as much diversity of physical and material characteristics as possible were included. The selected bridges are discussed in detail in Section 17.2.1.

The available documents for each of the selected bridges were then reviewed to get an overall knowledge of the structure at hand, including as-built plans, maintenance reports, and previous material tests of the bridge over the years. These were later complemented by results of visual inspections of the bridges in the field, which provided an opportunity to verify the accuracy of the plans, as well as fill any missing information for the initial modeling of the structures. Primitive FE model of each selected bridge was then developed based on the acquired data from the available documents as well as results of visual inspections.

It was necessary to choose the potential types of sensors and instrumentation plans for each selected bridge. Based on reviewing the available literature and technical requirements of the research program, two types of sensors were selected to be installed on each bridge, accelerometers, and extensometers. A complete description of sensors and instrumentation techniques is provided in Section 17.2.3.

Once the sensors had been selected, the field investigations were started and dynamic load tests were carried out on the selected bridges over a period of two years. Recorded measurements were then filtered and used to calibrate the developed FE models. A detailed report on calibration and verification of FE models could be found in published articles by the authors [3–11].

17.2 Monitoring program

17.2.1 An overview of selected bridges

Twelve masonry arch bridges from Iranian railway network were selected for analysis. All bridges were built more than 70 years ago and have been since then in operation. A view of each bridge is presented in Fig. 17.1. Overall, a total of 38 spans were instrumented with different sensors (some spans are instrumented in both faces of the span). Geometrical characteristics of tested spans for each bridge, along with superstructure characteristics are presented in Table 17.1. Geometrical characteristics are mainly determined based on archived as-built plans of bridges and in-site measurements. It should also be mentioned that all bridges are constructed from stone blocks, except for No. 1 and No. 2 bridges which are constructed from non-reinforced concrete blocks.

17.2.2 Test train formation

All of the selected bridges are currently in operation and accommodate mixed passenger and freight trains. As train formation might have an impact on the results, it was decided to consider various train formations in the dynamic load testing programs. Three 6-axle diesel locomotives (total weight of 120 tons) and five 4-axle freight wagons (total weight of 80 tons) are used to form the test train. Axle spacing and loads are presented in Fig. 17.2, schematically.

Overall, four different train formations were considered throughout the tests: 3D5W consisting of three locomotives and five freight wagons (nominal total weight of 760 tons), 3D consisting of three locomotives (nominal total weight of 360 tons), 2D consisting of two locomotives (nominal total weight of 240 tons), and 1D consisting of a single locomotive (nominal total weight of 120 tons). Also, tests were repeated with various crossing speeds to measure the impact of speed on results.

Test speeds were different for each bridge, as speed limitations were advised based on the topography and operational characteristics of each region. However, a speed range of 5–75 km/h was carried out for all of the tested bridges, with some bridges having higher speeds included in the tests. A total of 845 test runs were carried out on 38 spans of 12 selected bridges.

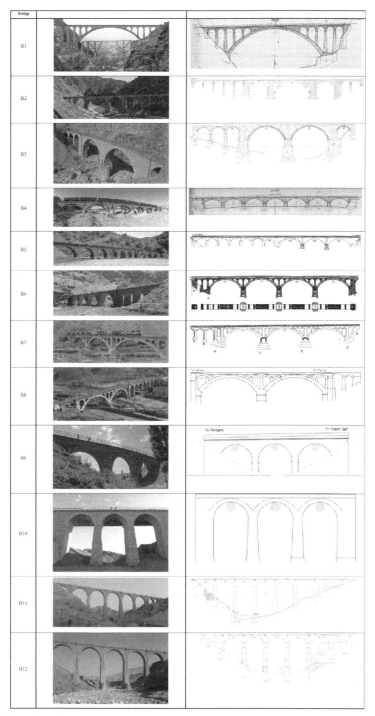

Fig. 17.1 A view of each bridge and its sensoring plan (*green dots* (gray in the print version) represent sensor locations). (For interpretation of the references to color in this figure legend, the reader is referred to the web version of this article.)

Table 17.1 Geometrical characteristics of selected bridges.

Bridge	Number and length of spans	Number and length of tested spans	Total length (m)	Span rise (m)	Rail	Ballast depth (cm)	Sleeper	Curved radius (m)
B1	1 × 66 m	1 × 66 m	75	18	U33	30	Steel	–
B2	7 × 10 m + 1 × 40 m + 2 × 21 m	1 × 40 m	200	10	U33	35	Steel	430, 300
B3	4 × 10 m + 2 × 30 m	2 × 10 m and 2 × 30 m	130	11.5	U33	35	Steel	300
B4	7 × 21 m	4 × 21 m	190	5.25	UIC60	30	Wood	–
B5	3 × 21.5 m + 7 × 10 m	1 × 10 m and 3 × 21 m	270	5	U33	35	Wood	–
B6	4 × 8 m + 4 × 25 m	1 × 8 m and 3 × 25 m	176	6	U33	35	Steel	–
B7	3 × 36 m + 7 × 8 m	2 × 36 m	270	11	U33	35	Steel	250, 250
B8	2 × 25 m + 7 × 8 m + 2 × 3.5 m	1 × 25 m	147	6.5	U33	35	Steel	250
B9	3 × 11 m	2 × 11 m	40	2.75	U33	35	Steel	250
B10	3 × 9.5 m	3 × 9.5 m	35	2.5	UIC60	35	Wood	–
B11	2 × 10 m + 1 × 20.68 m + 5 × 21.5 m	4 × 21.5 m	194	5.4	U33	30	Wood	–
B12	5 × 21.5 + 2 × 10 m	5 × 21.5 m	164	5.4	U33	30	Steel	–

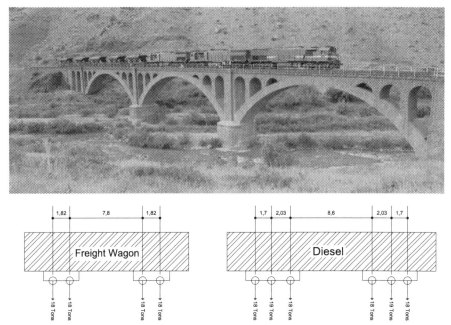

Fig. 17.2 (A) A photo of a test train on no. "7" bridge during the loading tests and (B) schematic characteristics of test train.

17.2.3 Sensors and instrumentation technique

Three types of sensors were employed in the test program, namely extensometers, accelerometers, and strain gauges.

17.2.3.1 Extensometers

An extensometer of "Deflected Cantilever Displacement Transducer" (DCDT for simplicity of reference) was used for the test program. This type of DCDT could measure displacements of up to 25 mm with an accuracy of 10 μm. As the aim was to measure the vertical displacement of a location on the bridge relative to the ground, the installment scheme presented in Fig. 17.3 was adopted.

In this installment method, DCDT was mounted on the desired location on the arch, and a heavyweight (a decommissioned steel or wooden sleeper) was placed exactly beneath it on the ground. A steel cable connected the hook of the DCDT to the weight. The cable was then manually pulled using the turnbuckles and DCDT output was monitored until a value of 12.5 mm (half the capacity of the sensor) was observed. The turnbuckle was then locked and DCDT output was reset to zero. As the cable was fully extended, any vertical movement in either upward or downward directions (up to 12.5 mm) was accurately measured by the DCDT. A sample record is presented in Fig. 17.4.

Railway bridge under increased traffic demands 361

(a)

(b)

Fig. 17.3 (A) DCDTsensor mounted in the middle of the span and (B) DCDT's cable fixed to the decommissioned steel sleeper placed beneath the span.

Extensometers were installed to measure the vertical displacement of keystones in arches. Recorded vertical displacements in various locations on the arch were then used for the purpose of FE model calibration and verification.

17.2.3.2 Accelerometers

Piezoresistive accelerometers with a capacity of 2 and 5 g were used throughout the test program. As the accelerometers were mono-directional, two were mounted, one in vertical and one in lateral directions, to measure the bridge accelerations in both directions. A photo of installed accelerometers is presented in Fig. 17.5. A sample record is also given in Fig. 17.6.

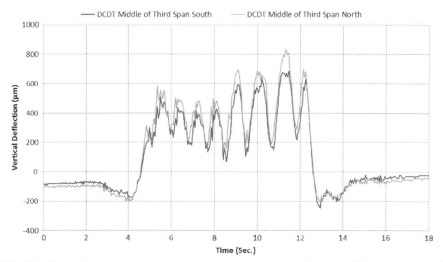

Fig. 17.4 Vertical deflection signature of northern and southern sides of middle span of one of the tested bridges, as a 3D5W train crossed the bridge with a speed of 63 km/h.

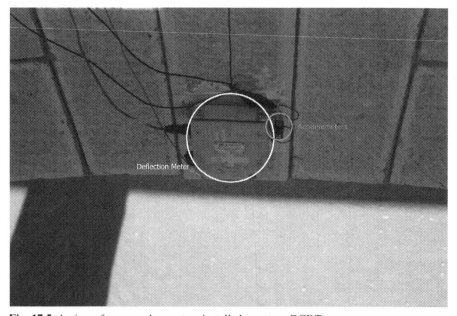

Fig. 17.5 A view of two accelerometers installed next to a DCDT sensor.

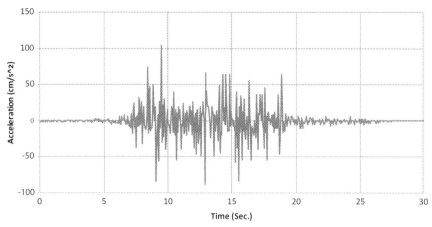

Fig. 17.6 A sample acceleration measurement for the middle span of one of the tested bridges, as a 3D5W train crossed the bridge with a speed of 46 km/h.

Free vibration segments of acceleration measurements were then used to extract the natural frequencies of the bridge. Also, as the excessive vibrations transferred from the track to the bridge structure could be detrimental, mitigation strategies were of high interest. For this reason, a more in-depth study of the recorded acceleration signatures was carried out to detect various frequency contents, which will be discussed in Section 17.4.2.

17.3 Processing data

Once the sensors were mounted on each bridge, dynamic load tests were carried out. Throughout the tests, a constant measuring frequency of 2 kHz was adopted. A number of test runs were completed for each bridge and raw data was recorded. Moving average and zero shifting were used to preprocess the raw data.

Moving average filter is a fast and practical tool to filter out any extreme noisy peaks in the records. This filter is easy to apply as it already is incorporated in many data processing packages. The range of moving average for acceleration and displacement data was set to 10 and 100 samples, respectively.

With the start of each test run, all the sensors were reset to zero and the measurements started. However, in some cases, some offset was observed in the initial (before train reaching the bridge) segment of the records. To rectify this error, the average value of the initial segment was calculated and reduced from all measured values, which is known as zero shifting.

The processed data could then be used for extensive analysis. Displacement measurements were used to extract the maximum displacement of the location on the bridge where the sensor was installed for various train formations or crossing speeds. The maximum measured displacement was also employed for calculating dynamic amplification factor of the span, which will be presented in Section 17.4.1.

However, for acceleration signatures, root mean square (RMS) was used as an index for analyzing acceleration signatures in various frequency bands. This statistical tool allows for studying a large stream of numbers by extracting a single representative value. This is especially useful for comparing acceleration values for various test runs with different train formations or crossing speeds.

To do so, first the PSD of the signals was computed. In almost all standard textbooks related to the signal analysis [12–14], it is mentioned that the double-sided auto-spectral density, PSD, denoted $S_{xx}(f)$, is the forward Fourier transform of the autocorrelation function ($R_{xx}(\tau)$).

$$S_{xx}(f) = \int_{-\infty}^{\infty} R_{xx}(\tau) \, e^{-j2\pi f \tau} d\tau. \tag{17.1}$$

The single-sided spectral density is denoted by $G_{xx}(f)$ [15].

$$G_{xx}(f) = 2 \, S_{xx}(f) \, for \, f > 0 \tag{17.2}$$

The interpretation of it indicates that sum of the area under the PSD in a specific frequency range is the mean square value of the signal. Consequently, it can be derived that:

$$x_{rms} = \sqrt{\int G_{xx}(f) df} \tag{17.3}$$

As mentioned, built on the processing procedure performed in the frequency domain, the root mean square (RMS) values were estimated in different frequency bands. All acceleration data gathered for the analysis and the RMS computation were accomplished in different frequency bands, as indicated in Fig. 17.7.

17.4 Analysis of results

17.4.1 Calculating DAF value

One of the primary objectives of the research program was to study the dynamic amplification factor (DAF) of masonry arch bridges. DAF is one of the main parameters in structural assessment of bridges, which is well researched for road bridges, RC, and steel railway bridges [16–22]. Yet DAF of masonry arch railway bridges is rather scarce. The aim of this part is to have an in-depth analysis of the results of the field tests to get a better image of the DAF of masonry arch railway bridges.

DAF is defined as follows:

$$R_{\text{Dynamic}} = (1 + DAF) \times R_{\text{Static}} \tag{17.4}$$

In which R_{Dynamic} is the dynamic response of the bridge and R_{Static} is the static response of the bridge. The response of the bridge may be calculated based on

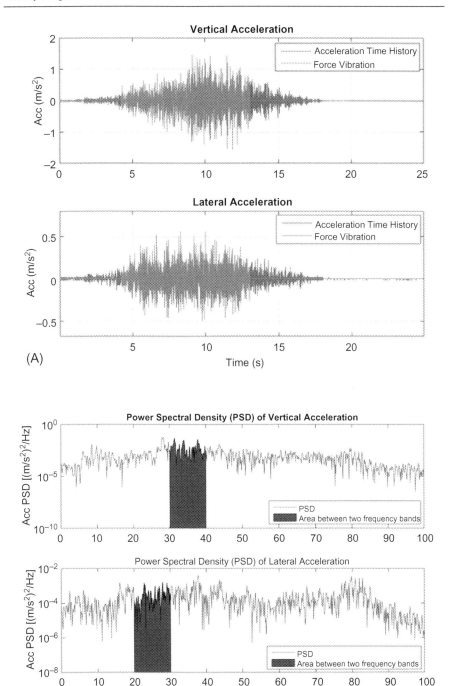

Fig. 17.7 The sample of (A) acceleration time history and forced vibration segment and (B) related PSD (with area between two frequency bands).

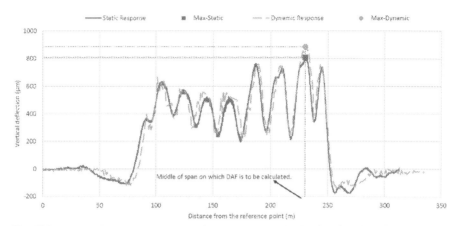

Fig. 17.8 Static and dynamic responses of a tested span, and selecting the values for calculating the DAF (static test crossing speed was 4 km/h, and dynamic test crossing speed was 65 km/h).

deflection or strain measurements. However, in this study, the vertical displacement of tested spans was used for this purpose, as shown by a sample record in Fig. 17.8.

17.4.1.1 Statistical approach for selecting a DAF value for a bridge

In some of the tested bridges, more than one span was instrumented. Also, for each train run, one DAF value was calculated for each instrumented location. This means that for one bridge, more than one DAF value was available. As these values were not exactly equal, a statistical approach was considered to select one representative DAF value for the bridge. Cumulative density function of all DAF values for one bridge was plotted, and DAF value corresponding to a pre-defined cumulative density function (CDF) level was considered as the DAF value for the bridge.

In probability theory, the CDF of a real-valued random variable X evaluated at x is the probability that X will take a value less than or equal to x. In mathematical terms:

$$F_X(x) = P(X \leq x) \tag{17.5}$$

In other words, CDF is the analysis of the frequency of occurrence of values of a phenomenon less than a reference value. CDF level could be determined based on the degree of importance of the bridge. A CDF level of 90% was considered throughout the project. A sample of the statistical procedure is shown in Fig. 17.9, and final results are presented in Table 17.2.

17.4.1.2 Correlation of DAF and structural parameters of the bridge

The relation between the DAF value and structural parameters of selected bridges and operational characteristics of the trains were of great interest. To find the extent of correlation between the two parameters, Pearson linear correlation coefficient was used.

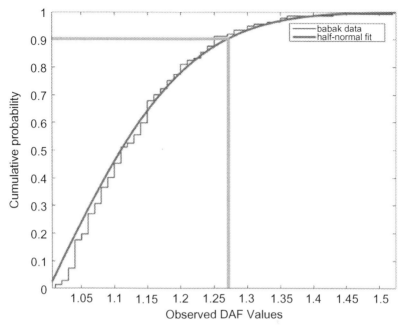

Fig. 17.9 Determining DAF value corresponding to a CDF level of 90% (five bridges).

Correlation coefficient (CC) is a linear statistical value used to determine whether two variables are related, and the extent of relation. For two sets of data (x and y), correlation coefficient (ρ) is determined using Eq. (17.6). CC is between -1 and 1, and the closer the value to 1 or -1, the stronger the correlation between the two parameters. In this respect, it is a great indicator to see whether DAF values are influenced by any structural characteristics of the bridge or operational characteristics of the trains.

Table 17.2 Corresponding DAF values for a CDF level of 90%.

Bridge	DAF
B1	1.27
B2	1.37
B3	1.53
B4	1.22
B5	1.27
B6	1.26
B7	1.17
B8	1.15
B9	1.15
B10	1.51
B11	1.41
B12	1.39

$$\rho = \frac{R_{xy}}{\sqrt{R_{xx} R_{yy}}} \tag{17.6}$$

where

$$R_{xx} = \sum x^2 - \frac{\left(\sum x\right)^2}{n} \tag{17.7}$$

$$R_{yy} = \sum y^2 - \frac{\left(\sum y\right)^2}{n} \tag{17.8}$$

$$R_{xy} = \sum xy - \frac{\left(\sum x\right)\left(\sum y\right)}{n} \tag{17.9}$$

17.4.1.3 Correlation of DAF and train formation

Fig. 17.10 shows experimental DAFs for each train formation for all tested spans (average values are determined for each speed for better visibility). No evident correlation is present between DAF and trains formation or crossing speed. CCs between DAF and speed for 3D5W, 3D, 2D, and 1D are −0.04, 0.11, 0.09, and 0.06, respectively. It is however observed that a majority of DAF values are smaller than 1.2.

Histogram of experimental DAF values for each train formation is presented in Fig. 17.11. For the heaviest train formation, namely 3D5W, almost 70% of all DAF values are less than 1.2, about 25% are in the range of 1.2–1.4, and roughly 5% are larger than 1.4. Overall, less than 10% of all experimental DAF values are over 1.4. Based on the observations of Figs. 17.10 and 17.11, it was concluded that the

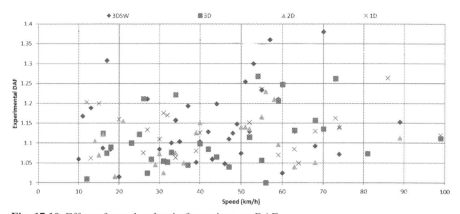

Fig. 17.10 Effect of speed and train formations on DAF.

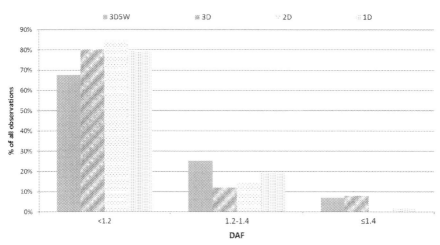

Fig. 17.11 Histogram of experimental DAF values for each train formation.

operational characteristics of train including its formation and crossing speed have a marginal impact on DAF values.

17.4.1.4 Correlation of DAF and span length

Fig. 17.12 presents the span length and DAF values. As evident, no perceptible trend could be traced between the two. Maximum, average, and minimum DAF values remain almost constant for various span lengths. The CC between the two is 0.09.

Euro code, AREMA, and ORE proposed equations for DAF consider span length as an input parameter, as presented in Fig. 17.12. However, according to results of this research, the proposed values of these guidelines are overestimated, especially for smaller spans of less than 20 m.

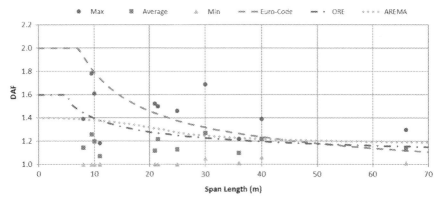

Fig. 17.12 Correlation of span length and DAF (CC = 0.09).

17.4.1.5 Correlation of DAF and rise/span ratio

Fig. 17.13 investigates the effects of rise/Span ratio on average DAF values. Rise to span ratio is a representative of the physical shape of the arch, and is less variant compared to span lengths. For each ratio, a large variety of DAF values are available as evident from Fig. 17.13. The CC value between these two parameters is 0.41, which is higher than that between the DAF and span length, which signifies that the rise to span length has a higher correlation with DAF than span length. Yet a CC value of 0.41 is still relatively small and it cannot be argued that rise to span ratio is a significant affecting parameter for DAF.

17.4.1.6 Correlation of DAF and natural frequency of the bridge

Natural frequencies are among the most important characteristic parameters of any structure. To get the modal parameters from recorded data, pick peaking method (abbreviated PP) was employed. PP method performs an approximate decomposition of the response signature into a set of single degrees of freedom systems, each representing a single mode. Due to its simplicity and speed, it is used for determination of modal characteristics of ambient vibration measurements. A sample of recorded ambient vibration of B4 is presented in Fig. 17.14.

Table 17.3 presents the experimental natural frequencies of tested bridges, along with the thresholds for the first natural frequencies based on UIC 778-3 [2]. As evident, the first vertical natural frequency of B9 is significantly lower than the proposed value of UIC 778-3. Also, the first vertical natural frequency of B4 is slightly higher than proposed value of UIC 778-3.

Correlation of DAF and first natural frequencies in lateral and vertical directions are presented in Fig. 17.15. According to Fig. 17.15, most natural frequencies are in the range of 2–6 Hz, while average DAF values vary between 1.05 and 1.35. No evident trend is detected between the natural frequencies and experimental DAF values.

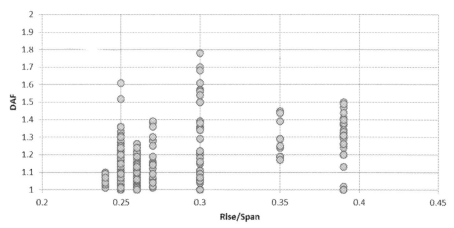

Fig. 17.13 Correlation of rise/span ratio and DAF (CC = 0.41).

Railway bridge under increased traffic demands

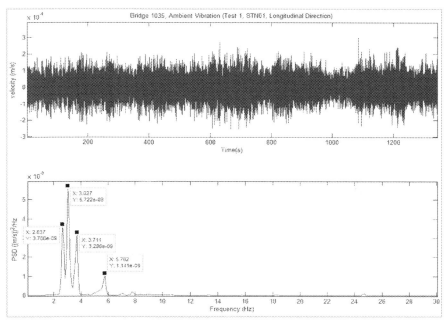

Fig. 17.14 A sample of ambient vibrations (*top*) and application of PP method to determine natural frequencies of the structure (*bottom*).

Table 17.3 Experimental natural frequencies of the tested bridges along with proposed values of UIC 778-3 [2].

Bridge name	First vertical natural frequency (Hz)			First lateral natural frequency (Hz)
	Experimental	UIC-lower limit	UIC-upper limit	
B3	3.5	3.15	7.44	2.5
B4	10.25	4	10.1	6.27
B5	8.8	3.8	9.55	3.9
B6	4.5	3.5	8.5	3
B7	5.9	2.8	6.5	1.95
B9	4	7.3	15.76	3
B10	9	8	16.93	3.2
B11	4	3.8	9.55	1.5
B12	4	3.8	9.55	1.5

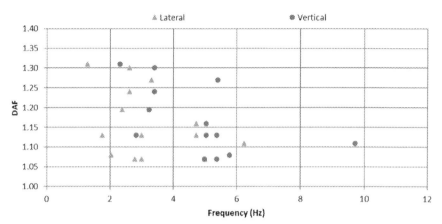

Fig. 17.15 Correlation of first lateral and vertical natural frequencies and average DAF values (CC of lateral = 0.37; CC of vertical = 0.53).

17.4.1.7 Correlation of DAF and combined modulus of elasticity of mortar and masonry

To study the correlation between DAF and material characteristics of the bridge, combined modulus of elasticity of mortar and masonry is used. This value is taken from the calibrated FE models, as the modulus of elasticity of masonry was used as the calibration parameter. Fig. 17.16 shows the correlation of calibrated modulus of elasticity with DAF values for each span. As could be seen, there is a moderate correlation between the two, and the CC is 0.56. For a detailed description of calibration of modulus of elasticity, reader is referred to Refs. [3–12].

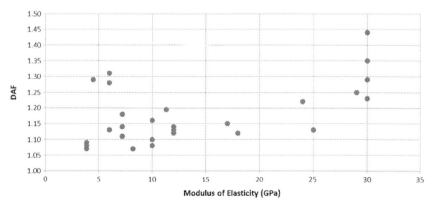

Fig. 17.16 Correlation of modulus of elasticity of masonry and experimental DAF values (CC = 0.56).

Railway bridge under increased traffic demands

17.4.1.8 Commentary on results

Fig. 17.17 and Table 17.4 allow for a comparison between the experimental DAF values with those proposed in some of the dominant guidelines. Root mean square deviation is also used to determine the difference between the two, which is as follows:

$$\text{RMSD} = \sqrt{\frac{1}{N}\sum_{i=1}^{N}\delta_i^2} \qquad (17.10)$$

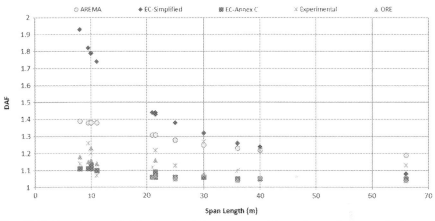

Fig. 17.17 Experimental DAF values and those proposed by various railway standards.

Table 17.4 DAF values of tested spans according to various standards and experimental results.

Bridge	Span length (m)	Experimental	EC-simplified	EC-annex C	ORE	AREMA
B1	66	1.27	1.08	1.05	1.04	1.19
B2	40	1.37	1.24	1.05	1.06	1.22
B3	10	1.53	1.79	1.13	1.23	1.38
B3	30	1.53	1.32	1.06	1.08	1.25
B4	21	1.22	1.44	1.06	1.06	1.31
B5	10	1.27	1.79	1.11	1.16	1.38
B5	21.5	1.27	1.44	1.06	1.07	1.31
B6	8	1.26	1.93	1.11	1.18	1.39
B6	25	1.26	1.38	1.06	1.06	1.28
B7	36	1.17	1.26	1.05	1.04	1.23
B8	25	1.15	1.38	1.06	1.05	1.28
B9	11	1.15	1.74	1.1	1.14	1.38
B10	9.5	1.51	1.82	1.11	1.15	1.38
B11	21.5	1.41	1.43	1.09	1.16	1.31
B12	21.5	1.39	1.43	1.09	1.16	1.31

In which N is the number of points being compared, and δ is the distance between observation i and the reference value, which in this case are guideline and experimental values, respectively. This statistical tool measures the average distance between two given sets of data. The RMSD values are presented in Table 17.5. The RMSD is also categorized for three types of spans to have a better look at the results, namely small spans (<15 m), medium spans (between 15 and 30 m), and large spans (larger than 30 m).

As Table 17.5 suggests, the worst results correspond to simplified method of EC, as the values based on this method are significantly overestimated. This is especially true for small and medium spans, whereas values for large spans are relatively more accurate. DAF values proposed by detailed EC method and ORE are relatively close to experimental values, yet slightly underestimated.

17.4.2 Frequency domain-based statistical analysis of acceleration records

As far as the whole data is concerned, the dominant acceleration RMS belongs to the 30–40 Hz frequency band in vertical and 20–30 Hz in lateral direction [23]. For both vertical and lateral accelerations, however, the frequency band of 20–60 Hz has the highest contents, as presented in Fig. 17.18.

17.4.2.1 Effect of train formation on transmitted accelerations to the bridge

Fig. 17.19 shows the average of RMS of different frequency bands for each train formation in both vertical and lateral directions.

According to Fig. 17.19, the effect of train formation on singular frequency bands is rather limited, yet is more perceptible for the frequency band of 0–100 Hz. In vertical direction, 3D train formation results in the highest transmitted frequency contents to the bridge, being 10.9%, 3.6%, and 22.8% higher than 1D, 2D, and 3D5W train formations, respectively.

This is due to the fact that the locomotives induce the highest vertical accelerations to the track; therefore, the higher the number of locomotives, the higher the

Table 17.5 RMSDs between experimental DAF values and those proposed by various railway standards.

Standard	EC-simplified	EC-annex C	ORE	AREMA
Whole range	0.42	0.11	0.10	0.16
Small span (<15 m)	0.65	0.09	0.06	0.22
Middle span (15–30 m)	0.25	0.11	0.09	0.14
Long span (>30 m)	0.09	0.14	0.14	0.07

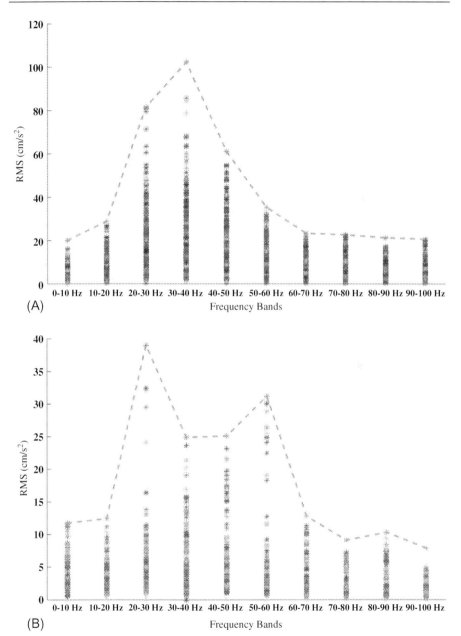

Fig. 17.18 Acceleration RMS and the push-up in different frequency bands in (A) vertical and (B) lateral directions.

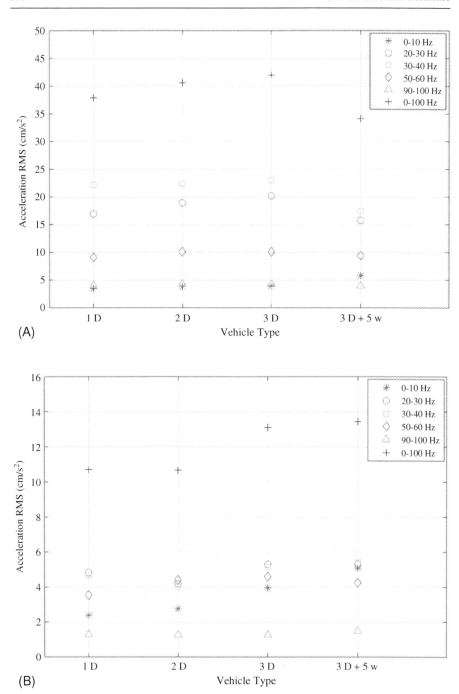

Fig. 17.19 Effect of train formation on average of RMS of different frequency bands in (A) vertical and (B) lateral directions.

transmitted vertical accelerations. On the other hand, wagons induce relatively lower vertical accelerations compared to locomotives. So, the 3D5W vertical acceleration signal consists of a short segment of high vertical accelerations (due to three locomotives) and a long segment of low vertical accelerations (due to five wagons). Hence the overall frequency content of transmitted vertical accelerations on the bridge for 3D5W train formation is relatively lower compared to the case of train formations consisting of just locomotives (1D, 2D, and 3D).

Yet in the lateral direction, a rather flat pattern is observable, which indicates that train formation has minimal effect on frequency contents of transmitted lateral acceleration to the bridge. When the frequency band of 0–100 Hz is considered, it can be observed that the longer the train formation, the higher the frequency content of transmitted lateral acceleration to the bridge. This indicates that the induced lateral accelerations caused by crossing of locomotives and wagons are almost similar for the tested bridges.

17.4.2.2 Effect of train crossing speed on transmitted accelerations to the bridge

The average push-up of speed in different frequency bands was determined and is presented in Fig. 17.20. It is worth mentioning that only the train crossing speed was considered for calculations and train formation was disregarded.

Two frequency bands of 20–30 Hz and 30–40 Hz have relatively higher contents compared to other frequency bands (90–100 Hz is presented as a sample). According to Fig. 17.20, highest frequency contents are observed for a crossing speed of 52 km/h, although the effect is observed in different frequency bands for vertical and lateral directions. Peak frequency contents for vertical and lateral directions are observed for 20–30 and 30–40 Hz frequency bands, respectively. The correlation of train crossing speed and RMS of various frequency bands for vertical and lateral directions are presented in Table 17.6.

17.4.2.3 Effect of span length on transmitted accelerations to the bridge

The recorded acceleration values are assorted based on span length, and RMS of acceleration values are calculated for frequency band of 0–100 Hz, which is shown in Fig. 17.21. As observed in this figure, no trend can be traced between the span length and RMS of transmitted accelerations to the bridge.

Fig. 17.22 presents the results for each frequency band. As evident, no perceptible trend between span length and frequency contents can be found. CCs between span length and RMS of vertical and lateral frequencies are calculated and presented in Table 17.7, which shows that no strong correlation is present between the two sets of parameters.

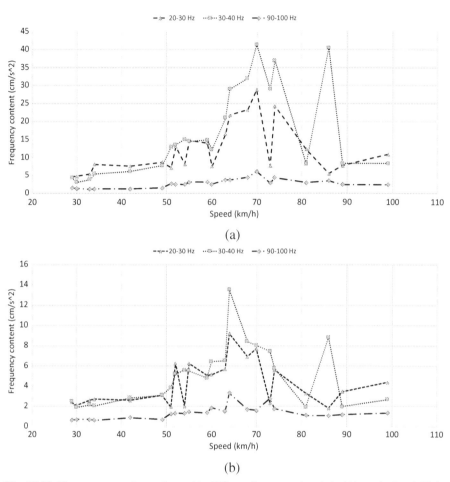

Fig. 17.20 The average push-up of speed in different frequency bands in (A) vertical and (B) in lateral direction.

17.4.2.4 Effect of bridge's natural frequency on transmitted accelerations to the bridge

Effect of first natural frequency in vertical and lateral direction on frequency contents in various bands is presented in Figs. 17.23 and 17.24. As observed, no clear trend exists between bridge's natural frequencies and RMS of various frequency bands. CCs between first vertical natural frequencies and RMS of vertical acceleration for various frequency bands are presented in Table 17.8, and those for first lateral natural frequency are presented in Table 17.9.

Table 17.6 CC between train crossing speed and various frequency bands in vertical and lateral directions.

Frequency band (Hz)	0–10	10–20	20–30	30–40	40–50	50–60	60–70	70–80	80–90	90–100	0–100
Vertical	0.21	0.30	0.13	0.24	0.23	0.19	0.17	0.17	0.22	0.16	0.17
Lateral	0.05	0.13	0.07	0.10	0.13	0.07	0.13	0.18	0.22	0.16	0.06

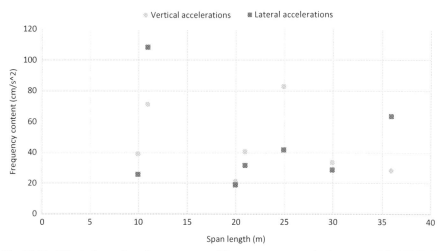

Fig. 17.21 Effect of span length on average of frequency contents in the range of 0–100 Hz in vertical and lateral directions.

17.5 Conclusion

Results of a comprehensive test program on masonry arch bridges in Iranian railway network were reported here. The main goal of the test program was to enhance the knowledge of structural performance of masonry arch bridges through dynamic load tests.

A total of 12 bridges with different characteristics were chosen, allowing for an in-depth analysis of the effects of various parameters on the structural behavior of masonry arch bridges. Moreover, four train formations and various crossing speeds were considered throughout the tests. Two types of sensors, including extensometers and accelerometers, were used to measure the response of the bridge to crossing trains. Recorded data were used to calculate the dynamic amplification factor of the bridges, as well as study the frequency contents of the transmitted accelerations to the tested bridges.

Based on the results, no significant correlation was observed between train formation and crossing speed, span length, span to rise ratio, first natural frequencies of the bridge, and updated modulus of elasticity of mortar and masonry (based on the FE model) and experimental DAF values. On the other hand, it was observed that 95% of DAF values were less than 1.4. Also, it was observed that among some of the dominant guidelines, the values proposed by Euro code (detailed method of appendix 3) and ORE were closest to the experimental results.

Frequency contents of transmitted vertical and lateral accelerations to the bridge were studied by calculating RMS of acceleration for ten frequency bands with a range of 10 Hz, covering a whole frequency range of 0–100 Hz. It was concluded that frequency content for various frequency bands was not identical, but rather significantly different.

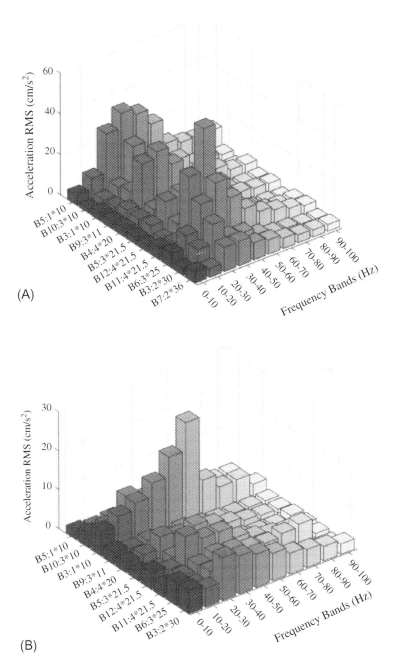

Fig. 17.22 The effect of span length on frequency contents in (A) vertical and (B) lateral directions for various frequency bands.

Table 17.7 CC between train span length and various frequency bands in vertical and lateral directions.

Frequency band (Hz)	0–10	10–20	20–30	30–40	40–50	50–60	60–70	70–80	80–90	90–100	0–100
Vertical	0.24	0.09	−0.11	−0.22	−0.25	−0.16	−0.20	−0.25	−0.18	−0.20	−0.19
Lateral	0.25	−0.08	0.14	0.00	−0.22	−0.29	−0.18	−0.05	−0.08	−0.15	−0.09

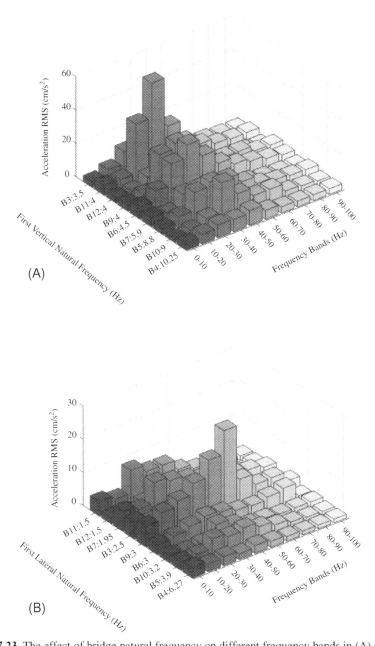

Fig. 17.23 The effect of bridge natural frequency on different frequency bands in (A) vertical and (B) lateral directions.

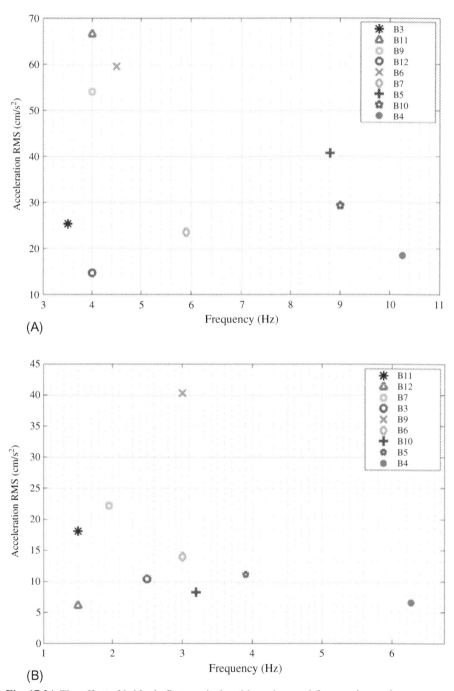

Fig. 17.24 The effect of bridge's first vertical and lateral natural frequencies on frequency contents in the range of 0–100 Hz in (A) vertical and (B) lateral directions (average values in the whole frequency range are presented).

Table 17.8 CC between first vertical natural frequency and various frequency bands in vertical direction.

Frequency band (Hz)	0–10	10–20	20–30	30–40	40–50	50–60	60–70	70–80	80–90	90–100	0–100
Vertical	−0.16	−0.17	−0.33	−0.23	−0.13	−0.25	−0.14	−0.17	−0.02	0.03	−0.26

Table 17.9 CC between first lateral natural frequency and various frequency bands in lateral direction.

Frequency band (Hz)	0–10	10–20	20–30	30–40	40–50	50–60	60–70	70–80	80–90	90–100	0–100
Lateral	−0.16	−0.16	−0.29	−0.38	−0.25	−0.15	−0.18	−0.16	−0.10	−0.13	−0.27

The correlation of RMS of acceleration values with structural parameters of the bridge and crossing trains was calculated as well. It was observed that two frequency bands of 20–30 Hz and 30–40 Hz had higher contents relative to other frequency bands. Moreover, it was observed that a crossing speed of 52 km/h resulted in high-frequency contents, specifically in the abovementioned ranges.

References

[1] G. Michele Calvi, M. Moratti, G.J. O'Reilly, N. Scattarreggia, R. Monteiro, D. Malomo, P. Martino Calvi, R. Pinho, Once upon a time in Italy: the tale of the Morandi bridge, Struct. Eng. Int. 29 (2) (2019) 198–217.
[2] UIC (Union Internationale des Chemins), Recommendations for the Inspection, Assessment and Maintenance of Masonry Arch Bridges, 2011. UIC 778-3, Paris.
[3] S. Ataei, M. Tajalli, A. Miri, Assessment of load carrying capacity and fatigue life expectancy of a monumental masonry arch bridge by field load testing: a case study of Veresk, Struct. Eng. Mech. 59 (4) (2016) 703–718.
[4] S. Ataei, A. Miri, M. Tajalli, Dynamic load testing of a railway masonry arch bridge: a case study of Babak Bridge, Sci. Iran. 24 (4) (2017) 1834–1842.
[5] S. Ataei, A. Miri, Investigating dynamic amplification factor of railway masonry arch bridges through dynamic load tests, Constr. Build. Mater. 183 (2018) 693–705.
[6] S. Ataei, A. Miri, M. Jahangiri, Assessment of load carrying capacity enhancement of an open spandrel masonry arch bridge by dynamic load testing, Int. J. Archit. Herit. 11 (8) (2017) 1086–1100.
[7] S. Mohammadzadeh, A. Miri, M. Nouri, Enhancing the structural performance of masonry arch bridges with ballast mats, ASCE J. Perform. Constr. Facil. 31 (5) (2017), 04017089.
[8] S. Mohammadzadeh, A. Miri, Assessing ballast cleaning as a rehabilitation method for railway masonry arch bridges by dynamic load tests, Proc. Inst. Mech. Eng. F J. Rail Rapid Transit 232 (4) (2017) 1135–1148.
[9] S. Mohammadzadeh, A. Miri, Ballast cleaning as a solution for controlling increased bridge vibrations due to higher operational speeds, J. Perform. Constr. Facil. 32 (5) (2018), 04018064.
[10] S. Ataei, A. Miri, M. Tajalli, Implementing relative deflection of adjacent blocks in model calibration of masonry arch bridges, J. Perform. Constr. Facil. 32 (4) (2018), https://doi.org/10.1061/(ASCE)CF.1943-5509.0001171.
[11] S. Ataei, A. Miri, M. Jahangiri, Assessing safety of a railway stone arch bridge by experimental and numerical analyses, J. Croat. Assoc. Civ. Eng. 69 (11) (2017) 1017–1029.
[12] J. Bendatm, A.G. Piersol, Random Data: Analysis and Measurement Procedures, fourth ed., Wiley Interscience, 2010.
[13] D.E. Newland, An Introduction to Random Vibrations, Spectral, and Wavelet Analysis, third ed., Dover Publications Inc., 2005.
[14] P.H. Wirsching, T.L. Paez, H. Ortiz, Random Vibrations: Theory and Practice, Wiley Interscience, 1995.
[15] A. Brandt, Noise and Vibration Analysis, Signal Analysis and Experimental Procedure, Wiley, 2011.
[16] Y. Zhang, C.S. Cai, X. Shi, C. Wang, Vehicle-induced dynamic performance of FRP versus concrete slab bridge, J. Bridg. Eng. 11 (2006) 410–419.
[17] L. Deng, C.S. Cai, Development of dynamic impact factor for performance evaluation of existing multi-girder concrete bridges, Eng. Struct. 32 (1) (2010) 21–31.

[18] H. Li, Dynamic Response of Highway Bridges Subjected to Heavy Vehicles, PhD thesis, Department of Civil and Environmental Engineering, Florida State University, 2005.
[19] L. Deng, Y. Yu, Q. Zou, C. Cai, State-of-the-art review of dynamic impact factors of highway bridges, J. Bridg. Eng. (2014), https://doi.org/10.1061/(ASCE)BE.1943-5592.0000672.
[20] D. Huang, Vehicle-induced vibration of steel deck arch bridges and analytical methodology, J. Bridg. Eng. (2012), https://doi.org/10.1061/(ASCE)BE.1943-5592.0000243.
[21] M. Samaan, J.B. Kennedy, K. Sennah, Impact factors for curved continuous composite multiple-box girder bridges, J. Bridg. Eng. (2007), https://doi.org/10.1061/(ASCE)1084-0702(2007)12:1(80).
[22] O. Hag-Elsafi, W.F. Albers, S. Alampalli, Dynamic analysis of the Bentley Creek Bridge with FRP deck, J. Bridg. Eng. (2012), https://doi.org/10.1061/(ASCE)BE.1943-5592.0000244.
[23] Iranian Railways Technical Report. Analysis of Dynamic Load Tests on 12 Masonry Arch Bridges, Tehran, 2015 (in Persian).

Structural health monitoring strategy for damage detection in railway bridges using traffic induced dynamic responses

Andreia Meixedo[a], Diogo Ribeiro[b], João Santos[c], Rui Calçada[a], and Michael Todd[d]
[a]CONSTRUCT—LESE, Faculty of Engineering (FEUP), University of Porto, Porto, Portugal,
[b]CONSTRUCT-LESE, School of Engineering, Polytechnic of Porto, Porto, Portugal,
[c]LNEC, National Laboratory for Civil Engineering, Lisbon, Portugal, [d]Department of Structural Engineering, University California San Diego, San Diego, CA, United States

18.1 Introduction

The high dependency of modern societies on structural and mechanical systems leads to an active field of research that aims to reduce the costs of visual inspection and maintenance. Particularly, the maintenance of bridges is central to the structural integrity and cost-effectiveness of any transportation system [1]. However, in the case of railway infrastructures, their intense use by frequent and heavy traffic makes the task of detection and possible repair of damaged sections problematic. Moreover, many of these infrastructures are currently nearing the end of their life cycle.

Because these systems cannot be economically replaced, techniques for damage detection that make use of structural monitoring in real-time, are being developed and implemented so that these infrastructures can continue to be safely used if their operation is extended beyond the design basis service life. These circumstances demand that the onset of damage in new systems can be detected at the earliest possible time to prevent failures that can have serious life safety and economic consequences [2].

Structural Health Monitoring (SHM) represents a promising strategy in this ongoing challenge to achieve sustainable infrastructure since it has the potential to identify a structural change before it becomes critical. SHM for damage detection involves the collection of reliable data on the baseline condition of a bridge, the observation of its evolution over time, and the characterization of the degradation. By permanently installing a number and variety of sensors, which continuously measure structural responses, it is possible to obtain a real-time representation of the structure's current state. However, this information is only useful by assuring that reliable SHM systems, methods for data analysis, and statistics tools are put into practice.

In this context, the present research work aims at developing and validating an SHM strategy for damage detection in railway bridges using traffic-induced dynamic responses. To achieve this goal, an unsupervised data-driven strategy is implemented, consisting of multivariate statistical techniques. The signals resulting from train crossings correspond to a large mass traveling at significant speeds, thus generating features that can obscure information associated with damage. The set of techniques implemented herein allows removing all the train-related features to expose, with high sensitivity, those generated by damage. The effectiveness of the proposed methodology is validated in a long-span steel-concrete composite bowstring arch railway bridge tuned with a permanent structural monitoring system. An experimentally validated finite element model was used, along with experimental values of temperature, noise, and train loadings and speeds, to realistically simulate baseline and damage scenarios.

After this introduction, in Section 18.2, a literature review on SHM for damage detection is conducted. Section 18.3 presents the case study, detailing the monitoring system installed and the simulation of different structural conditions. In Section 18.4, the strategy for damage detection is implemented and validated. Finally, Section 18.5 presents the main conclusions drawn from this research work.

18.2 Literature review on SHM for damage detection

The assessment of damage usually requires a comparison between two states and, consequently, each SHM approach requires a baseline system. How the training set is defined will depend on which level of damage detection is aimed and, therefore, the data set can be established based on normal conditions only or a combination of normal and damaged conditions of the structure. Hence, SHM for damage detection can be seen as a four-step process (Fig. 18.1): (i) operational evaluation, (ii) data acquisition, (iii) feature extraction and (iv) feature discrimination.

The first step to developing an SHM strategy is to perform an operational evaluation. This phase attempts to provide answers to four questions, which are mentioned in Fig. 18.1, regarding the implementation of a damage detection investigation [3,4]. By providing answers to these questions, the operational evaluation process begins to set limitations on what will be monitored and how the monitoring will be accomplished.

Obtaining accurate measurements of a system's dynamic response is essential to SHM. There are many different sensors and data acquisition systems that can be applied to the SHM problem and the one employed will be application-specific. In this sense, Fig. 18.1 details the several considerations that one should make during the data acquisition step.

Identifying features that can accurately distinguish a damaged structure from an undamaged one is the focus of most SHM technical literature. Fig. 18.1 summarizes the main ideas that support the third step of the process, i.e., the feature extraction.

Fundamentally, feature extraction refers to the process of transforming the measured data into some alternative form where the correlation with the damage is more readily observed [5]. Often in SHM, the feature extraction process is based on fitting some model, either physics-based or data-based, to the measured response data. The parameters of these models, quantities derived from the parameters, or the

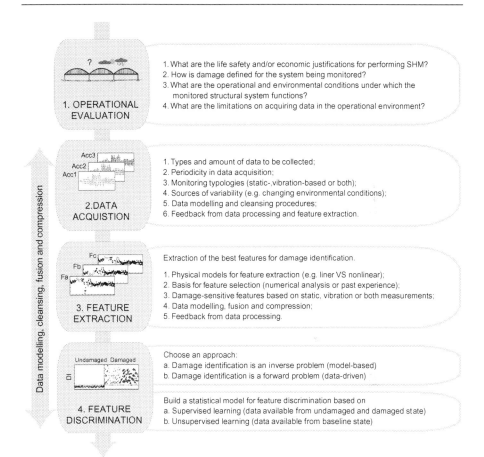

Fig. 18.1 SHM for damage detection as a four-step process.

predictive errors associated with these models, become then the damage-sensitive features [6].

Modal or model-based features are the most common in the literature [7–11] due to the advantage of being directly associated with the mass and, more importantly, with structural stiffness, which is expected to change in the presence of damage. Regardless of these advantages, OMA-based information can also be considered not sensitive to early damage due to the need of identifying high order modes shapes, which proved to be very challenging for real structure monitoring [12]. Symbolic data [13], wavelet components [14] and basic signal statistics are also examples of techniques successfully applied as extractors of damage-sensitive features for both static and dynamic monitoring. In applications comprising acceleration measurements, autoregressive (AR) models, wavelets, and principal component analysis (PCA) have been widely reported [12,15–21].

Once an operational evaluation stage has passed and a sensor network has been designed, the SHM system can begin to deliver data. At this stage, one is now faced with the challenge of making an accurate assessment of the damage condition of a given structure based on any extracted features. Feature discrimination, which comprises the choice and implementation of algorithms to process the data and carry out the identification, is arguably the most crucial component of an intelligent SHM strategy for damage detection. Before even choosing the algorithm, it is necessary to choose between two complementary approaches to the problem, as described in Fig. 18.1: (a) model-based or (b) data-driven.

The inverse approach (model-based) combines an initial model of the structure and measured data to improve the model or test a hypothesis. In practice, the model is commonly based on finite element analysis. Once the model is built, it is updated based on measured data from the real structure, such as acceleration and force responses, often in the form of a modal database, although frequency response function data may also be directly used [22,23]. The goal is to adjust the built model in such a way as to make it conform better with data from the real structure. Although, it is important to be aware that the updating step brings up an important point; it is very difficult to build an accurate model of a structure since the information will be lacking in many areas. For example, the material properties may not be known with great accuracy, especially in civil engineering where each structure is unique.

In turn, the forward approach (data-driven) does not require the development of numerical or analytical models to be fitted with in situ data; instead, it is based on the discipline of machine learning or, often more specifically, the pattern recognition aspects of machine learning. The idea is that one can learn relationships from data. In the context of SHM, this means that one can learn to assign a damage state or class to a given measurement vector from the structure or system of interest. The measurement vectors must be formed from measurements that are sensitive to the damage; in the normal terminology of pattern recognition, they are referred to as features, as previously discussed. Once features have been defined, the mapping between the features and the diagnosis can be constructed. In the forward approach, one can still make effective use of the law-based models as a means of establishing good features for damage identification [24].

The portion of the SHM process that is less documented in the technical literature is the development of statistical models for discrimination between features from the undamaged and damaged structures. Statistical model development is concerned with the implementation of algorithms that operate on the extracted features to quantify the damage state of the structure. The functional relationship between the selected features and the damage state of the structure is often difficult to define. Therefore, the statistical models are derived using machine learning techniques. The machine learning algorithms used in statistical model development for feature discrimination usually fall into two categories (i) supervised learning and (ii) unsupervised learning, see Fig. 18.1.

When training data is available from both undamaged and damaged structures, supervised learning algorithms can be used; group classification and regression analysis are primary examples of such algorithms. In the case of group classification, the output of the algorithm is a discrete class label. In its most basic form, this algorithm might simply assign a "damage" or "not damage" label to features. This type of algorithm is useful in the sense that the algorithms can be trained to give the probability of class membership. Using a regression algorithm, the outputs are one or more continuous variables. This problem is often nonlinear and is particularly suited to neural networks or other machine learning algorithms [3].

Since data obtained from damaged civil engineering structures is rare or inexistent, unsupervised learning algorithms have been increasingly observed in the literature. Damage detection methods are the primary class of algorithms used in this situation. This type of algorithm is a two-class problem that indicates if the acquired data comes from normal operating conditions or not [3]. There are many damage detection techniques, e.g., outlier analysis, kernel density estimation, and auto-associative neural networks [25,26]. All techniques fit a probability distribution to the normal condition data, then assess the probability of the test data having been generated by the same mechanism. It is important to notice that supervised and unsupervised learning come usually associated with a forward damage identification approach.

Inherent in the data acquisition, feature extraction, and feature discrimination portions of the SHM strategy are data cleansing, fusion, and compression procedures, as well as data modeling (see Fig. 18.1). Data cleansing is the process of selectively choosing data to pass on to, or reject from, the feature selection process. On the other hand, data fusion focuses on reducing the volume of data, while preserving its most relevant information. The fusion process may combine features from a single sensor, features from spatially distributed sensors, or even heterogeneous data types. In all situations, the objective of a data fusion process is to reach a new type of information with less volume and greater or similar ability to characterize the measured phenomena, when compared to that achieved when using any of the original information sources alone [27]. The Mahalanobis distance has been thoroughly used in this context due to its capacity to describe the variability in multivariate data sets [7,28]. Data compression, in turn, is the process of reducing the dimensionality of the data, or the features extracted from the data, to facilitate efficient storage of information and to enhance the statistical quantification of these parameters.

The ability to perform robust data modeling is one of the biggest challenges facing SHM when attempting to transition this technology from research to field deployment and practice on in situ structures [2]. As it applies to SHM, data modeling is the process of separating changes in sensor reading caused by damage from those caused by varying operational and environmental conditions, such as temperature or trains crossing at different speeds [29]. Two approaches are generally found in the literature and in the practice of feature modeling [12]: (i) input-output, based on regression methods such as multiple linear regression (MLR) [30,31] or (ii) output-only, based on latent variable methods such as PCA [8,12]. The first removes the effects of the environmental and operational variations (EOVs), establishing relationships between measured actions (e.g., temperature, traffic, wind) and measured structural responses. When monitoring systems do not include the measurement of EOVs, latent variable methods can be employed. These methods are able to suppress independent actions using only structural measurements.

18.3 Railway bridge over the Sado River

18.3.1 Bridge description

A bowstring-arch railway bridge over the Sado River was selected as the case study used throughout this work. It is located on the southern line of the Portuguese railway network that establishes the connection between Lisbon and the Algarve (Fig. 18.2). The bridge is prepared for conventional and tilting passenger trains with speeds up to 250 km/h, as well as for freight trains with a maximum axle load of 25 t. Even though the bridge accommodates two rail tracks, only the upstream track is currently in operation.

The bridge has a total length of 480 m, divided into three continuous spans of 160 m each. The bridge deck is suspended by three arches connected to each span of the deck by 18 hangers distributed over a single plane on the axis of the structure. The superstructure is composed of a steel-concrete composite deck, while the substructure, which includes the piers, the abutments, and the pile foundations, is built with reinforced concrete. The deck is fixed on pier P1, whereas on piers P2, P3, and P4 only the transverse movements of the deck are restrained, while the longitudinal movements are constrained by seismic dampers.

18.3.2 Monitoring system

The structural health condition of the railway bridge over the Sado River has been controlled with a comprehensive autonomous online monitoring system, as detailed in Fig. 18.3, since the beginning of its life cycle. This monitoring system was defined based on an operational evaluation and allowed the acquisition of data necessary to implement the strategy for damage detection (steps 1 and 2 of Fig. 18.1).

To identify each train that crosses the bridge and compute its speed, two pairs of optical sensors were installed at both ends of the bridge. The structural temperature

Fig. 18.2 Overview of the bridge over the Sado River.

Structural health monitoring strategy for damage detection

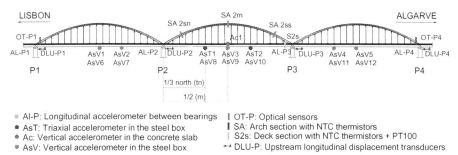

Fig. 18.3 The SHM system of the railway bridge over the Sado River.

action is measured using PT100 thermometers and NTC thermistors. Three sections of the arch were instrumented with 12 NTC thermistors. Additionally, four NTC thermistors were fixed to the steel box girder and three PT100 thermometers were embedded in the concrete slab. To control the behavior of the bearing devices, the responses from longitudinal displacement transducers were obtained from four sensors, each adjacent to a bearing device. The set of sensors also includes one vertical piezoelectric accelerometer fixed at the midspan of the concrete slab, two triaxial force balance accelerometers at the thirds of the midspan steel box girder, and 12 vertical force balance accelerometers fixed along each span of the steel box girder. Four longitudinal MEMS DC accelerometers were also installed at the top of each pier. Data acquisition is carried out continuously, at a sampling rate of 2000 Hz, by a locally deployed industrial computer to save the time history during the passage of the trains.

18.3.3 Numerical modeling and validation

A simulation of healthy and damage scenarios was conducted to test and validate the strategy implemented in Section 18.4 since damage scenarios were not observed experimentally during the period of this research. After an effective validation of the strategy, it can be applied straight to experimental data from different types of bridges.

For this purpose, a 3D finite element (FE) numerical model of the bridge was developed in ANSYS software and fully validated with experimental data (Fig. 18.4A). Among the modeled structural elements, those defined as beam finite elements consist of piers, sleepers, ballast-containing beams, rails, arches, hangers, transverse stiffeners, diaphragms, and diagonals. Shell elements were used to model the concrete slab and the steel box girder, while the pads, the ballast layer, and the foundations were modeled using linear spring-dashpot assemblies. The mass of the nonstructural elements and the ballast layer was distributed along with the concrete slab. Concentrated mass elements were used to reproduce the mass of the arches' diaphragms and the mass of the sleepers, which were simply positioned at their extremities. The connection between the concrete slab and the upper flanges of the steel box girder, as well as the connection between the deck and the track, were performed using rigid links. Special attention was paid to the bearings supports, as they can strongly influence the

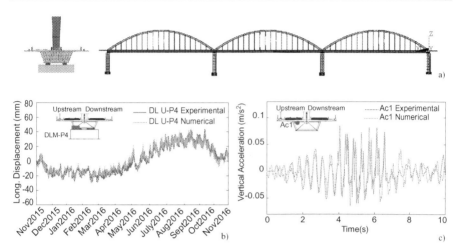

Fig. 18.4 Numerical modeling and validation: (A) 3D FE numerical model of the bridge over the Sado River, (B) static validation of the displacements measured on pier P4, and (C) dynamic validation of vertical accelerations at the concrete slab (Ac1) with the AP at 216 km/h.

performance of the bridge. Hence, to simulate the sliding behavior of the bearings, nonlinear contact elements were applied. Moreover, constraint elements located between the bearings were used to restrict the transversal movement in each pier, and the longitudinal and transversal movements in the case of the first pier.

To validate the static behavior of the numerical model, the response of the structure to the action of temperature was studied. The structural static behavior of the bridge was simulated in the FE model by running a time-history analysis using experimental data as input. The simulation procedure consisted of using the temperatures acquired every hour on-site over the course of 1 year. Fig. 18.4B presents a very good agreement between the numerical and experimental displacements of pier P4 for the temperature measured on-site between November 2015 and November 2016.

Regarding the dynamic behavior, numerical simulations were conducted considering the Alfa Pendular (AP) train as a set of moving loads crossing the bridge over the Sado River at a speed of 216 km/h. Fig. 18.4C shows a very good agreement between the experimental and numerical responses, in terms of the vertical accelerations acquired on the concrete slab at the second midspan (Ac1). Before the comparison, the time series were filtered based on a low-pass digital filter with a cut-off frequency equal to 15 Hz.

A detailed description of the numerical model and its validation can be found in Meixedo et al. [32].

18.3.4 Simulation of different structural conditions

The dynamic numerical simulations implemented in the present research work aimed at replicating the structural quantities measured in the exact locations of the accelerometers installed on-site (Fig. 18.3) during the passage of a train in the bridge.

To correctly reproduce these structural responses, the temperature action was introduced as input in the numerical model. The measurements of the optical sensors' setup were used to obtain the train speed and axle configuration, as well as the type of train.

The dynamic analyses mentioned hereafter were carried out for two of the passenger trains that typically cross the bridge over the Sado River, namely, the AP train and the Intercity (IC) train. Their frequent speeds on the bridge are 220 km/h for the AP train and 190 km/h for the IC train. The nonlinear problem was solved based on the Full Newton-Raphson method and the dynamic analyses were performed by the Newmark direct integration method, using a methodology of moving loads [9]. The integration time step (Δt) used in the analyses was 0.005 s.

Fig. 18.5A summarizes the 100 simulations of the baseline (undamaged) condition that aim at reproducing the responses of the bridge taking into account the variability of temperature, speed, type of train, and loading schemes (LS) [33].

On the other hand, the damage scenarios were chosen based on possible vulnerabilities identified for the type of structural system, taking into account its materials, behavior, loadings, and connections [34]. As shown in Fig. 18.5B, damage scenarios were simulated according to different groups:

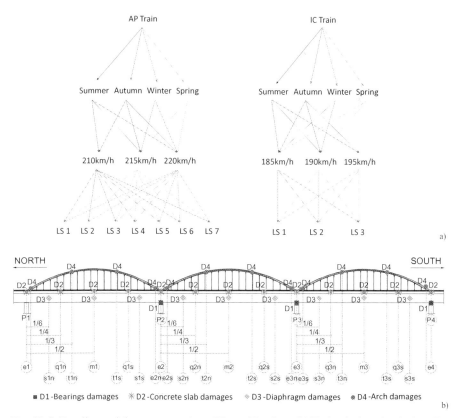

Fig. 18.5 Baseline and damage scenarios: (A) combination of 100 simulations for the baseline condition, (B) types of damages and their location on the bridge over the Sado River.

(i) damage in the bearing devices (type D1);
(ii) damage in the concrete slab (type D2);
(iii) damage in the diaphragms (type D3);
(iv) damage in the arches (type D4).

Each scenario was simulated considering only one damage location. Regarding the group of type D1, four severities of damage were included, namely, increases of the friction coefficient from a reference value of 1.5% to 1.8%, 2.4%, 3.0%, as well as to a full restrain of the movements between the pier and the deck. The remaining damage scenarios consisted of 5%, 10%, and 20% stiffness reductions in the chosen sections of the bridge (Fig. 18.5B) on the concrete slab (type D2), the diaphragms (type D3), and arches (type D4). These structural changes were simulated by reducing the modulus of elasticity of concrete (type D2) and of steel (types D3 and D4). A total of 114 damage scenarios were simulated for AP train crossings.

To obtain the most reliable reproduction of the real SHM data, the noise measured on-site by each accelerometer was added to the corresponding numerical output. These noise distributions were acquired while no trains were traveling over the bridge and under different ambient conditions. Each simulation was corrupted with different noise signals acquired at different days, thus ensuring the most representative validation for the techniques developed herein.

The time-series illustrated in Fig. 18.6 are examples of simulated responses for baseline and damage conditions, acquired from the accelerometer Ac1.

The variations associated with different train types, loading schemes, and train speeds are shown in Fig. 18.6A1 and A2. A clear distinction between the bridge responses for the IC (Fig. 18.6A1) train and the AP train (Fig. 18.6A2) passages can be observed, thus displaying the necessity of taking into account different train types for implementing damage detection strategies. Contrariwise, Fig. 18.6A1 allows observing that different LS generate smaller changes in the dynamic responses. The train speed also has an important influence on the structural response induced by trains crossing the bridge, as shown in Fig. 18.6A2.

The influence of damage scenarios in the signal obtained for the train crossings appears to be much smaller than that observed for changes in operational and environmental conditions, even when regarding sensors adjacent to the damages and for the biggest magnitudes considered (20% stiffness reductions). This conclusion can be easily observed in Fig. 18.6B1 and B2, where the bridge responses considering friction increments in the bearing devices of pier P2 and stiffness reductions in the concrete slab are, respectively, presented.

18.4 Strategy for damage detection using train induced dynamic responses

18.4.1 Overview

To address steps 3 and 4 of Fig. 18.1, an unsupervised data-driven strategy for damage detection in bridges, based on traffic-induced dynamic responses, which aims at being as effective as robust, is presented in Fig. 18.7, and comprises the following main operations:

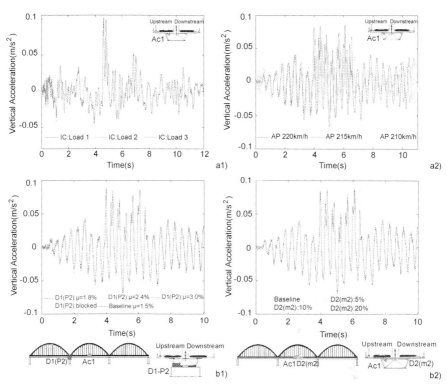

Fig. 18.6 Numerical simulations of sensor Ac1: (A1) baseline time-series using different LS of the IC train at 190 km/h, (A2) baseline time-series using different speeds of the AP train with LS5, (B1) damage time-series considering friction increase D1 (P2), and (B2) damage time-series considering stiffness reduction D2 (m2).

(i) damage-sensitive feature extraction from the acquired structural responses and feature modeling to remove EOVs, through the implementation of a double PCA (a latent-variable method);
(ii) data fusion using the Mahalanobis distance to merge multisensor features without losing damage related information;
(iii) feature discrimination to classify the extracted features in two categories, healthy or damaged, by applying an outlier analysis.

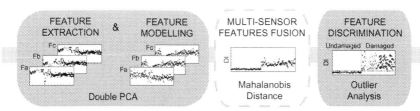

Fig. 18.7 Flowchart of the SHM strategy for damage detection.

18.4.2 Feature extraction and modeling—Double PCA

Feature extraction and feature modeling are addressed in this research work by implementing a double PCA. PCA is a multivariate statistical method that produces a set of linearly uncorrelated vectors called principal components, from a multivariate set of vector data [35].

The first operation intends to extract damage-sensitive features from the dynamic responses of the bridge. Considering an n-by-m matrix X with the original time series, where n is the number of measurements (i.e., 2112 in this case study) and m is the number of sensors (i.e., 23), a transformation to another set of m sensors, Y, designated principal components or scores, can be achieved by the following equation:

$$Y = X \cdot T \tag{18.1}$$

where T is an m-by-m orthonormal linear transformation matrix that applies a rotation to the original coordinate system. The covariance matrix of the measurements, C, is related to the covariance matrix of the scores, Λ, as follows:

$$C = T \cdot \Lambda \cdot T^T \tag{18.2}$$

in which T and Λ are matrixes obtained by the singular value decomposition of the covariance matrix C. The columns of T are the eigenvectors and the diagonal matrix Λ comprises the eigenvalues of the matrix C in descending order. Hence, the eigenvalues stored in Λ are the variances of the components of Y and express the relative importance of each principal component in the entire data set variation [7].

To allow data compression, four statistical parameters, namely the root mean square (RMS), the standard deviation, the Skewness, and the Kurtosis, are afterward extracted from the scores, Y. Thereby, the information presented in a matrix of 2112-by-23 is transformed into a matrix of 4-by-23. A total of 92 features are thus extracted from the 23 acceleration measurements. This operation is implemented for each of the 214 structural conditions.

To illustrate the feature extraction procedure, the four statistical parameters obtained for two of the twenty-three sensors, AL-P1 and AsV3, are represented in Fig. 18.8. The eight features are divided according to the structural condition in two main groups: baseline (first 100 simulations) and damage (subsequent 114 simulations). A comparison between the values of the four features from each sensor, across all 214 scenarios, allows concluding that each statistical parameter is describing distinct trends in the analyzed data. Also, the information obtained from each feature is different depending on the sensor location. The main changes in the amplitudes of the features are induced by the type and speed of the trains. In addition, for each speed value, the changes observed in the amplitude of the statistical parameters are generated by changes in the structural temperature values (chosen for autumn, spring, summer, or winter). The different LS (the seven symbols in a row in the case of the AP and three symbols in a row in the case of the IC) considered for each train type and speed, and each temperature, are the operational factors with the smallest influence on the feature variability regarding the baseline simulations.

Structural health monitoring strategy for damage detection

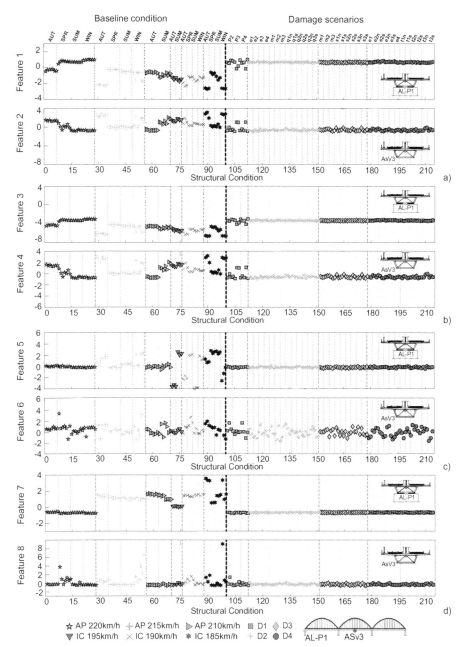

Fig. 18.8 Feature extraction of the responses from sensors AL-P1 and AsV3 for all 214 structural conditions: (A) RMS of the PCs, (B) standard deviation of the PCs, (C) Skewness of the PCs, and (D) Kurtosis of the PCs.

The analysis of the features shown in Fig. 18.8, as well as the time series presented in Fig. 18.6, allow drawing some conclusions about the difficulty in distinguishing undamaged and damage scenarios since the variations caused by environmental and operational effects result in similar or greater changes in the parameters.

Assuming that environmental conditions have a linear effect on the identified features, the implementation of a double PCA to the continuous monitoring results may efficiently remove environmental and operational effects, without the need to measure these actions [7,28].

Considering now an n-by-m matrix X with the features extracted from the dynamic responses, where n is the number of simulations for the baseline condition (i.e., 100 in this case study) and m is the number of features from all the sensors (i.e., 92), a transformation to another set of m parameters, Y, can be achieved by applying Eq. (18.1).

As demonstrated by Santos et al. [12], the PCA is able to cluster meaningful information related to EOVs in the first components, while variations related to other small-magnitude effects, such as early damage, may be retained in latter components. Since the purpose of the present research work is to detect damage, which has generally a local character, the feature modeling operation consists of eliminating the most important principal components (PCs) from the features and retaining the rest for subsequent statistical analysis. Bearing this in mind, the matrix Λ from Eq. (18.2) can be divided into a matrix with the first e eigenvalues and a matrix with the remaining m-e eigenvalues. Defining the number of e components remains an open question with regard to the representation of the multivariate data; although several approaches have been proposed, there is still no definitive answer [36]. In this work, the value of e (or the number of PCs to discard) is determined based on a rule of thumb in which the cumulative percentage of the variance reaches 80% [36,37]. After choosing e, the m-e components of the matrix Y can be calculated using Eq. (18.1) and a transformation matrix $\widehat{T}$ built with the remaining m-e columns of T. Those m-e components can be remapped to the original space using the following:

$$F_{\mathrm{PCA}} = X \cdot \widehat{T} \cdot \widehat{T}^{T} \tag{18.3}$$

where F_{PCA} is the n-by-m matrix of double PCA-based features, expected to be less sensitive to environmental and operational actions and to be more sensitive to the damage scenarios.

Since the cumulative percentage of the variance of the sum of the first six principal components was higher than 80% for different structural conditions, these six PCs were discarded during the modeling process (i.e., $e = 6$).

Fig. 18.9 shows the series of eight features across the 214 scenarios obtained for the AL-P1 and AsV3 accelerometers, after the application of the double PCA. The direct comparison of these action-free damage-sensitive features with those shown before the feature modeling (Fig. 18.8) allows observing that the feature modeling enabled removing the variations generated by the temperature, as well as by the type and speed of the train, but not those generated by damage. Moreover, the feature's sensitivity to the damage scenarios was increased.

Structural health monitoring strategy for damage detection 403

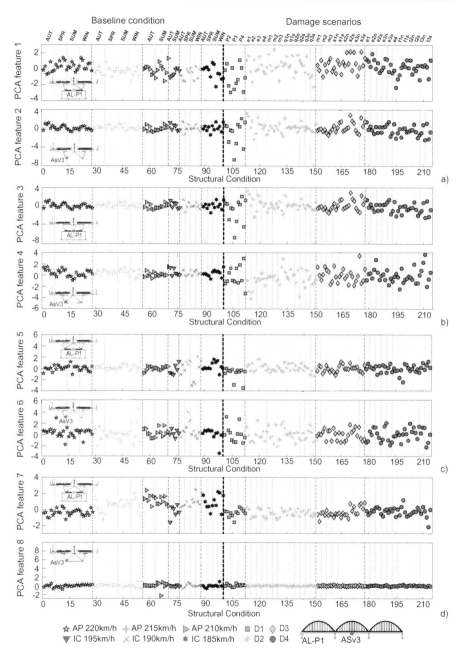

Fig. 18.9 Feature modeling of the features from sensors AL-P1 and AsV3 for all 214 structural conditions: (A) PCA of the RMS-based features, (B) PCA of the standard deviation-based features, (C) PCA of the Skewness-based features, and (D) PCA of the Kurtosis-based features.

18.4.3 Multisensor features fusion

To improve the features' discrimination sensitivity, data fusion was performed. A Mahalanobis distance was implemented to the modeled features, allowing for an effective fusion of the multisensor information. The outcome was a damage indicator, DI, for each train crossing. The analytical expression of the Mahalanobis distance for each simulation i, denoted as DI_i, is the following

$$DI_i = \sqrt{(x_i - \bar{x}) \cdot S_x^{-1} \cdot (x_i - \bar{x})^T} \tag{18.4}$$

where x_i is a vector of m features representing the potential damage/outlier, $\bar{x}$ is the matrix of the means of the features estimated in the baseline simulations, and S_x is the covariance matrix of the baseline simulations.

Hence, to detect all damage scenarios, a data fusion of the double PCA-based features from all the 23 sensors located on the bridge was implemented. It leads to a single vector *214-by-1* that represents all the data acquired through the 23 sensors. As a result, a clear distinction between simulations of the baseline condition and damage scenarios was achieved, as presented in Fig. 18.10.

18.4.4 Feature discrimination—Outlier analysis

Feature discrimination is addressed herein applying an outlier analysis that allows for automatic classification of each DI into healthy or damaged. A statistical confidence boundary CB based on the Gaussian inverse cumulative distribution (ICDF) function considering a mean μ and standard deviation σ of the baseline feature vector, and for a level of significance α is implemented. The inverse function can be defined in terms of the Gaussian cumulative distribution function as follows

$$CB = invF(1 - \alpha) \tag{18.5}$$

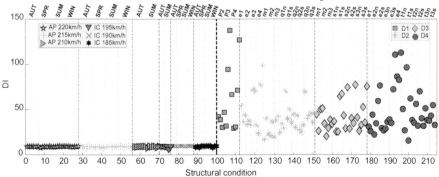

Fig. 18.10 DI values obtained with double PCA-based features for all 214 structural conditions considering the responses from all sensors.

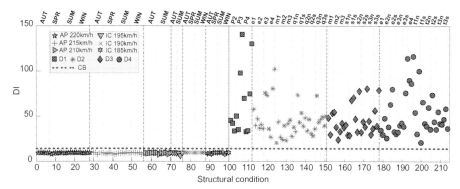

Fig. 18.11 Automatic damage detection using *DI* values for all 214 structural conditions considering the responses from all sensors and a *CB* determined for a significance level of 1%.

where

$$F(x|\mu,\sigma) = \frac{1}{\sigma\sqrt{2\pi}} \int_{-\infty}^{x} e^{-\frac{1}{2}\left(\frac{x-\mu}{\sigma}\right)^2} dy, \quad \text{for } x \in \mathbb{R} \qquad (18.6)$$

Thus, a feature is considered an outlier when its *DI* is equal or greater than *CB*. A significance level of 1% was defined, as it is commonly observed in several SHM works addressing damage identification [2,38].

Fig. 18.11 corroborates the effectiveness of the methodology in distinguishing baseline from damage scenarios. The strategy does not display either Type I (false positive) nor Type II errors (false negative).

18.5 Conclusions

This research presents an innovative data-driven SHM strategy for conducting unsupervised damage detection in railway bridge vibration response from traffic-induced excitation, applying multivariate statistical techniques. The strategy consists of fusing sets of acceleration measurements to improve sensitivity and combines:

(i) double PCA for feature extraction and modeling;
(ii) Mahalanobis distance for data fusion;
(iii) outlier analysis for feature discrimination.

The effectiveness of the presented strategy was validated on a bowstring-arch railway bridge through the simulation of several structural conditions using only experimentally obtained actions as input, namely temperature, noise, train loadings, and speeds. Damage severities of 5%, 10%, and 20% stiffness reductions in the concrete slab, diaphragm, and arches were simulated, as well as friction increases in the movements of the bearing.

The damage-sensitive features were extracted by implementing a PCA to the bridge accelerations induced by train crossings in different locations along the bridge.

Statistical parameters were extracted from the principal components to allow data compression. The study of damage-sensitive features obtained from different structural conditions, allowed concluding the supremacy of the environmental and operational variations when compared with damage, proving the importance of feature modeling. Moreover, the information obtained from each feature is different depending on the sensor location and the statistical parameter. PCA was once again implemented to model the features, allowing to successfully remove the environmental and operational effects without losing sensitivity to damage.

To enhance sensitivity, the fusion of the 92 features extracted from all the sensors was implemented and a single damage indicator for each train crossing was defined and obtained. This step proved to be crucial to achieve the highest possible level of information fusion and to obtain a clear distinction between undamaged and damaged conditions.

To automatically detect the presence of damage, an outlier analysis was performed based on a CB computed for a significance level of 1%. The robustness and effectiveness of the proposed strategy were demonstrated by automatically detecting the damage scenarios as different from those belonging to undamaged structural conditions. Using features modeled based only on measurements of structural responses, no false detections occurred.

Acknowledgments

This work was financially supported by the Portuguese Foundation for Science and Technology (FCT) through the PhD scholarship SFRH/BD/93201/2013. The authors would like to acknowledge the support of the Portuguese Road and Railway Infrastructure Manager (Infraestruturas de Portugal, I.P), the Portuguese National Laboratory for Civil Engineering (LNEC), the SAFE-SUSPENSE project—POCI-01-0145-FEDER-031054 (funded by COMPETE2020, POR Lisboa and FCT) and the Base Funding—UIDB/04708/2020 of the CONSTRUCT—Instituto de I&D em Estruturas e Construções—financed by national funds through the FCT/MCTES (PIDDAC). The authors are also sincerely grateful to the European Commission for the financial sponsorship of H2020 MARIE SKŁODOWSKA-CURIE RISE Project, Grant No. 691135 "RISEN: Rail Infrastructure Systems Engineering Network."

References

[1] C.H. Carey, E.J. O'Brien, J. Keenahan, Investigating the use of moving force identification theory in bridge damage detection, Key Eng. Mater. 569–570 (2013) 215–222, https://doi.org/10.4028/www.scientific.net/KEM.569-570.215.
[2] C.R. Farrar, K. Worden, Structural Health Monitoring: A Machine Learning Perspective, Wiley, 2013.
[3] C.R. Farrar, K. Worden, An introduction to structural health monitoring, Philos. Trans. R. Soc. A Math. Phys. Eng. Sci. 365 (1851) (2007) 303–315, https://doi.org/10.1098/rsta.2006.1928.
[4] H. Sohn, C.R. Farrar, F.M. Hemez, D.D. Shunk, D.W. Stinemates, B.R. Nadler, J.J. Czarnecki, A review of structural health monitoring literature: 1996–2001, Los Alamos (USA), 2004.

[5] K. Worden, J.M. Dulieu-Barton, An overview of intelligent fault detection in systems and structures, Int. J. Struct. Health Monit. 3 (1) (2004) 85–98.
[6] C.R. Farrar, S.W. Doebling, D.A. Nix, Vibration-based structural damage identification, Philos. Trans. R. Soc. A Math. Phys. Eng. Sci. 359 (1778) (2001) 131–149, https://doi.org/10.1098/rsta.2000.0717.
[7] A. Yan, G. Kerschen, P. De Boe, J. Golinval, Structural damage diagnosis under varying environmental conditions—part I : a linear analysis, Mech. Syst. Signal Process. 19 (2005) 847–864, https://doi.org/10.1016/j.ymssp.2004.12.002.
[8] A. Alvandi, C. Cremona, Assessment of vibration-based damage identification techniques, J. Sound Vib. 292 (2006) 179–202, https://doi.org/10.1016/j.jsv.2005.07.036.
[9] V. Alves, A. Meixedo, D. Ribeiro, R. Calçada, A. Cury, Evaluation of the performance of different damage indicators in railway bridges, Procedia Eng. 114 (2015) 746–753, https://doi.org/10.1016/j.proeng.2015.08.020.
[10] A. Meixedo, V. Alves, D. Ribeiro, A. Cury, R. Calçada, Damage identification of a railway bridge based on genetic algorithms, in: Maintenance, Monitoring, Safety, Risk and Resilience of Bridges and Bridge Networks—Proceedings of the 8th International Conference on Bridge Maintenance, Safety and Management, IABMAS 2016, Foz Do Iguaçu, Brazil, 2016.
[11] V.N. Alves, M. Oliveira, D. Ribeiro, R. Calçada, A. Cury, Model-based damage identification of railway bridges using genetic algorithms, Eng. Fail. Anal. 118 (August) (2020), https://doi.org/10.1016/j.engfailanal.2020.104845, 104845.
[12] J.P. Santos, C. Crémona, A.D. Orcesi, P. Silveira, Multivariate statistical analysis for early damage detection, Eng. Struct. 56 (2013) 273–285, https://doi.org/10.1016/j.engstruct.2013.05.022.
[13] A. Cury, C. Cremona, Assignment of structural behaviours in long-term monitoring: application to a strengthened railway bridge, Struct. Health Monit. 11 (4) (2012) 422–441, https://doi.org/10.1177/1475921711434858.
[14] D. Posenato, P. Kripakaran, I.F.C. Smith, Methodologies for model-free data interpretation of civil engineering structures, Comput. Struct. 88 (7–8) (2010) 467–482, https://doi.org/10.1016/j.compstruc.2010.01.001.
[15] E. Figueiredo, G. Park, C.R. Farrar, K. Worden, J. Figueiras, Machine learning algorithms for damage detection under operational and environmental variability, Struct. Health Monit. 10 (6) (2010) 559–572, https://doi.org/10.1177/1475921710388971.
[16] O.R. De Lautour, P. Omenzetter, Damage classification and estimation in experimental structures using time series analysis and pattern recognition, Mech. Syst. Signal Process. 24 (2010) 1556–1569, https://doi.org/10.1016/j.ymssp.2009.12.008.
[17] A. Datteo, G. Busca, G. Quattromani, A. Cigada, On the use of AR models for SHM: a global sensitivity and uncertainty analysis framework, Reliab. Eng. Syst. Saf. 170 (2018) 99–115, https://doi.org/10.1016/j.ress.2017.10.017.
[18] R. Azim, M. Gül, Damage detection of steel girder railway bridges utilizing operational vibration response, Struct. Control Health Monit. (2019) 1–15, https://doi.org/10.1002/stc.2447. August.
[19] A. Meixedo, J. Santos, D. Ribeiro, R. Calçada, M. Todd, Damage detection in railway bridges using traffic-induced dynamic responses, Eng. Struct. 238 (2021), https://doi.org/10.1016/j.engstruct.2021.112189, 112189.
[20] A. Meixedo, J. Santos, D. Ribeiro, R. Calçada, M. Todd, Online unsupervised detection of structural changes using train–induced dynamic responses, Mech. Syst. Signal Process. 165 (2022), https://doi.org/10.1016/j.ymssp.2021.108268, 108268.
[21] A. Meixedo, D. Ribeiro, J. Santos, R. Calçada, M. Todd, Real-time unsupervised detection of early damage in railway bridges using traffic-induced responses, Structural Health

Monitoring Based on Data Science Techniques. Structural Integrity, vol. 21, Springer Science and Business Media Deutschland GmbH, 2022, pp. 117–142.
[22] M.I. Friswell, Damage identification using inverse methods, in: Dynamic Methods for Damage Detection in Structures, Springer Wien, New York, 2008.
[23] L. Colombo, C. Sbarufatti, M. Giglio, Definition of a load adaptive baseline by inverse finite element method for structural damage identification, Mech. Syst. Signal Process. 120 (2019) 584–607, https://doi.org/10.1016/j.ymssp.2018.10.041.
[24] S.W. Doebling, C.R. Farrar, M.B. Prime, D.W. Shevitz, Damage identification and health monitoring of structural and mechanical systems from changes in their vibration characteristics: a literature review, Report: LA, 1996.
[25] D. Posenato, F. Lanata, D. Inaudi, I.F.C. Smith, Model-free data interpretation for continuous monitoring of complex structures, Adv. Eng. Inform. 22 (2008) 135–144, https://doi.org/10.1016/j.aei.2007.02.002. 22.
[26] I. Gonzalez, R. Karoumi, BWIM aided damage detection in bridges using machine learning, J. Civ. Struct. Heal. Monit. 5 (5) (2015) 715–725, https://doi.org/10.1007/s13349-015-0137-4.
[27] C. Haynes, M. Todd, Enhanced damage localization for complex structures through statistical modeling and sensor fusion, Mech. Syst. Signal Process. 54–55 (2015) 195–209, https://doi.org/10.1016/j.ymssp.2014.08.015.
[28] W.H. Hu, C. Moutinho, E. Caetano, F. Magalhães, Á. Cunha, Continuous dynamic monitoring of a lively footbridge for serviceability assessment and damage detection, Mech. Syst. Signal Process. (2012), https://doi.org/10.1016/j.ymssp.2012.05.012.
[29] C. Farrar, H. Sohn, K. Worden, Data normalization: a key for structural health monitoring, 2001. Technical Report, Los Alamos National Laboratory; 836(LA-UR-01-4).
[30] F. Cavadas, I.F.C. Smith, J. Figueiras, Damage detection using data-driven methods applied to moving-load responses, Mech. Syst. Signal Process. 39 (1–2) (2013) 409–425, https://doi.org/10.1016/j.ymssp.2013.02.019.
[31] B. Peeters, G. de Roeck, One-year monitoring of the Z24 bridge environmental effects versus damage events, Earthq. Eng. Struct. Dyn. 30 (2) (2001) 149–171.
[32] A. Meixedo, D. Ribeiro, J. Santos, R. Calçada, M. Todd, Progressive numerical model validation of a bowstring-arch railway bridge based on a structural health monitoring system, J. Civ. Struct. Heal. Monit. 11 (2) (2021) 421–449, https://doi.org/10.1007/s13349-020-00461-w.
[33] R. Pimentel, D. Ribeiro, L. Matos, A. Mosleh, R. Calçada, Bridge Weigh-in-Motion system for the identification of train loads using fiber-optic technology, Structures 30 (November 2020) (2021) 1056–1070, https://doi.org/10.1016/j.istruc.2021.01.070.
[34] J. Santos, Smart Structural Health Monitoring Techniques for Novelty Identification in Civil Engineering Structures (PhD Thesis), Instituto Superior Técnico—University of Lisbon, 2014.
[35] D. Ribeiro, J. Leite, A. Meixedo, N. Pinto, R. Calçada, M. Todd, Statistical methodologies for removing the operational effects from the dynamic responses of a high-rise telecommunications tower, Struct. Control Health Monit. 28 (4) (2021), https://doi.org/10.1002/stc.2700, e2700.
[36] W.K. Härdle, L. Simar, Applied Multivariate Statistical Analysis, fourth ed., Springer, 2015.
[37] I.T. Jolliffe, Principal Component Analysis, second ed., Springer, New York, 2002.
[38] J.P. Santos, C. Crémona, L. Calado, P. Silveira, A.D. Orcesi, On-line unsupervised detection of early damage, Struct. Control Health Monit. (2015), https://doi.org/10.1002/stc.

Improved dynamic resilience of railway bridges using external dampers

19

Sarah Tell[a], Andreas Andersson[a], Amirali Najafi[b], Billie F. Spencer, Jr.[b], and Raid Karoumi[a]
[a]Royal Institute of Technology (KTH), Stockholm, Sweden, [b]University of Illinois at Urbana-Champaign, Urbana-Champaign, IL, United States

19.1 Dampers for structural vibration mitigation

The vertical bridge deck acceleration is generally the decisive design limit in dynamic analyses of railway bridges for high-speed lines since this criterion is most probable to be exceeded first [1]. The maximum value of the vertical bridge deck acceleration is reached at resonance, which occurs when the forcing frequency of the applied load coincides with the natural frequency of the structure.

Railway bridges are especially prone to resonant behavior, due to the repetitive and equidistant nature of the train-induced loads. At resonance, the response of the bridge deck may reach unacceptable levels, which could impair both the riding comfort for passengers and the stability of the track [2–4]. Even small changes in the geometry of the bridge or the train could converge their natural frequencies, resulting in a possibility that the critical forcing speed occurs within the design interval. However, a significant reduction of the dynamic response could be obtained by applying a relatively moderate increase in the overall damping of the bridge.

Vibration reduction using damping devices is common within civil engineering, in which damping is mechanically inserted into the system. Due to its reliability and robustness, fluid viscous damper (FVD) retrofitting has attracted much attention recently. A FVD generally consists of a cylinder housing with a piston that transfers a compressive silicone fluid through orifices to different chambers located on opposite sides of the piston head [5]. As a consequence of the viscosity of the damper fluid, the motion of the FVD will dissipate energy from the controlled structure and generate output damper forces that are out-of-phase with the input displacement [6]. The output damper force p_d is proportional to the damping coefficient c_d and is expressed as a function of the relative velocity across the damper $\dot{u}(x_d)$ as follows [5]

$$p_d(t) = c_d |\dot{u}(x_d, t)|^r \operatorname{sgn}[\dot{u}(x_d, t)] \qquad (19.1)$$

where r is a constant exponent which describes the nonlinear behavior of the device. Since FVDs do not impose any restoring forces on the system, the structure must resist all static loads to which it is exposed [7].

In contrast to the passive control achieved by FVDs, semiactive and active dampers could be applied to insert mechanical forces into the primary structure. Such an example is the magnetorheological damper (MRD), which contains a ferromagnetic mixture of particles and fluid that is able to rapidly and reversibly change states from liquid to semisolid [8]. This MR fluid, an accumulator, and a piston-embedded magnetic coil are enclosed in a cylinder housing [9]. During operation, an electric current is supplied from a set of wires to the coil that generates a magnetic field. Due to this, the particles will rearrange in a column-like manner perpendicular to the magnetic field and further restrict the motion of the piston. By controlling the input voltage, the resistance within the damper and, thus, the output force could be adjusted. In the absence of a magnetic field, the MRD will continue to operate in a fail-safe manner similar to that of a linear viscous damper [8,10]. A common approach for modeling the restoring force $p_d(t)$ of an MR damper is by means of a Bouc-Wen element [11,12] in parallel with spring with stiffness k_0 and a dashpot with damping coefficient c_0 as

$$\begin{cases} p_d(t) = c_0 \dot{u}(x_d, t) + k_0 u(x_d, t) + \alpha z(t) \\ \dot{z}(t) = A\dot{u}(x_d, t) - \gamma |\dot{u}(x_d, t)| |z(t)| |z(t)|^{r-1} - \beta \dot{u}(x_d, t) |z(t)|^r \end{cases} \quad (19.2)$$

where α is a current dependent variable and $z(t)$ is the evolutionary variable. Further, A, β, γ, and r are parameters controlling the hysteretic behavior of the damper.

Several studies [13–18] have covered the possibility of reducing vertical deck accelerations using FVDs and MRDs connected between the superstructure of the bridge and an auxiliary beam. To minimize the response of the bridge deck, the dampers are installed to the midpoint where the deflection reaches its maximum. The results from these studies show that the dynamic response could be substantially reduced using this retrofit method.

In terms of installation, maintenance, and surveillance, however, the midpoint of the bridge could be difficult to access, e.g., due to environmental conditions. Thus, a generalization of the system is proposed [19,20] to account for phenomena that should be considered when installing dampers to arbitrary points along the bridge deck. For some bridges, the neutral axis of the deck is located close to the top surface, resulting in an eccentricity from the supports (see Fig. 19.1). During deflection of the bridge, this eccentricity may cause the beam edges to rotate when the friction of the moveable bearing is overcome. As a consequence, all points located along the bridge in the vicinity of the supports will receive an extra motional contribution in the horizontal direction. This means that the FVDs can be placed closer to the supports where the movement otherwise would be too small for the dampers to operate.

This chapter is based on theoretical studies of FVDs and hybrid testing of MRDs, with damping devices installed at arbitrary positions along the bridge deck. Some examples showing the influence of supplemental damping devices on the railway bridge response are also given.

Improved dynamic resilience of railway bridges

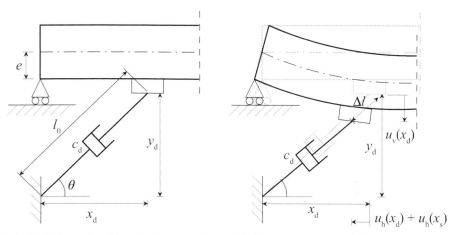

Fig. 19.1 Example of installation setup for a fluid viscous damper on a bridge.

19.2 Equation of motion for a bridge with viscous damper

By assuming that: (1) the strains and displacements are small, (2) the material response and contact conditions between all constituent parts are linear and (3) the modes of vibration are well-separated and restricted to vertical motion, it is possible to describe the bridge as a Bernoulli-Euler beam and solve the governing equation of motion with mode superposition [21,22]. The dynamic response can, thus, be expressed as a weighted sum of N natural vibration modes capable of describing the deformed shape of the structure. Typically, the first bending mode is the one mainly contributing to the response of railway bridges, but several modes of vibration could also have an impact [23]. This means that for a certain mode of vibration, the ordinary differential equation of motion of the bridge described with generalized coordinates $q_n(t)$ of the nth mode of vibration $\phi_n(x)$ at an arbitrary location x at any time instant t is given by

$$m_n \ddot{q}_n(t) + c_n \dot{q}_n(t) + k_n q_n(t) = p_n(t) \tag{19.3}$$

where m_n, c_n, and k_n are the modal mass, modal damping, and modal stiffness of the bridge respectively. By separation of variables, the vertical displacement $u_v(x,t)$ of the bridge is expressed as

$$u_v(x,t) = \sum_{n=1}^{N} \phi_n(x) q_n(t) \tag{19.4}$$

Due to the eccentricity (denoted e) shown in Fig. 19.1, the mass inertia in the axial direction must be included. The horizontal displacement of the bridge is

$$u_h(x,t) = e \sum_{n=1}^{N} \phi'_n(x) q_n(t) \tag{19.5}$$

Consequently, the modal mass of a bridge of length L can be described as

$$m_n = \int_0^L m(x) \left[\phi_n(x)^2 + e^2 \phi'_n(x_s)^2 \right] dx \tag{19.6}$$

where x_s corresponds to the location of the roller bearing. Further, the modal stiffness is given by

$$k_n = \int_0^L EI(x) \phi''_n(x)^2 dx \tag{19.7}$$

where E is the equivalent modulus of elasticity of the deck and I is the area moment of inertia. The modal damping is proportional to the mass and stiffness

$$c_n = 2\xi_n \sqrt{k_n m_n} = 2\xi_n m_n \omega_n \tag{19.8}$$

In which ξ_n is the modal damping ratio and ω_n is the circular natural frequency of mode n. The modal force $p_n(t)$ is described as a group of concentrated axle forces that are acting on the bridge at certain equidistant and repetitive intervals. Hence, it could be expressed as the sum of R amplitude functions as follows

$$p_n(t) = \sum_{r=1}^{R} P_r \varepsilon_r(t) \delta(t) \phi_n(vt - d_r) \tag{19.9}$$

where v is the train speed, P_r is the load from the rth axle and d_r is the distance from the rth axle to the first point of the beam. Further, $\delta(t)$ is the Dirac delta function and ε_r is defined by means of the Heaviside step function $H(t)$ as

$$\varepsilon_r(t) = H[t - d_r/v] - H[t - (L + d_r)/v] \tag{19.10}$$

Railway bridges are generally considered as lightly damped systems. Given this condition and the presumption that the additional damping from the FVDs does not change the mode shapes of the bridge, the damping coefficient from the FVDs could be included in the modal equation of motion. The displacement along the line of action of the dampers can be described as a coordinate corresponding to the change in length of the dampers at any time instant. Due to the relatively large eccentricity, the displacement at the roller support is significantly larger than at the fixed support. From

Fig. 19.1, this coordinate is derived in terms of the vertical and horizontal displacements from Eqs. (19.3) and (19.5)

$$l_0 + \Delta l(t) = \sqrt{[x_d + u_h(x_d) + u_h(x_s)]^2 + [y_d + u_v(x_d)]^2} \tag{19.11}$$

The change in length of the dampers is found by removing the terms corresponding to the initial length of the damper on both sides of the equation, as well as omitting the negligible second-order terms of the derived solution

$$\Delta l(t) = u(x_d, t) = [u_h(x_d, t)\cos\theta + u_h(x_s, t)\cos\theta + u_v(x_d, t)\sin\theta] \tag{19.12}$$

which results in

$$\dot{u}(x_d, t) = \sum_{n=1}^{N} [e\phi_n'(x_d)\cos\theta + e\phi_n'(x_s)\cos\theta + \phi_n(x_d)\sin\theta]\dot{q}_n(t)$$

$$= \sum_{n=1}^{N} \psi_n(\theta, x_d)\dot{q}_n(t) \tag{19.13}$$

The modal damping coefficient from the FVDs for one isolated mode of vibration is the sum of J damping devices placed at position $x_{d,j}$ with inclination θ_j

$$c_{d,n} = \sum_{j=1}^{J} c_{d,j} \psi_n(\theta_j, x_{d,j})^2 \tag{19.14}$$

The total modal damping coefficient from all FVDs is included in the total modal damping ratio of the combined bridge-damper system

$$\xi_{tot,n} = \frac{c_n + c_{d,n}}{2m_n\omega_n} \tag{19.15}$$

Since each mode of vibration is described as an SDOF system, the displacement at each time step could be solved by means of Duhamel's integral

$$q_n(t) = \int_{t=0}^{T} H(\tau)p(t-\tau)d\tau = \frac{1}{m_n\omega_{d,n}} \int_{t=0}^{T} e^{-\xi_{tot,n}\omega_n t}\sin(\omega_{d,n}t)p(t-\tau)d\tau \tag{19.16}$$

Consequently, the vertical bridge deck acceleration at an arbitrary location x along the bridge deck is found to be

$$\ddot{u}_v(x, t) = \frac{1}{\Delta t^2} \sum_{n=1}^{N} \phi_n(x) q_n''(t) \tag{19.17}$$

19.3 Real-time hybrid simulation and testing

Real-time experimental evaluation could be employed to properly account for the rate-dependent nature of supplemental energy dissipation devices [24]. A hybrid simulation is a reduced full-scale experiment in the sense that substructuring is applied to connect a physical component (e.g., a damping device) to a numerical model (e.g., a bridge). By a loop of action-reaction communication between the physical and numerical components, the real-time response of the reference system is obtained.

A layout for a hybrid simulation setup is illustrated in Fig. 19.2. The maximum vertical bridge deck acceleration of a railway bridge equipped with a supplemental MRD is simulated. Firstly, the modal entities of the mass, damping, and stiffness of the bridge, as well as the modal force from the train are calculated numerically in MATLAB and sent to a SIMULINK model. Thereafter, the real-time communication between the numerical model and the experimental component is initiated. The resulting displacement along the line of action of the damper is transferred as input to the experimental load actuator acting on the full-scale MRD shown in Fig. 19.3. Finally, the restoring force from the MRD is returned as an input force p_d to the SIMULINK model and the loop of action and reaction continues throughout the simulated train passage. A model-based compensation algorithm is implemented on the actuator signal to avoid the instability of the real-time experimental setup. Both the control and the data acquisition system are managed with ControlDesk from dSPACE.

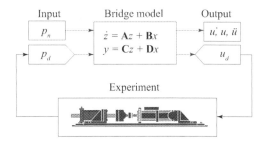

Fig. 19.2 The setup of the real-time hybrid experiment.

Fig. 19.3 A photo of the MRD used in the experiment.

The real-time hybrid simulation system is further described in Phillips and Spencer [24–26] and Tell et al. [27].

To implement the modal entities within the equation of motion of the system to the hybrid simulation, a reformulation to the state-space form is required. This is done by converting the second-order differential equation into two first-order differential equations

$$\mathbf{M}\ddot{\mathbf{q}}(t) + \mathbf{C}\dot{\mathbf{q}}(t) + \mathbf{K}\mathbf{q}(t) = \mathbf{P}(t) \Rightarrow \left\{ \mathbf{z} = \begin{bmatrix} \mathbf{q} \\ \dot{\mathbf{q}} \end{bmatrix}, \ \mathbf{x} = \mathbf{P} \right\}$$
$$\Rightarrow \begin{cases} \dot{\mathbf{z}}(t) = \mathbf{A}\mathbf{z}(t) + \mathbf{B}\mathbf{x}(t) \\ \mathbf{y}(t) = \mathbf{C}\mathbf{z}(t) + \mathbf{D}\mathbf{x}(t) \end{cases} \quad (19.18)$$

using the system and input matrices $\mathbf{A}$ and $\mathbf{B}$, together with the output matrix $\mathbf{C}$ and feedthrough matrix $\mathbf{D}$ shown in Eqs. (19.19) and (19.20). $\mathbf{M}$, $\mathbf{C}$, and $\mathbf{K}$ could either be scalar (for one mode of vibration) or matrices containing an arbitrary number of modes. The modal matrices for the bridge are denoted ϕ and the modal coordinates at the location of the bridge-damper connection are defined as ψ. The input consists of a modal load vector $\mathbf{P}$ for each mode and the output is the displacement, velocity, and acceleration of the bridge deck as well as the displacement at the position of the damper.

$$\mathbf{A} = \begin{bmatrix} 0 & \mathbf{I} \\ -\mathbf{M}^{-1}\mathbf{K} & -\mathbf{M}^{-1}\mathbf{C} \end{bmatrix}, \ \mathbf{B} = \begin{bmatrix} 0 & 0 \\ \mathbf{M}^{-1} & -\psi \mathbf{M}^{-1} \end{bmatrix} \quad (19.19)$$

$$\mathbf{C} = \begin{bmatrix} \psi & 0 \\ \phi & 0 \\ 0 & \phi \\ -\mathbf{M}^{-1}\mathbf{K} & -\mathbf{M}^{-1}\mathbf{C} \end{bmatrix}, \ \mathbf{D} = \begin{bmatrix} 0 & 0 \\ -\mathbf{M}^{-1} & -\psi \mathbf{M}^{-1} \end{bmatrix} \quad (19.20)$$

19.4 Response of the bridge-damper systems

A parametric study is conducted to show the influence of the FVDs in combination with the eccentricity on the total modal damping ratio ($c_d = 1$ MNs/m) for the first bending mode for different values of θ and x_d. For the present case, a simply supported steel-concrete composite girder bridge is analyzed with $L = 42$ m, $EI = 122$ GNm^2, $m = 18,400$ kg/m, $\xi = 0.5\%$ and $e = 2.2$ m. The result is shown in Fig. 19.4, in which the increase in the total modal damping ratio clearly emphasizes that the eccentricity allows for damper installation closer to the moveable bearing.

To find the vertical bridge deck acceleration for the railway bridge, train passages of certain high-speed load models (HSLM-A in EN 1991-2) are simulated. The design criterion of the maximum allowed vertical deck acceleration is 3.5 m/s^2 for bridges with ballasted tracks according to EN 1990. As seen in Fig. 19.5, the maximum vertical bridge deck acceleration is exceeded for all HSLM trains included in the analysis.

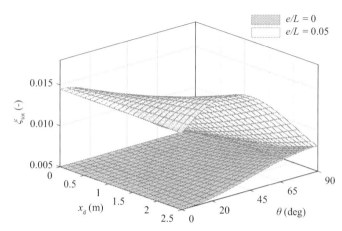

Fig. 19.4 The total damping ratio of the first vertical bending mode of vibration as a function of x_d and θ.

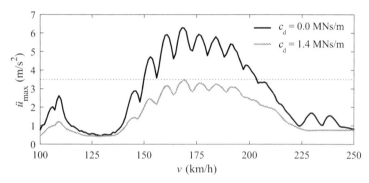

Fig. 19.5 Envelope of the vertical bridge deck acceleration due to HSLM-A, without and with viscous dampers.

By adding supplemental FVDs, it is possible to reduce the vertical bridge deck acceleration below the stated design limit. Using a single FVD, the required damping coefficient for reducing the acceleration level below the limit is found to be 1.4 MNs/m.

In the real-time hybrid simulation, a train load corresponding to HSLM-A4 is simulated to cross the railway at different speeds. The procedure is further described in Tell et al. [27]. The peak deck acceleration for each speed is presented in Fig. 19.6 for different currents of the MRD, which are held constant during the entire train passage. The dampers are installed horizontally ($\theta = 0$) at a distance of $x_d = 2$ m from the moveable bearing. To fulfill the Eurocode design criteria when using a single MR-damper, a current of at least 1 amp is required.

A parametric study of the damper position has also been performed, to minimize the vertical deck acceleration. A single MR-damper with a current of 2 amp was used.

Improved dynamic resilience of railway bridges

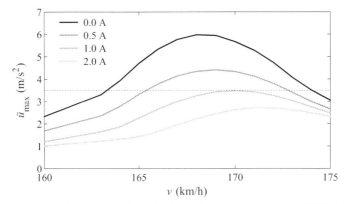

Fig. 19.6 Envelope of the vertical bridge deck acceleration for a simulated HSLM-A4 on the bridge.

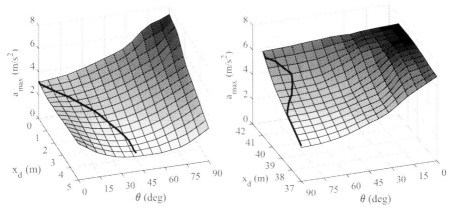

Fig. 19.7 Peak deck acceleration for different damper positions and 2.0 amp current, left: at the roller support, right: at the fixed support.

The resulting peak deck acceleration during train passage is shown in Fig. 19.7 as a function of the mounting distance x_d and the angle θ. The left figure shows results when the damper is mounted near the roller support. If mounted directly at the support the movement is restricted only to the horizontal component, i.e., $\theta = 0$. When extending the damper to 5 m from the support, the optimal angle is about 45 degrees. The difference in acceleration between the two cases is about a factor 2, but extending the damper to 5 m from the support would require a significantly larger mounting and an increased risk of loss in performance due to the additional flexibility of the supporting structure. It is instead deemed more reliable to install more dampers but horizontally at the support. The right figure shows similar results but with the damper mounted near the fixed support. Exactly at the support, $L = 42$ m, the displacement is zero and the damper is not activated. Since the longitudinal component at the fixed support is much smaller than at the roller support, the optimal angle for some distance away from the fixed support corresponds to a near vertical damper. This is however not practically feasible since it would require additional support and potentially obstruct the passage under the bridge. There would also be a risk of additional flexibility that may reduce the performance of the damper. In conclusion, considering the results from the parametric study and the practical implications it is deemed best to mount the dampers horizontally at the roller support.

19.5 Laboratory testing of an FVD

Experimental testing has also been performed with a fluid viscous damper. The target for the design of the damper was a damping coefficient $c_d = 1000$ kNs/m and a peak load capacity of 100 kN. A view of the damper in the test rig is shown in Fig. 19.8. The damper is intended for a bridge with a natural frequency in the range of 2–3 Hz and a set of harmonic load tests were performed in the range of 1–10 Hz with stroke lengths ranging from 0.5 to 8.0 mm, arranged in six test series. When setting up the harmonic tests the target force was estimated assuming a perfectly viscous damper according to Eq. (19.21).

$$F_{\max} = \delta_{\max} c_d 2\pi f \tag{19.21}$$

An example of results from the experimental testing is illustrated in Fig. 19.9. The experimental results revealed that the damper may not work ideally viscoelastic and also produced some nonlinear behavior at low amplitudes. The experimental results were fitted with a linear Kelvin-Voigt model consisting of a dashpot with a damping coefficient c_d in parallel with a spring with stiffness k_d.

Part of the results from the experimental testing is summarized in Tables 19.1 and 19.2, for the case of 2 and 3 Hz load frequency respectively. The target peak displacement and resulting force were calculated based on Eq. (19.21). For the tests at 2 Hz, the experimental displacements were consistently lower, possibly due to settings in the experimental system. The fitted parameters of the Kelvin-Voigt model indicate that the damping performance may be less at low displacement amplitudes. The variation in estimated stiffness k_d may be due to potential nonlinearities in the damper or that the damper does not completely conform to a Kelvin-Voight model.

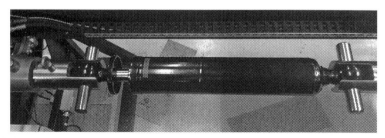

Fig. 19.8 View of a 100 kN passive FVD during laboratory testing.

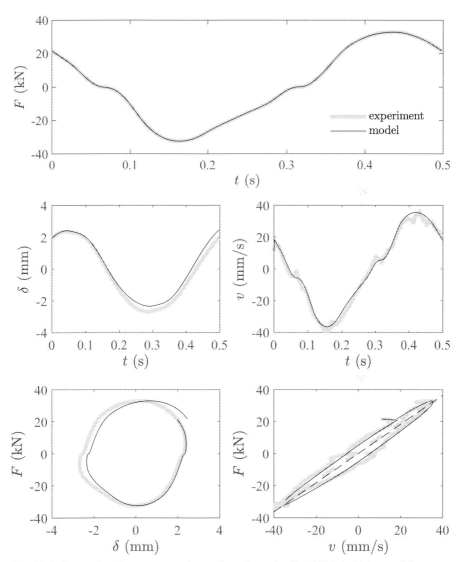

Fig. 19.9 Comparison between experimental results and a fitted Kelvin-Voigt model.

19.6 Comparison of damper performance

A comparison between different damper models on the bridge performance is studied. An FVD damper with only a linear damping coefficient is compared to a visco-elastic damper (VED) having both a primary and secondary stiffness component, modeled with a generalized Zener model as reported in [28]. The results are further compared with the MR-Damper in either Passive mode (0 amp) or Active mode (1 amp) current.

Table 19.1 Peak results from harmonic load tests at 2 Hz with the FVD damper.

Target		Experimental		Kelvin-Voigt model	
δ_{max} (mm)	F_{max} (kN)	δ_{max} (mm)	F_{max} (kN)	c_d (MNs/m)	k_d (MN/m)
0.5	6.3	0.4	4.4	0.5	5.6
1.0	12.6	0.9	11.2	0.9	4.8
2.0	25.1	1.8	23.0	1.0	3.1
3.0	37.7	2.7	32.9	0.9	2.2
4.0	50.3	3.7	43.0	0.9	1.7
5.0	62.8	4.7	53.6	0.9	1.4

Table 19.2 Peak results from harmonic load tests at 3 Hz with the FVD damper.

Target		Experimental		Kelvin-Voigt model	
δ_{max} (mm)	F_{max} (kN)	δ_{max} (mm)	F_{max} (kN)	c_d (MNs/m)	k_d (MN/m)
0.5	9.4	0.6	6.2	0.6	9.2
1.0	18.8	1.0	14.2	0.9	6.8
2.0	37.7	1.9	30.3	0.9	4.0
3.0	56.5	2.8	42.5	0.8	3.4
4.0	75.4	3.8	56.8	0.8	3.2

Table 19.3 Input parameters for different damper models.

Device	N_{damp}	c_d (MNs/m)	k_s (MN/m)	k_p (MN/m)	I_s (A)	A (−)	r (−)
FVD	1.2	1.00	–	–	–	–	–
VED	1.8	0.65	50	2.5	–	–	–
MRD-P	5.1	0.77	–	0	0	240	2
MRD-I	1.0	0.31	–	0	1	240	2

The reference input for the different models is given in Table 19.3, where N_{damp} is the factor for the number of dampers required to meet the design limit of 3.5 m/s^2 of peak bridge deck acceleration.

Simulations of the time-history response from HSLM-A9 at resonance speed are presented in Fig. 19.10. The results show that all dampers are able to mitigate the vibrations to the required amplitude. For the FVD and VED, the required damping is about the same, 1.2 MNs/m compared to $1.8 \times 0.65 = 1.2$ MNs/m. The same response is obtained for an MR-damper with a current of 1 amp. However, if the

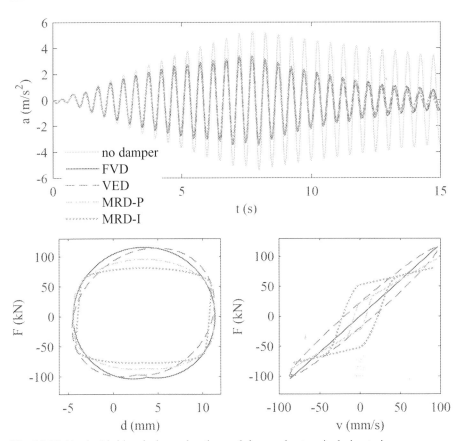

Fig. 19.10 Vertical bridge deck acceleration and damper hysteresis during train passage.

MR-damper is set to 0 amp, it would require five times more dampers to obtain the required performance.

The energy dissipated by each device is shown in the hysteresis loops, in which the relations between the force and the displacement, as well as force and velocity, are shown. The dissipated energy is given by the are inside each loop. For the FVD and the MRD-P, the area in the force-velocity hysteresis is practically zero. For the MRD-I, the combined force-displacement and force-velocity hysteresis approximately correspond to the total area of the FVD as well as the MRD-P in force-displacement. However, the combined area for the VED exceeds the energy dissipation for all other damping devices that were studied. The reason for this is the stiffness components, which are not included in any of the other models. The parameters of the MRD are dependent on the input current I_s. If the input current is 0 amp, it means that the parameters related to the Bouc-Wen element are reduced. Hence, an increased damping coefficient is required for the MRD-P compared to the MRD-I to compensate for this loss.

19.7 Conclusions

The work presented in this chapter is to a large extent based on the research presented in [20,28]. The concept of external dampers to mitigate vibrations from railway bridges is illustrated with both theoretical models and experimental testing. For bridges with relatively large eccentricity between the neutral axis and the support, vertical bending of the main girder results in a noticeable longitudinal movement at the roller support. This movement is facilitated for vibration mitigation with external dampers. Since the movement in this direction is much less at the fixed support, dampers should not be installed there. Hybrid testing using a magneto-rheological damper was performed in a laboratory with successful results. For practical applications, however, a passive viscous damper may be more cost-efficient and require less maintenance. Experimental testing was performed with a purpose-built viscous damper. The results showed that the target damping coefficient was not met at all load amplitudes and that the response was rather nonlinear for low displacement amplitudes.

Acknowledgments

The work with the real-time hybrid simulations was performed at the Smart Structures Technology Laboratory at the University of Illinois at Urbana-Champaign and was partly funded by the EU-project RISEN (H2020-MSCA-RISE-RISEN).

References

[1] M. Zacher, M. Baeßler, Dynamic Behaviour of Ballast on Railway Bridges, Taylor & Francis Group, 2009.
[2] L. Frýba, A rough assessment of railway bridges for high speed trains, Eng. Struct. 23 (5) (2001) 548–556.
[3] L. Frýba, Dynamics of bridges under moving loads. Past, present and future, in: Dynamics of High-Speed Railway Bridges: Selected and Revised Papers From the Advanced Course on 'Dynamics of High-Speed Railway Bridges', Porto, Portugal, 20–23 September 2005, 2008.
[4] H. Xia, N. Zhang, W.W. Guo, Analysis of resonance mechanism and conditions of train-bridge system, J. Sound Vib. 297 (3–5) (2006) 810–822.
[5] M.C. Constantinou, M.D. Symans, Experimental and Analytical Investigation of Seismic Response of Structures With Supplemental Fluid Viscous Dampers, National Center for Earthquake Engineering Research, State University of New York at Buffalo, Buffalo, US, 1992.
[6] G.W. Housner, L.A. Bergman, T.K. Caughey, A.G. Chassiakos, R.O. Claus, S.F. Masri, R. E. Skelton, T. Soong, B. Spencer, J.T. Yao, Structural control: past, present, and future, J. Eng. Mech. 123 (9) (1997) 897–971.
[7] P. Duflot, D. Taylor, Experience and practical considerations in the design of viscous dampers, in: Footbridge Vibration Design, CRC Press, 2009, pp. 189–202.
[8] T.T. Soong, B.F. Spencer, Supplemental energy dissipation: state-of-the art and state-of-the-practice, Eng. Struct. 24 (3) (2002) 243–259.
[9] D.Q. Truong, K.K. Ahn, MR fluid damper and its application to force sensorless damping control system, in: Smart Actuation and Sensing Systems—Recent Advances and Future Challenges, BoD–Books on Demand, 2012, pp. 383–425.

[10] M.D. Symans, M.C. Constantinou, Semi-active control systems for seismic protection of structures: a state-of-the-art review, Eng. Struct. 21 (6) (1999) 469–487.
[11] R. Bouc, Forced vibration of mechanical systems with hysteresis, in: Proceedings of the 4th Conference on Nonlinear Oscillations, Prague, Czechoslovakia, 1967, p. 315.
[12] Y.K. Wen, Method for random vibration of hysteretic systems, J. Eng. Mech. Div. 102 (2) (1976) 249–263.
[13] J. Lavado, A. Doménech, M.D. Martínez-Rodrigo, Dynamic performance of existing high-speed railway bridges under resonant conditions following a retrofit with fluid viscous dampers supported on clamped auxiliary beams, Eng. Struct. 59 (2014) 355–374.
[14] M. Luu, M.D. Martínez-Rodrigo, V. Zabel, C. Könke, H∞ optimization of fluid viscous dampers for reducing vibrations of high-speed railway bridges, J. Sound Vib. 333 (9) (2014) 2421–2442.
[15] M. Luu, M.D. Martínez-Rodrigo, V. Zabel, C. Könke, Semi-active magnetorheological dampers for reducing response of high-speed railway bridges, Control. Eng. Pract. 32 (2014) 147–160.
[16] M.D. Martínez-Rodrigo, J. Lavado, P. Museros, Dynamic performance of existing high-speed railway bridges under resonant conditions retrofitted with fluid viscous dampers, Eng. Struct. 32 (3) (2010) 808–828.
[17] M.D. Martínez-Rodrigo, P. Museros, Optimal design of passive viscous dampers for controlling the resonant response of orthotropic plates under high-speed moving loads, J. Sound Vib. 330 (7) (2011) 1328–1351.
[18] P. Museros, M.D. Martínez-Rodrigo, Vibration control of simply supported beams under moving loads using fluid viscous dampers, J. Sound Vib. 300 (1–2) (2007) 292–315.
[19] S. Rådeström, M. Ülker-Kaustell, A. Andersson, V. Tell, R. Karoumi, Application of fluid viscous dampers to mitigate vibrations of high-speed railway bridges, Int. J. Rail Transp. 5 (1) (2017) 47–62.
[20] S. Tell, Vibration Mitigation of High-Speed Railway Bridges: Application of Fluid Viscous Dampers (Lic. Thesis), KTH Royal Institute of Technology, Stockholm, Sweden, 2017.
[21] A. Chopra, Dynamics of Structures: International Edition, Pearson Higher Ed, 2015. 2015.
[22] R.W. Clough, J. Penzien, Dynamics of Structures, MaGraw-Hill, 1993.
[23] P. Museros, E. Alarcón, Influence of the second bending mode on the response of high-speed bridges at resonance, J. Struct. Eng. 131 (3) (2005) 405–415.
[24] B.M. Phillips, B.F. Spencer, Model-based feedforward-feedback tracking control for real-time hybrid simulation, 2011. Report NSEL-028, Newmark Structural Engineering Laboratory, University of Illinois at Urbana-Champaign, US.
[25] B.M. Phillips, B.F. Spencer, Model-based framework for real-time dynamic structural performance, 2012. Report NSEL-031, Newmark Structural Engineering Laboratory. University of Illinois at Urbana-Champaign, US.
[26] B.M. Phillips, B.F. Spencer, Model-based multiactuator control for real-time hybrid simulation, J. Eng. Mech. 139 (2) (2013) 219–228.
[27] S. Tell, A. Andersson, R. Karoumi, A. Najafi, B.F. Spencer, Advances in engineering materials, structures and systems: innovations, mechanics and applications, in: A. Zingoni (Ed.), Proceedings of the 7th International Conference on Structural Engineering, Mechanics and Computation, SEMC 2019, Cape Town, South Africa, September 2–4, 2019, 2019, pp. 1810–1813.
[28] S. Tell, Vibration Mitigation of High-Speed Railway Bridges, Application of Damping Devices in Theory and Practice (Doctoral thesis), KTH Royal Institute of Technology, Stockholm, Sweden, 2021.

Responses of mast structure and overhead line equipment (OHLE) subjected to extreme events

Chayut Ngamkhanong[a,b], Sakdirat Kaewunruen[a], Rui Calçada[c], and Rodolfo Martin[d]
[a]School of Engineering, University of Birmingham, Birmingham, United Kingdom, [b]Department of Civil Engineering, Faculty of Engineering, Chulalongkorn University, Bangkok, Thailand, [c]CONSTRUCT—LESE, Faculty of Engineering (FEUP), University of Porto, Porto, Portugal, [d]Evoleo Technology Pty Ltd., Porto, Portugal

20.1 Introduction

Nowadays, because of the rapid population growth, both passenger and freight train journeys have been increasing [1]. Therefore, additional capacity is needed to accommodate the economic growth [2]. The electric train has become an efficient railway system and is allowed to run more frequently and quickly. Overhead line equipment (OHLE) is the equipment used to supply power to electric trains and consists of masts, gantries, and wires found along electrified railways. This is now the preferred means of powering trains throughout the world. OHLE is an important asset of the railway infrastructure and is one of the most vulnerable components because of its poor dynamic behavior [3].

The OHLE is normally supported by side mast structures. The structure of the mast could be either a cantilever mast column for one or two tracks (Fig. 20.1) or a steel portal frame structure (Fig. 20.2) for more than two tracks because of the longer length where the cantilever mast is not feasible. Note that the major factor influencing the OHLE vibration characteristics is the ground condition or the structure–support interaction, as the soil, can be disturbed or loosened by construction, operation, or other human activities [4]. Moreover, the soil could become unstable undergo long-term deformation and gradually collapse because of the void propagation leading to poor ground conditions, particularly where the shallow work is done.

This chapter reviews the previous works relevant to the structural responses of the mast structure and OHLE and its support conditions from the civil engineering aspect. Firstly, this chapter presents the maintenance criteria based on the lateral deflection of the contact wire, which has been used in the UK and Australia as a construction tolerance. The vibration characteristics of the cantilever mast structure are then presented using finite element modeling (FEM). Furthermore, the responses of OHLE are

Fig. 20.1 Cantilever mast structure.

Fig. 20.2 Portal frame mast structure.

presented under harsh environments such as earthquakes, wind, and hurricanes. They are followed by the ground vibration which could affect the structural responses and lead to the failure of OHLE. The understanding of OHLE behaviors and structural limits under extreme cases will improve the design standard and thinking approach to effectively improve the ground support condition of OHLE.

20.2 Maintenance criteria

OHLE is generally inspected routinely, as the contact wire, which is made of pure copper or a copper alloy, could wear over time because of the train passage with the pantograph leading to the possibility of wire breakage. Sometimes, failure can be catastrophic when extreme events occur. The wire or other parts can be displaced from the gantry, resulting in the loss of contact between the train and the OHLE. Consequently, the train cannot be operated until the broken equipment or the cut wires have been repaired or replaced. In general, the contact wire runs along a zigzag path (also called "stagger"), as shown in Fig. 20.3, above the track to avoid wearing the groove in the pantograph. The contact wire is normally tensioned to resist the swaying movement, and the movement of the overhead wire could possibly come from the movement of the support. Extreme events and human activities could knock the electric system down because of the loss of contact between the OHLE and the train. To minimize the risk of the pantograph losing contact with the contact wire (Fig. 20.4), maintenance criteria have been identified in many countries such as the UK and Australia [2,7]. The maintenance criteria limit the lateral deflection of the contact wire, which is assumed to be the construction tolerance. The 50-mm construction tolerance of the contact wire is widely used as the allowable maximum displacement at the contact wire in the transverse direction.

Fig. 20.3 Zigzag path on overhead wire [5].

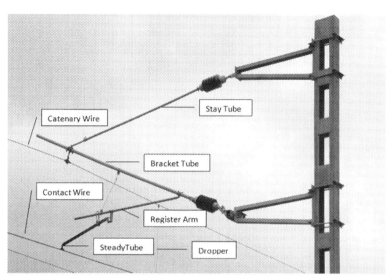

Fig. 20.4 OHLE support components [6].

The maintenance index can be calculated by using the ratio between the overhead contact wire displacement over the allowable displacement (50 mm). It is assumed that the contact wire will lose contact when the maintenance index reaches 1; thus, replacement and maintenance are needed.

Note that the rotation of the mast foundation can be alternatively used as an indicator of the maintenance limit proposed by Haiderali [4]. The tri-axial tilt node and the tilt beam were installed on top of the mast structure foundation to record the rotation in parallel and perpendicular to the track. The OHLE real-time foundation monitoring was conducted in the railway line between Manchester Victoria and Euxton Junction in the UK, which was two of the busiest lines. The sensors were installed in areas with a high risk and a low risk of ground disturbance and coal mining. The allowable pile tile and deflection were determined and linearly linked to the mast structure, which was assumed to be rigid in the finite element modeling. However, in this study, the high risk of mining subsidence did not show a significant tilt rotation and reached the maintenance level.

20.3 Vibration characteristics of OHLE

Even though an electric train has less vibrations than a conventional diesel train, the OHLE and their equipment could still experience severe vibrations due to extreme events. Note that the single cantilever mast structure is a slender structure by nature and is easy to sway. Structural engineers must consider and design this structure as a slender column, taking into account the impact of the second-order effect. The vibration characteristics including the modal shape and the corresponding frequencies were studied by Ngamkhanong et al. [8] using finite element analysis [9]. In this previous study, they presented a three-dimensional mast structure and its support with the consideration of various structure–support properties. The first 10 modes were presented: 4 modes for twisting, 3 modes for transverse bending, and 3 modes for longitudinal bending were observed when the structure was placed on the full fixed support. The examples of mast support are given in Fig. 20.5A and B, while the simplified spring support of mast structure is presented in Fig. 20.5C. Mode shapes and natural frequencies are presented in Table 20.1. The first mode of vibration presents the first twisting mode, while the second and the fifth modes show the first and the second modes of bending about the transverse direction. The third and the fourth modes represent the first and the second modes of bending about the longitudinal direction, respectively.

Note that the cantilever mast structure is very sensitive to vibrations, as the crossing phenomenon and switching modes can be observed when the support condition is changed (Fig. 20.6). Moreover, when the stiffness is reduced to 10,000 kNm/rad, the frequency of the first bending about the x-axis (mode 2) becomes greater than that

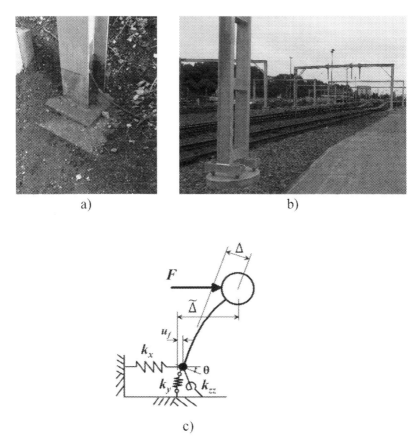

Fig. 20.5 Support of (A) cantilever mast and (B) frame mast. (C) Schematic load to structure with rotational flexibility at support [5].

Table 20.1 Mode shapes and natural frequencies of cantilever mast structure under fixed support conditions [8].

Mode	Mode shape and natural frequency (Hz)	Mode	Mode shape and natural frequency (Hz)
1	1.07 (1st twisting)	6	48.46 (2nd twisting)
2	5.63 (1st bending abt X)	7	83.43 (3rd twisting)
3	7.76 (1st bending abt Z)	8	117.4 (3rd bending abt Z)
4	35.04 (2nd bending Z)	9	144.52 (4th twisting)
5	42.63 (2nd bending abt X)	10	188.64 (3rd bending abt X)

of the first bending mode about the z-axis (mode 3). The crossing phenomenon is observed at the point where the structure has a rotational stiffness of around 17,000 kNm/rad. There is a crossover between mode 2 and mode 3 at 3.5 Hz. According to the obtained results, the rotational stiffness rarely affects the natural frequencies and the mode shape in a higher mode. As for the lower frequencies, the first mode of bending about the x-axis intersects the first mode of bending about the z-axis at 3.5 Hz. Thus, it can be concluded that the cantilever mast structure is sensitive to vibrations, and its characteristics could be influenced by the ground or support conditions (Fig. 20.6).

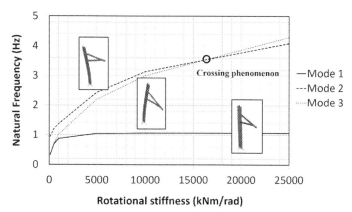

Fig. 20.6 Natural frequencies of cantilever mast structure and its support stiffness [8].

20.4 OHLE under harsh environment

The behavior of OHLE and its support under extreme events and human activities has been studied by many researchers. Most of them used the finite element modeling technique to capture the responses. Earthquakes, wind, and hurricanes have been previously presented in the open literature. Note that not only does a harsh environment lead to a failure of the OHLE system, but human activities such as maintenance activities in the surrounding area and a train running past also could potentially cause slight damage to the OHLE.

20.4.1 Earthquakes

The responses of OHLE under strong far-field earthquakes recorded in the PEER ground motion database, which had magnitudes of 6.5–8.0 Mw and occurred over a distance of 150 km, were studied by Ngamkhanong et al. [5]. This study used finite element software to analyze the seismic responses by using a nonlinear time history analysis. The stronger earthquakes were applied in the transverse direction of the railway track, as the electric system could be cut off when the displacement at the contact wire was relatively high in comparison to that in the other directions. This was confirmed by the push-over analysis, which revealed that the transverse displacement at the contact wire was the highest when the load was applied in the transverse direction.

Fig. 20.7 shows the displacement response of the mast structure at the contact wire with the support stiffness of 10,000 kNm/rad in the transverse direction subjected to the Chi-Chi and Kobe earthquakes. For the Kobe earthquake, the maximum displacement reached the maintenance limit, and thus, because of the resonance effect generated by the Kobe earthquake, the OHLE system could be cut off and maintenance was needed.

Overall, as expected, the worst condition was observed when the structure sat on poor support, as can be seen in Fig. 20.8. The maximum displacement could be seen in

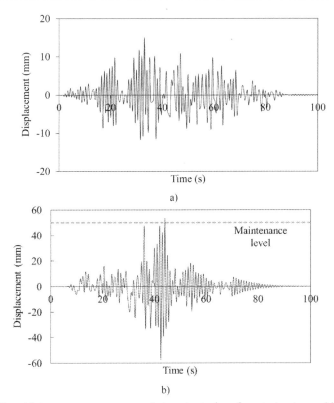

Fig. 20.7 Time history responses measured at contact wire of mast structure subjected to (A) Chi-Chi Taiwan earthquake and (B) Kobe Japan earthquake [5].

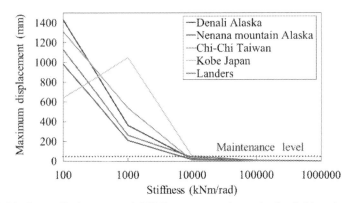

Fig. 20.8 Maximum displacement of OHLE at contact wire under far-field earthquakes [5].

the case when the rotational stiffness was between 100 and 1000 kNm/rad. As is known, the lower the support stiffness is, the higher is the maximum displacement. However, for the Kobe earthquake, the worst case was not observed as for the others. The maximum response significantly increased when the poorer support was taken. This was because each earthquake-generated ground motions in different dominant frequencies, thereby resulting in the different levels of structural vibrations depending on the dominant frequency of the structure. The resonance effect, which amplifies the intensity and produces a large oscillation by more than double, occurs when the frequency of the ground motion matches the natural frequency of the structure. It is recommended that the structural support be designed carefully, as the higher rotational stiffness at the support can significantly increase the natural frequencies of the mast structure while the earthquake frequencies are relatively low to prevent the resonance effects.

20.4.2 Wind and hurricanes

Presently, a considerable number of failure events of a free-standing structure, such as a billboard, lattice tower, or a lighting pole, have been observed to be caused by strong wind [10–13], as seen in many pieces of evidence (Fig. 20.9). Most of the failures were caused by the slenderness of the structure and its support failure. Sometimes, wind action could fail the railway system indirectly. For example, the trees next to the train lines might fall and hit the overhead contact wires or knock down the mast structure [14], thereby leading to the failure of the electrical power system.

The wind and hurricane effects were studied by Ngamkhanong et al. [15] by using the classical linear static analysis. The intensities of wind action were varied. The wind level scaled from 20 to 300 km/h, which fell into hurricane category 5. The wind speed and the types of damage caused by hurricanes are presented in Table 20.2. The wind force computed by using the EURO code was applied to the mast structure. Based on BS EN 50341-1:2012 [17], the wind action on the overhead line was

Fig. 20.9 Example of failure of slender structures under strong wind.

Table 20.2 Hurricane wind scale and types of damage [16].

Category	Wind speed	Types of damage due to hurricane winds
1	74–95 mph 64–82 kt 119–153 km/h	*Very dangerous winds will produce some damage*: Well-constructed frame homes could have damage to the roof, shingles, vinyl siding, and gutters. Large branches of trees will snap, and shallowly rooted trees may be toppled. Extensive damage to power lines and poles will probably result in power outages that could last a few to several days
2	96–110 mph 83–95 kt 154–177 km/h	*Extremely dangerous winds will cause extensive damage*: Well-constructed frame homes could sustain major roof and siding damage. Many shallowly rooted trees will be snapped or uprooted and block numerous roads. Near-total power loss is expected with outages that could last from several days to weeks
3	111–129 mph 96–112 kt 178–208 km/h	*Devastating damage will occur*: Well-built framed homes may incur major damage or removal of roof decking and gable ends. Many trees will be snapped or uprooted, blocking numerous roads. Electricity and water will be unavailable for several days to weeks after the storm passes
4	130–156 mph 113–136 kt 209–251 km/h	*Catastrophic damage will occur*: Well-built framed homes can sustain severe damage with loss of most of the roof structure and/or some exterior walls. Most trees will be snapped or uprooted and power poles downed. Fallen trees and power poles will isolate residential areas. Power outages will last weeks to possibly months. Most of the area will be uninhabitable for weeks or months
5	157 mph or higher 137 kt or higher 252 km/h or higher	*Catastrophic damage will occur*: A high percentage of framed homes will be destroyed, with total roof failure and wall collapse. Fallen trees and power poles will isolate residential areas. Power outages will last for weeks to possibly months. Most of the area will be uninhabitable for weeks or months

calculated by multiplying the acting wind pressure with the projected area and structural factors. The wind action calculated from the wind velocity was presented as a static pressure or force acting on the face of the structure [17–20]. The wind actions had characteristic values depending on the type and the location of the structure.

The hurricane scale was developed in 1971 by civil engineer Herbert Saffir and meteorologist Robert Simpson, who at the time was a director of the US National Hurricane Center. The US National Weather Service, Central Pacific Hurricane Center, and the Joint Typhoon Warning Center define sustained winds as average winds over

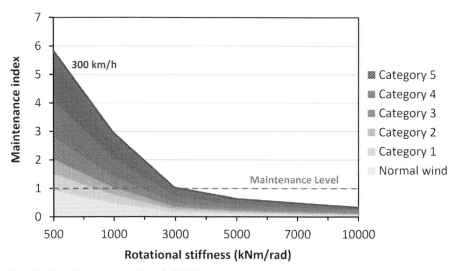

Fig. 20.10 Maintenance index of OHLE under wind action and hurricane [22].

a period of 1 min, measured at the same 33 ft (10.1 m) height; this definition was used for this scale [16,21]. These velocities were then used to calculate the static force to apply to the cantilever mast. The hurricane's sustained wind speed could be classified into five categories based on the potential property damage, as shown in Table 20.2.

After the application of the wind forces to the mast structure, the responses and the maintenance index of OHLE were obtained [22]. The maintenance index of OHLE under wind actions is presented in Fig. 20.10. We found that for the normal wind actions, the failure of OHLE could not be observed even when the support condition was not well designed. However, the wind speed of approximately 130 km/h, which fell into hurricane category 1, could knock down the OHLE with a support stiffness of 500 kNm/rad, as the maintenance index had reached 1. Fig. 20.10 suggests that the rotation stiffness of 5000 kNm/rad of support had to be taken to prevent the failure from a hurricane. Note that the well support designed as fixed could significantly prevent an OHLE failure even when the strongest hurricane occurred.

20.4.3 Ground vibrations from trains

A train passing by generates ground vibrations that are transmitted as waves through the track support system into the surrounding areas. The intensity of the ground vibrations is generally influenced by the train speed, vibration source, vibration path, and the receiver. The vibration waves propagate through the soil layers and reach a nearby structure, creating movement and annoyance to the people around. Note that railway vibrations are a serious global concern, as they can affect property and cause annoyance to people in the surrounding areas [23]. The effect of ground-borne vibrations on the building in the surrounding areas has been studied in the previous literature

[24–27]. Even though ground-borne vibrations might not cause damage to the structure, they may annoy the people in the building [28,29].

The effects of ground-borne vibrations generated by trains on cantilever masts were studied previously [15,30]. Ground-borne vibrations generated by train speeds in the range of 100–300 km/h were applied to the mast structure. The ground vibrations were calculated under the consideration of semi-empirical models for predicting low-frequency vibrations under soft ground conditions [31–33]. The vibrations were applied from different angles depending on the location of the train at a certain point in time, as shown in Fig. 20.11.

The semi-empirical model was based on vibration measurements. Furthermore, the factors discussed by Madshus et al. [32] that were of primary importance for the low-frequency railway-induced vibrations on the ground and their effect in the surrounding areas included the following: (1) ground conditions, (2) train type, (3) line quality and embankment design, (4) train speed, (5) distance from track to structure, and (6) building foundation and structure.

According to the results reported by Ngamkhanong and Kaewunruen [30] and Ngamkhanong et al. [15] (Fig. 20.12), the ground-borne vibrations could not fail the OHLE system, as the maintenance indices were all less than 1 (<50 mm of displacement at the contact wire) even if the very low stiffness of the structure–support interaction was taken into account. The displacement hardly occurred when the fixed support was applied. Even though the resonance could amplify the intensity of the

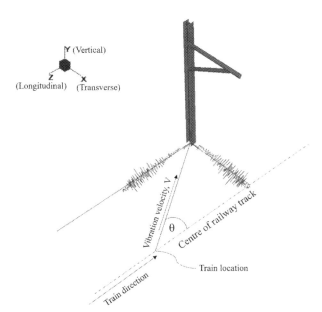

Fig. 20.11 Location of the train traveling on the railway track and ground vibration distribution to the cantilever mast structure [15,30].

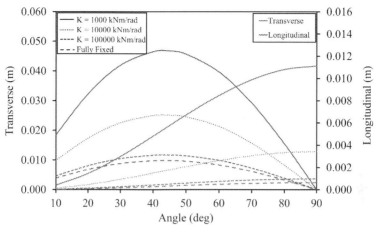

Fig. 20.12 Maximum displacement at contact wire subjected to ground-borne vibration from train passage [15,30].

ground motion as expected, the displacement of OHLE at the contact wire was still in the acceptable range, and thus, maintenance was not needed.

20.5 Summary

This chapter focused on the responses of a cantilever mast structure, which is a support of the OHLE system for single or double railway tracks. The cantilever mast structure and the OHLE system are two of the most vulnerable components of the rail infrastructure. It was found that the cantilever mast structure was very sensitive to vibrations because of its shape. From the previous studies, it was interesting to note that the structure–soil interaction and the support connection of the mast structure play a significant role in the vibration characteristics of the mast structure as the poor support can significantly reduce the overall structural performance and capacity. Interestingly, the ground subsidence caused by shallow coal mining can cause damage to OHLE. The OHLE responses were indirectly measured by the foundation deflection and rotation. The numerical simulations were performed together with the real-time foundation monitoring on the tilt degree and settlement. As expected, a hurricane can knock down the OHLE, while a normal wind cannot fail the OHLE. Earthquakes also have an impact on OHLE even when their epicenter is more than 150 km away because of their low frequency and long wavelength. Ground-borne vibrations caused by high-speed trains are generally not sufficiently strong to cause damage to the contact wire despite the occurrence of the resonance effect. The outcome reported in this chapter will help civil and track engineers to effectively and efficiently inspect OHLE structures and their support by using the structural responses from the wind actions. Moreover, this chapter contributes significantly to our understanding of the responses of OHLE to extreme events and human-induced vibrations, allowing engineers to create a more resilient design for this vital part of the infrastructure.

References

[1] A. Baxter, A Guide to Overhead Electrification Equipment, Network Rail, London, United Kingdom, 2015.

[2] RailCorp, Design of Overhead Wiring Structures & Signal Gantries. Engineering Manual—Civil, 2011.

[3] A. Beagles, D. Fletcher, M. Peffers, P. Mak, C. Lowe, Validation of a new model for railway overhead line dynamics, Proc. Inst. Civ. Eng. 169 (2016) 339–349.

[4] A.E. Haiderali, Mitigation of ancient coal mining hazards to overhead line equipment structures, Proc. Inst. Civ. Eng. Transp. (2019), https://doi.org/10.1680/jtran.18.00143.

[5] C. Ngamkhanong, S. Kaewunruen, C. Baniotopoulos, Far-field earthquake responses of overhead line equipment (OHLE) structure considering soil-structure interaction, Front. Built Environ. 4 (35) (2018), https://doi.org/10.3389/fbuil.2018.00035.

[6] M.K. Jain, Over Head Equipment—Cantilever, 2014. https://www.railelectrica.com/traction-distribution/over-head-equipment-cantilever.

[7] Network Rail, NR/L2/CIV/073: Design of Overhead Line Structures, Network Rail, London, UK, 2015.

[8] C. Ngamkhanong, S. Kaewunruen, C. Baniotopoulos, M. Papaelias, Crossing phenomena in overhead line equipment (OHLE) structure in 3D space considering soil-structure interaction, IOP Conf. Ser. Mater. Sci. Eng. 245 (2017), 032047.

[9] G+D Computing Pty Ltd., Using Strand7: Introduction to the Strand7 Finite Element Analysis System, Sydney, Australia, second, G+D Computing Pty Ltd, Sydney, NSW, Australia, 2002.

[10] C.W. Letchford, Wind loads on rectangular signboards and hoardings, J. Wind Eng. Ind. Aerodyn. 89 (2001) 135–151.

[11] Z. Li, D. Wang, X. Chen, S. Liang, J. Li, Wind load effect of single-column-supported two-plate billboard structures, J. Wind Eng. Ind. Aerodyn. 179 (2018) 70–79.

[12] R. Ramalingam, Failure analysis of lattice tower like structures, IOP Conf. Ser. Earth Environ. Sci. 80 (1) (2017), 012024.

[13] Y. Tamura, S. Cao, Climate change and wind-related disaster risk reduction, in: Proceedings of the APCWE-VII, Taipei, Taiwan, 2009.

[14] Network Rail, How Storms and Flooding Affect the Railway, 2017. https://www.networkrail.co.uk/storms-affect-railway-team-orangeprepares/.

[15] C. Ngamkhanong, S. Kaewunruen, R. Calçada, R. Martin, Condition monitoring of overhead line equipment (OHLE) structures using ground-bourne vibrations from train passages, in: H. Rodrigues, A. Elnashai (Eds.), Advances and Challenges in Structural Engineering. GeoMEast 2018. Sustainable Civil Infrastructures, Springer, Cham, 2019, https://doi.org/10.1007/978-3-030-01932-7_2.

[16] Tropical Cyclone Weather Services Program, Tropical Cyclone Definitions, National Weather Service, 2006.

[17] BSI, BS EN 50341-1:2012 Overhead Electrical Lines Exceeding AC 1 kV—Part 1: General requirements—Common Specifications, 2012.

[18] BSI, BS EN 1991-1-4:2005 Eurocode 1: Actions on Structures—Part 1–4: General Actions—Wind Actions, 2005.

[19] BSI, BS EN 1993-3-1 Eurocode 3: Design of Steel Structures—Part 3–1: Towers, Masts and Chimneys—Towers and Masts, 2006.

[20] BSI, BS EN 1993-3-2 Eurocode 3: Design of Steel Structures—Part 3–2: Towers, Masts and Chimneys—Chimneys, 2006.

[21] Federal Emergency Management Agency, Guide for All-Hazard Emergency Operations Planning, Federal Emergency Management Agency (FEMA), Washington, DC, 1996.

[22] C. Ngamkhanong, S. Kaewunruen, R. Calçada, R. Martin, Failure of overhead line equipment (OHLE) structure under hurricane, in: H. El-Naggar, K. El-Zahaby, H. Shehata (Eds.), Innovative Solutions for Soil Structure Interaction. GeoMEast 2019. Sustainable Civil Infrastructures, Springer, Cham, 2020, https://doi.org/10.1007/978-3-030-34252-4_6.

[23] D.P. Connolly, G.P. Marecki, G. Kouroussis, I. Thalassinakis, P.K. Woodward, The growth of railway ground vibration problems—a review, Sci. Total Environ. 568 (2016) 1276–1282.

[24] G. Kouroussis, L. Van Parys, C. Conti, O. Verlinden, Prediction of ground vibrations induced by urban railway traffic: an analysis of the coupling assumptions between vehicle, track, soil, and buildings, Int. J. Acoust. Vib. 18 (4) (2013) 163–172.

[25] H. Mouzakis, K. Vogiatzis, Ground-borne noise and vibration transmitted from subway networks to a typical Athenian multi-storey reinforced concrete building ICSV 2016, in: 23rd Int. Conf., 2016.

[26] C. Zou, Y. Wang, J.A. Moore, M. Sanayei, Train-induced field vibration measurements of ground and over-track buildings, Sci. Total Environ. 575 (2017) 1339–1351.

[27] C. Zou, Y. Wang, P. Wang, J. Guo, Measurement of ground and nearby building vibration and noise induced by trains in a metro depot, Sci. Total Environ. 536 (2015) 761–773.

[28] P. Lopes, J.F. Ruiz, P.A. Costa, L.M. Rodríguez, A.S. Cardoso, Vibrations inside buildings due to subway railway traffic. Experimental validation of a comprehensive prediction model, Sci. Total Environ. 568 (2016) 1333–1343.

[29] S.A. Suhairy, Prediction of Ground Vibration From Railways, 2000. SP Report, Boras, Sweden.

[30] C. Ngamkhanong, S. Kaewunruen, The effect of ground borne vibrations from high speed train on overhead line equipment (OHLE) structure considering soil-structure interaction, Sci. Total Environ. 627 (2018) 934–941, https://doi.org/10.1016/j.scitotenv.2018.01.298.

[31] L.G. Kurzeil, Ground borne noise and vibration from underground rail systems, J. Sound Vib. 66 (1979) 363–371.

[32] C. Madshus, B. Bessason, L. Harvik, Prediction model for low frequency vibration from high speed railways on soft ground, J. Sound Vib. 193 (1) (1996) 195–203.

[33] C. Madshus, A.M. Kaynia, High-speed railway lines on soft ground: dynamic behaviour at critical train speed, J. Sound Vib. 231 (3) (2000) 689–701.

Reliability quantification of the overhead line conductor   21

Sakdirat Kaewunruen[a], Chayut Ngamkhanong[b], and Jiabao Jiang[c]
[a]Department of Civil Engineering, School of Engineering, University of Birmingham, Edgbaston, Birmingham, United Kingdom, [b]Department of Civil Engineering, Faculty of Engineering, Chulalongkorn University, Bangkok, Thailand, [c]China Construction Seventh Engineering Division Corp Ltd., Zhengzhou, China

21.1 Introduction

The overhead lines are continuous structures consisting of three main components: towers (supports), insulators, and conductors. The major role of the conductor is to carry the current while the insulator provides the safe distance of the conductor from support structures to ensure that the current does not move to the support structure. It is very important to ensure that the overhead line equipment meets the standard for safety and reliability [1]. It is noted that the overhead line conductor must remain safe from the effects of weather and loading on a line and stay clear from the ground [1,2]. It should be noted that the support structure vibration may influence the overhead line conductor movement. In the overhead line system, most of the research studies focus on the reliability of tower structures which are the supports of the overhead line conductor. The influences of extreme events have been taken into consideration mostly for supporting structures [3,4]. It is evidenced from the previous studies that the vibration responses of support structures due to extreme events may cause some unprecedented damage and failure of the overhead line system [5–10]. However, it has been evidenced in the overhead line structure of the railway system in which the structure is much smaller and more sensitive to vibrations than the overhead transmission line [8]. As for the overhead transmission line, the span length is much wider and the larger lattice structures of supports are normally constructed to ensure the safety and integrity of the system to the overhead conductor and weather-related loads [1]. However, it is widely found that the strong wind causes damage to the overhead line and transmission systems [11,12].

Presently, the effects of weather-related loads have been included in the regulatory codes as they can significantly influence the overhead line structures and power transmission systems. For instance, the temperature change is also an important factor that can elongate the conductor resulting in increasing sag of the conductor which is likely to decrease the clearance to the ground [2]. More importantly, ice and wind may cause failure to this system as seen in many evidence. The shape of the catenary also changes with ice and wind loading, and time. These factors cause a significant increase in tension of conductor leading to the risk of failure. Consequently, it is important to ensure that the tension of conductor is within the strength tension limits which are defined by the percentage of rated breaking strength in tension that is set as an allowable strength to not be exceeded [13]. The rated strength is varied depending on the load conditions. This provides the basis of the design of conductor element. Importantly, the incorrect design of conductor and initial sag may lead to the failure of overhead line transmission due to the breakage of conductor since the additional loads such as ice, wind, etc., may add up to the conductor. To ensure adequate vertical and horizontal clearance under all weather and electrical loadings, and to ensure that the breaking strength of the conductor is not exceeded, the behavior of the conductor catenary under all conditions must be known before the line is designed.

Even though conductors are very important to overhead line systems, the reliability of conductors has not been fully investigated in the past. Previous studies mainly focused on the electrical aspects such as the power outage [14,15] and tower supports, whereas the structural and mechanical aspects of the conductor have not been fully studied. The overhead transmission or distribution conductors are usually a very flexible element in length and uniform in weight characteristics, formed as a catenary between the support towers [16]. Methods to determine the tension in overhead line conductors have been studied previously using the conductor curve equations. A conductor during the suspension period can be approximately formed as a catenary using a curve of the hyperbolic cosine function [17]. The catenary is a plane curve for tension calculation, whose shape corresponds to a hanging homogeneous flexible chain supported at its ends and sagging under gravity. The approximation for the sag curve can also use a parabolic equation based upon a MacLaurin expansion of the hyperbolic cosine. It is noted that the parabolic equation is not recommended for the long-span conductor [2] while it can well represent the sag curve for short-span conductor [18]. This chapter adopts the parabolic equation to evaluate the tension of the conductor and quantify the reliability of the conductor due to harsh environments.

This chapter analyses the reliability of the overhead line conductor, which is designed based on British standards [19] and is widely used in the United Kingdom. The probabilistic theory of random variables associated with the uncertainties of basic resistance and load variables is applied. The reliability indices of the overhead line conductor and its failure probability are developed to ensure the reliability of the structure. This chapter uses two analysis methods: Mean Value First-Order Second Moment Reliability Method (MVFOSM) and First-Order Moment Reliability Method (FORM). Sensitivity analyses of the reliability indices for tension according to the requirements of the limit state functions are investigated. The effects of ice and wind load, which the conductor may strongly experience, are taken into account. While several parameters such as sag, conductor unit weight, designed rated tensile strength are considered to evaluate the uncertainties on the reliability of overhead line conductor. The new findings will suggest the values for the standard design to improve the reliability and safety of overhead line conductors.

21.2 Concept of reliability analysis

Structural performance is determined by load and resistance parameters. The theory of structural reliability requires the relationship between the load actions (E) and structural resistance (R) to guarantee the safety of a structure. This limit state is defined as Z, being calculated as R minus E. Thus, the limit state function $Z(X)$ is a safe state of a structure when it is positive, that is, when $Z(X) > 0$. The structured state is unlikely safe when this state function $Z(X)$ is negative, which is $Z(X) < 0$. It is thus obvious that $Z(X) = 0$ is a critical point, and acts as the basic limitation of quantification of reliability.

In the structural reliability problem, the uncertain factors of materials are defined as the basic random variables, and the space of the basic random variables is divided into failure and safety regions by limit state function. Normally, the uncertain structure can be determined by random variables: $x_1, x_2, x_3, x_4, x_5, \ldots, x_n$. The reliable probability (safety region) of the structure can be expressed as:

$$p_s = P(Z > 0) = \iint_{Z<0} \cdots \int fx(x_1, x_2, \ldots, x_n) dx_1 dx_2 \ldots dx_n \qquad (21.1)$$

Further, the failure probability (failure region) of the structure can be expressed as:

$$p_f = P(Z < 0) = \iint_{Z<0} \cdots \int fx(x_1, x_2, \ldots, x_n) dx_1 dx_2 \ldots dx_n \qquad (21.2)$$

where Z represents the critical limit state function of the structure, which is a function of random variables $x_1, x_2, x_3, x_4, x_5, \ldots, x_n$. The limit state function can be plotted as shown in Fig. 21.1. The probability of failure is calculated by integrating the joint probability density of random variables in the failure region, shown in Fig. 21.1 [20]. Under the assumption of normal distribution, the probability of structural failure is equal to the probability of load action over resistance.

The reliability index β is used to compute structure failure probability. It should be noted that, as for the limit state form, the load action and resistance are normally

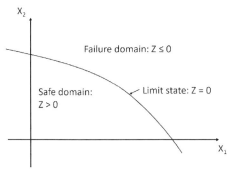

Fig. 21.1 Limit state function and safe and failure domains.

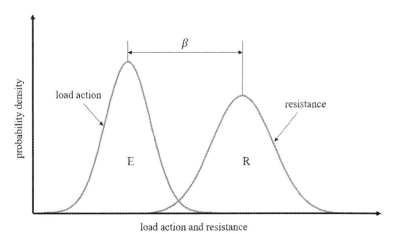

Fig. 21.2 Failure probability in a normal distribution [21].

derived from the probabilistic models based on statistical distributions. The relationship can be represented by distribution curves as seen in Fig. 21.2, which mainly describes the distributions of load action and resistance. It should be noted that the overlap area can be called the interference region. This region indicates the probability of failure in which the load action distribution exceeds the structural capacity. The structure is reliable when the interference region is small. Thus, it is important to move these distribution curves away from each other to reduce the probability of failure or increase the reliability index. Thus, this figure suggests that two methods to improve the reliability index are to increase the resistance or reduce load action. The reliability index (β) is inversely proportional to the probability of failure (P_f) by Eq. (21.3).

$$p_f = \phi(-\beta) = \phi\left(-\frac{\mu_z}{\sigma_z}\right) \tag{21.3}$$

where $\phi(-\beta)$ = standard normal distribution function; β reliability index, β can be expressed as: $=\frac{\mu_z}{\sigma_z}=\frac{\mu_R-\mu_E}{\sqrt{\sigma_R^2+\sigma_E^2}}$; μ_z mean value of the random variables; σ_z standard deviation of the random variables.

The relationship between failure probability and the reliability index is shown in Table 21.1.

Table 21.1 Relationship between reliability index and structure failure probability.

P_f	10^{-1}	10^{-2}	10^{-3}	10^{-4}	10^{-5}	10^{-6}	10^{-7}
β	1.3	2.3	3.1	3.7	4.2	4.7	5.2

As shown in Table 21.1, the larger the reliability index, the lesser the failure probability of the structure. To solve the reliability problem, several methods are proposed to solve reliability problem analytically. In this study, Mean Value First-Order Second Moment Reliability Method (MVFOSM) and First Order Reliability Method (FORM) are considered.

21.2.1 Mean value first-order second moment reliability method (MVFOSM)

Mean Value Order Second Moment Reliability Method (MVFOSM), also known as the mean value method, is the simplest and lowest cost reliability method. This method does not consider the actual distribution of basic variables, but directly according to its normal distribution or lognormal distribution. In the analysis, the Taylor series is used to expand at the center point index [22]. The reliability index formula is shown in Eq. (21.4).

$$\beta = \frac{\mu_z}{\sigma_z} = \frac{g(\mu x_1, \mu x_2, \ldots, \mu x_n)}{\sqrt{\sum_{i=1}^{n} \left(\frac{\partial g}{\partial x_i}\right)^2 \sigma x_i^2}} \tag{21.4}$$

It should be noted that, in the case of a highly nonlinear performance function, the mean value is normally in the reliability region rather than in the critical boundary. In addition, with the different critical equations (same physical meaning, but the mathematical expression is different), the reliability index may be different leading to the drawbacks of this method.

21.2.2 First order reliability method (FORM)

The design point method is proposed to overcome the drawbacks of MVFOSM by approximating the limit state function at the optimal point (design point), which is the minimum distance point on the limit state surface. This method can be divided into two methods: First Order Reliability Method (FORM) and Second-Order Reliability Method (SORM).

In 1974, Hasofer and Lind first proposed the reliability index more scientifically and introduced the concept of design point, which meant the second-moment model further was developed [23]. The design point method can consider the non-normal random variables. Under the condition of similar calculation workload, the reliability index β can be approximately calculated with higher accuracy than MVFOSM, and the design value of "design point" meeting the limit state equation can be obtained [23]. In FORM, the limit state function is linearized by the first-order Taylor expansion at the design point.

The steps for the iterative calculation of reliability index by the equivalent normal method are as follows:

1. Calculate $cos\theta$ value using Eq. (21.5).

$$\cos\theta_{x_i} = \frac{-\frac{\partial g}{\partial x}\sigma_{x_i}}{\sqrt{\sum_{i=1}^{n}\left(\frac{\partial g}{\partial x_i}\sigma_{x_i}\right)^2}} \quad (21.5)$$

2. Find random variable design point mean value, assume a β and random variable mean value, normally use initial basic random variable mean value to calculate at the first iteration.

$$x_i = \mu_{x_i} + \beta\sigma_{x_i}\cos\theta_{x_i} \quad (21.6)$$

3. Put the random variables design point mean value $x_1, x_2, \ldots, x_n$. in the critical equation.
4. Recompute reliability index β to change the random variable design point mean value to make the critical equation results close to zero.
5. In this chapter, if the critical equation results are less than 0.0001, the iteration is stopped. Otherwise, restart at (2) to continue the iteration until the accuracy requirements are met

21.2.3 Second-order reliability method (SORM)

The Second-Order Reliability Method (SORM), as its name implies, approximates the limit state function $f(z) = 0$ by the second-order Taylor expansion at the design point. This method is equivalent to FORM except for the limit state function which is approximated by second-order so the limit state becomes nonlinear and more accurate. The equation is shown as

$$\begin{aligned}f(z) &\approx \Delta f(z^*)(z-z^*) + \frac{1}{2}(z-z^*)^T f(z-z^*) \\ &= \|\Delta f(z^*)\|\left((\beta - \alpha z) + \frac{1}{2\|\Delta f(z^*)\|}(z-z^*)^T f(z-z^*)\right)\end{aligned} \quad (21.7)$$

where α can be expressed as $-\triangle f(z^*)/\|\triangle f(z^*)\|$

$z^* =$ design point value of the random variable.

In SORM, the Hessian matrix at the design point, z^*, need to be computed. When performance function is not explicitly known, usually, a finite difference scheme is used to compute the Hessian matrix [20].

Different reliability methods mean different limit state functions $f(z)$. In MVFORM, the nonlinear function $f(z)$ is expanded by the Taylor series at the "center point" of the mean value of random variables, and the linear term is retained. In

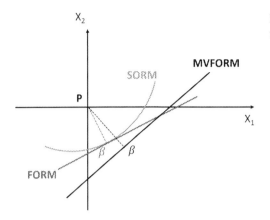

Fig. 21.3 Comparison among three methods.

FORM, the nonlinear function $f(z)$ is expanded by the Taylor series at the "design point" of the mean value of random variables and retains the linear term. The difference between FORM and SORM is that SORM considers the quadric term after expanding the nonlinear function $f(z)$. The three different limit state functions $f(z)$ in the plane coordinate system maps to three different curves, as shown in Fig. 21.3. The distances of the point P to the functions $f(z)$ are the reliability index β obtained by each method.

As shown in Fig. 21.3, the reliability index value β calculated by MVFORM has less accuracy in comparison to FORM and SORM. When the nonlinear degree of the structural function near the design point is high, the accuracy of results by using the first-order second-moment method is not enough. However, when the critical functions do not contain many non-linear terms like this case, FORM is similar to those from SORM. For this reason, SORM is not considered in this analysis.

21.3 Load calculations

21.3.1 Tension critical equation in conductor

The critical tension in the overhead conductor is derived based on the parabolic equation for the conductor with a span length of less than 400 m. Due to the conductor being regarded as a flexible cable, it only bears tangential tension at any point. Because of the different weights of the conductor at different points, the tension in the conductor along the line is also different as the tension changes with the length of the wire [24]. The tension in the conductor can be approximated by the parabolic equation of the sag curve. The arrangement of sag of conductor supported by two end towers is shown in Fig. 21.4.

According to the arrangement of the conductor, sag (f) can be calculated using the parabolic equation in Eq. (21.8).

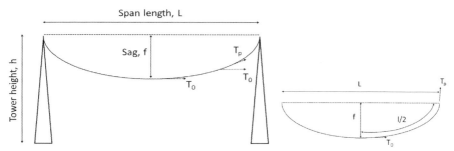

Fig. 21.4 Sag in overhead line conductor.

$$f = pL^2/8T_0 \tag{21.8}$$

where f = sag of the conductor (m); p = unit weight of the conductor (KN/m); L = span length between two tower structures (m); T_0 = tension of the conductor (KN).

To calculate the conductor length (l), Eq. (21.9) can be expressed as a function of sag. The conductor slack which is the difference between conductor length and span length ($l-L$) can be also calculated using this equation.

$$l = L + 8f^2/3L \tag{21.9}$$

It can be also back calculated to obtain sag length in terms of slack as shown.

$$f = \sqrt{3L(l-L)/8} \tag{21.10}$$

In the conductor design, two points are noteworthy and noted as critical. Firstly, the tension of the conductor at the point in the catenary where the conductor slope is equal to 0. From Eq. (21.11), the tension in the conductor can be calculated as the function of conductor unit weight per length. It should be noted that the tension of the conductor is constant throughout the length in the horizontal component. Secondly, the tension of the conductor at the ends of the level span. The tension of the conductor is equal to the tension of the conductor at the horizontal slope plus the conductor weight per unit length multiplied by sag length as presented in Eq. (21.12).

$$T_0 = \frac{pL^2}{8f} \tag{21.11}$$

$$T_p = T_0 + pf = \frac{pL^2}{8f} + pf \tag{21.12}$$

The critical limit state equation of the conductor in tension condition is shown as

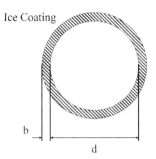

Fig. 21.5 Ice coating on the conductor.

$$F(z) = R - E = R - T_p = R - \frac{pL^2}{8f} - pf \qquad (21.13)$$

where $F(z)$ = critical state of the conductor; R = resistance of the conductor; E = load action of the conductor.

21.3.2 Conductor unit load actions

21.3.2.1 Ice and wind load

The design criteria of overhead transmission line generally take into account the effects of weather-related loads including ice, wind load, and ice and wind load. The conductor must be able to withstand the extreme external load without excessive sags, lateral movement, and tension force. It is important to note that, in winter, ice, which is accompanied by wind, is built up and accumulated on the conductor and may increase the sag of the conductor. The conductor can be coated up and stuck by ice as shown in Fig. 21.5. This situation can significantly increase the effective weight of the conductor per unit length. Also, the tension of the conductor is risen dramatically. Ice can be physically built up in different forms such as glaze ice, rime ice, wet snow, etc. It should be noted that ice load is taken into consideration for designing the structure and foundation.

Ice loading on the conductor can be calculated using Eq. (21.14). It should be noted that ice coating will add up to the self-weight of the conductor resulting in the increase in resultant weight in the downward direction. It is calculated from the density of ice multiplied by the volume of ice per unit length as presented in equation,

$$w_i = \pi b(b+d)\rho g \qquad (21.14)$$

where w_i = conductor weight per unit length due to ice (m/s); b = ice thickness (m); d = diameter of conductor (m); ρ = density of ice (kg/m^3); g = gravity force (9.81 m/s^2).

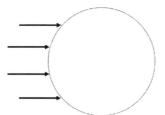

Fig. 21.6 Wind acting horizontally on the conductor.

Wind force acting on the conductor can change the conductor self-weight per unit length horizontally in the direction of the airflow as presented in Fig. 21.6. The relationship between wind pressure and wind velocity is shown in Eq. (21.15).

Basic wind pressure:

$$w_0 = v^2/1600 \tag{21.15}$$

where $w_0 =$ basic wind pressure standard value (m^2/s^2); $v =$ wind velocity (m/s).

The wind force per unit length is calculated by wind pressure per unit area multiplied by projected area per unit length. Horizontal wind unit-load:

$$w_w = \alpha w_0 \mu_z \beta_c dLB \sin\theta \tag{21.16}$$

where $w_w =$ horizontal wind unit load (kN/m); $\alpha =$ wind pressure asymmetrical coefficient; $\mu_z =$ wind pressure height variable coefficient; $\beta_c =$ conductor wind load adjustment coefficient; $B =$ wind load increasing factor (for ice thickness over 10 mm, $B = 1.2$); θ is the angle of wind direction and conductor.

The resultant of the conductor unit load lead due to the ice and wind load can be seen in Fig. 21.7. It is noted that the conductor self-weight and ice represent the total

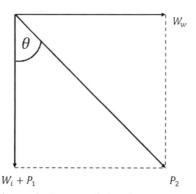

Fig. 21.7 Total weight of the conductor per unit length.

vertical load while the wind load represents the horizontal load on the conductor. The combination load per unit length can be calculated using Eq. (21.17).

$$p_2 = \sqrt{(w_i + p_i)^2 + w_w^2} \tag{21.17}$$

where p_1 is the weight of conductor per unit length (kN/m); p_2 is the total weight of conductor per unit length considering ice and wind load (kN/m).

21.3.2.2 Thermal elongation

If the conductor temperature increases from a reference temperature, T_{ref}, to another temperature, T, the conductor length, L, changes in proportion to the product of the conductor's effective thermal expansion coefficient, α, and the change in temperature, $T - T_{ref}$, as shown below in Eq. (21.18).

$$l_T = l_{Tref}\left(1 + \alpha\left(T - T_{ref}\right)\right) \tag{21.18}$$

It is clearly seen that sag and conductor length is changed due to the change in conductor tension. The conductor tension due to the increased temperature can be estimated considering the conductor sag during transmission. This results in ground clearance reduction.

21.3.3 Torsion critical equation in conductor

Torsion is one of the failure modes of wire. This also considered the reliability of the conductor due to torsion. It should be noted that the critical torsional deflection is considered as 30 degrees as the critical angle that the conductor can withstand [25]. The torsional angle of the conductor can be calculated according to torque applied, length of the conductor, and its properties as presented in Eq. (21.19).

$$\theta = \frac{TL}{GJ} = 32TL/G\pi d^4 \tag{21.19}$$

where θ = twist angle of the conductor (rads); T = torque of conductor (Nm); G = modulus of rigidity (Pa); J = polar second moment of area (m^4); L = length of conductor (m).

The critical state equation of conductor in torsion condition is presented as

$$F(z) = R - S = \theta_R - \theta_L = \theta_R - \frac{TL}{GJ} = \theta_R - 32TL/G\pi d^4 \tag{21.20}$$

where $F(z)$ = the critical state of the conductor; θ_R = conductor twist angle resistance; θ_E = conductor angle twist caused by imposed torque.

Table 21.2 Characteristics of 373-AL1 conductor.

Code	Old code	Area (mm²)	No. of wires	Diameter (mm) Wire	Diameter (mm) Cond.	Mass per unit length	Rated strength (kN)	DC resistance
373-AL1	MOTH	373.1	19	5.00	25.0	1025.3	59.69	0.0770

21.3.4 Assumptions and statistical parameters

The support towers with a span length of 260 m are considered in this study. The conductor used in this study is the aluminum conductor type 373-AL1 which is widely used in the United Kingdom. It should be noted that the conductor complies with BS EN 50182. It is a round wire concentric lay stranded overhead conductor whose properties are based on the British standard [19] (Table 21.2). The characteristics of the conductor are presented in Table. This study assumes that the overhead line conductor and its towers are in the area prone to strong wind and low temperature. It is assumed that the conductor is coated by 12.5 mm thickness of the ice and the horizontal wind velocity is 30 m/s. The statistical parameters and dimensions of conductor and tower arrangements for reliability analysis are presented in Table 21.3. It is assumed that these data produce a governing limit state over the service life of the conductor.

As for resistance, the recommended limits in tension uses the percentage of rated tensile or breaking strength that are not to be exceeded upon installation and during service life [13,26]. The tension limits depend on conditions that are recommended in previous studies. These limits can be used as an upper bound for the sag-tension calculation to predict the behavior of the conductor.

In general, the target reliability of any structure has been proposed in many works of literature to design those structures to meet the standards to ensure the safety level

Table 21.3 Statistical parameter for model uncertainties.

Basic variables	Symbol	Distribution	Mean value	Unit	Standard deviation	Coefficient of variation
Wind velocity	v	Constant	30	m/s	/	/
Ice thickness	b	Constant	12.5	mm	/	/
Conductor self-weight	p_1	Normal	1.0253	kg/m	0.020506	0.02
Diameter	d	Normal	0.025	m	0.00025	0.01
Modulus of rigidity	G	Normal	2.1×10^{10}	N/m²	4.2×10^8	0.02
Span length	L	Constant	260	m	/	/
Sag of conductor	f	Normal	7.6	m	0.076	0.01
Tower height	h	Constant	15	m	/	/

[27]. These indices are designed based on the return period in the function of failure probabilities of both element and structure as a lower bound. It should be noted that the target reliability indices of buildings, bridges, and other structures are relatively high as these structures may cause human casualties due to the failure of structures. However, the target reliability index of the overhead line conductor has never been fully presented in any standards. It is recommended that the failure probability of transmission lines should be between 0.0001 and 0.001 to avoid failure which is equal to the reliability index of 2.3–3.1 [28]. This is because their occasional failure does not usually lead to the loss of human lives.

21.4 Results and discussions

In this chapter, the results can be divided into two main parts. Firstly, the reliability of conductor 373-AL1 is analyzed considering the mean values of geometric and material properties of conductor presented in Table 21.2. The reliability indices in the tension of conductor due to normal, ice and wind, and high temperature are analyzed. Secondly, the sensitivity analysis is presented considering the uncertainties via standard deviation. The results are divided into three main cases: self-weight, ice and wind load, and thermal elongation.

21.4.1 Reliability index

21.4.1.1 Self-weight

The tension limit of the conductor is generally set as a percentage of the conductor rated strength. As for normal conditions with no ice and winds, the tension limits are set much lower than those with ice and winds. The resistance can be 15%–25% rated tensile strength. This study assumes the conductor resistance as 15% rated tensile strength. The reliability index of the conductor due to the normal condition with the normal conductor properties and dimensions are presented in Table 21.4.

21.4.1.2 Ice and wind load

In this study, the heavy loading districts based on NESC loading areas are assumed [26]. This study considers the ice coating on the conductor which is accompanied by wind. The radial thickness of ice coating is 12.5 mm with a horizontal wind speed of 30 m/s. It should be noted that for ice and wind loads, the tension limit is 60% rated tensile strength at 60 °F. The reliability index of the conductor due to ice and wind load with the normal conductor properties and dimensions are presented in Table 21.5.

Table 21.4 Reliability index of overhead conductor considering self-weight only.

Methods	MVFOSM	Form
Reliability index β	2.1759	2.1355

Table 21.5 Reliability index with maximum ice and wind load.

Methods	MVFOSM	Form
Reliability index β	2.7650	2.7023

21.4.1.3 Thermal elongation

This part assumes the temperature of conductor increases from 60 °F (15 °C) to 167 °F (75 °C) resulting in the change in length and tension of the conductor. It should be noted that conductor sag increases due to the increased temperature for transmission conductors. This is due to the fact that the allowable ground clearance is minimized by the high-temperature condition. In this case, sag change can be calculated using Eq. (21.20). It is found that sag increases from 7.6 m at 60 °F to 7.73 m at 167 °F. It is assumed that the tension limit is 20% rated tensile strength [26]. The reliability index of the conductor due to high temperature with the normal conductor properties and dimensions are presented in Table 21.6.

21.4.1.4 Torsion

This study also considers the torsion failure of the conductor to see whether this failure mode is more critical in comparison to tension mode. It should be noted that snow and wind can produce an eccentric load on the conductor causing the conductor to rotate. It has been observed from fields that the torque applied on the conductor due to snow can be roughly between 0.025 and 0.05 Nm/m [29]. This can be calculated analytically by assuming the conductor length of the conductor. This study assumes that torque of 130 Nm is applied as a load action at mid-span to produce the twist angle. While the torsional resistance of the conductor is approximate based on the torsion test [25]. The allowable torsion angle of 250 degrees is assumed at the torque that can initially destroy the aluminum strands. The reliability index calculated according to these assumptions is shown in Table 21.7.

From Table 21.1, the reliability index due to torsion obtained from MVFOSM and FORM are between 3.6308 and 3.5317 which are much greater than those in tension. It is clear that torsion failure is not critical compared to tension failure. This implies that the conductor is more reliable and hardly failed in torsion. Thus, the sensitivity analysis is conducted considering only the tension of the conductor in the next section.

Table 21.6 Reliability index at high temperature without external load.

Methods	MVFOSM	Form
Reliability index β	2.8110	2.7440

Table 21.7 Reliability index due to torsion.

Methods	MVFOSM	Form
Reliability index β	3.6308	3.5317

21.4.2 Sensitivity analysis

The overall results show that both methods provide quite close results, especially the random variables located around the mean value. While FORM methods tend to present better results rather than MVFOSM when the random variable is located away from the mean value as reliability indices seem more conservative and accurate.

21.4.2.1 Self-weight

Fig. 21.8 shows that the reliability index is slightly influenced by conductor sag This assumes that the conductor weight is not added up by any external factors such as ice, wind, etc., while the resistance is based on the 20% of rated tensile strength. Overall, increasing sag height at the middle can increase the reliability index of the conductor. It should be noted that the mean value of conductor sag is 7.6 m where the reliability indices obtained by MVFOSM and FORM are 2.176 and 2.136. These are slightly lower than the target reliability index. It is suggested that the sag of the conductor should be monitored to ensure its safety and reliability. If the level of the lowest point in the conductor increases due to uncertainties in installation and design, this can reduce the reliability index to even worse than the standard. The influences of

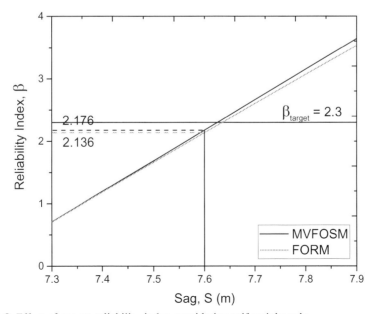

Fig. 21.8 Effect of sag on reliability index considering self-weight only.

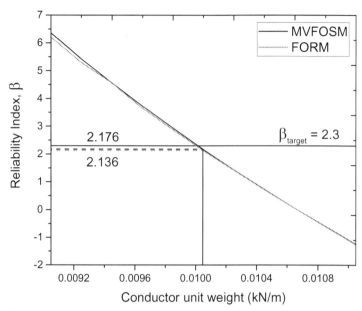

Fig. 21.9 Effect of conductor unit weight on reliability index considering self-weight only.

conductor unit weight on the reliability index of the conductor are presented in Fig. 21.9. It is clearly observed that conductor weight has a significant effect on the conductor. To improve the reliability of the conductor, the conductor weight should be decreased to reduce the tension of the conductor to meet the target reliability index.

As for resistance, it is varied by the rated tensile strength between 15% and 25% for self-weight only according to the suggested values from previous studies. Fig. 21.10 illustrates the effects of rated tensile strength of conductor on reliability index to improve the standard design and criterion. Rate tensile strength can have a significant influence on the reliability index of the conductor. It should be noted that, for this case, rated tensile strength must not be less than 18% as the resistance may be less than the load action obtained by self-weight leading to failure of the conductor. However, it is suggested that the resistance of this conductor should be increased to safely reach the target reliability. It is found that increasing by 1%–2% can make the conductor much more reliable than the suggested value.

21.4.2.2 Ice and wind load

When a conductor is covered by ice coating and exposed to wind, the effective conductor weight per unit length increases. This leads to increased tension and may exceed the resistance. The reliability index of the conductor with the consideration of ice and wind load is presented in Figs. 21.11 and 21.12. This includes the uncertainties of sag and conductor weight on reliability index while the tensile resistance is set as 60% of rated tensile strength equalizing to 35.81 kN. It is found that the mean value of sag and weight of conductor meets the target reliability as seen in Figs. 21.11

Reliability quantification of the overhead line conductor

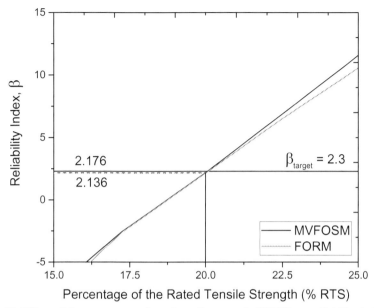

Fig. 21.10 Effect of the percentage of rated tensile strength (%RTS) on reliability index considering self-weight only.

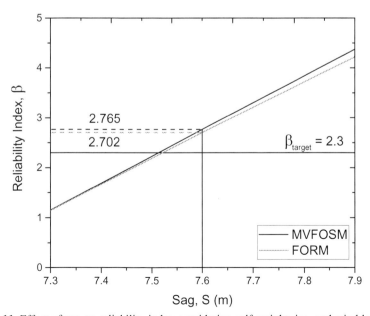

Fig. 21.11 Effect of sag on reliability index considering self-weight, ice, and wind load.

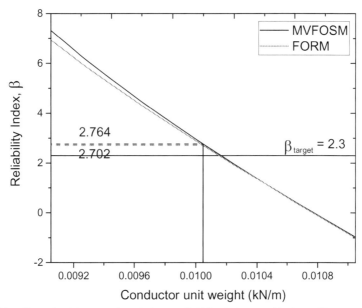

Fig. 21.12 Effect of conductor unit weight on reliability index considering self-weight, ice, and wind load.

and 21.12. It should be noted that the weight of the conductor, presented in Fig. 21.12, does not include the additional load from ice. Even though the ice weight is added upon the conductor and wind acts horizontally, the designed values of the conductor and recommended resistance can hold the conductor within the acceptable range.

Fig. 21.13 shows the effects of rated tensile strength of conductor on reliability index considering ice and wind load. The normal resistance suggested is 60% of rated tensile strength which is equal to 35.81 kN. It is found that the suggested resistance reaches the target reliability while the percentage of rated tensile strength should not be reduced to 59%. It is important that a percent reduction can make a significant impact on the reliability of the conductor in tension. It is suggested that an increasing percentage of rated tensile strength can significantly improve the reliability of the conductor. However, the suggested value at 60% of rated tensile strength is a proper value to make it reliable.

21.4.2.3 Thermal elongation

This study considers that the temperature of conductor increases from 60 °F (15 °C) change to 167 °F (75 °C) when it is transmitted. The conductor elongates with the increasing temperature of the conductor. This leads to a significant increase in length and sag of the conductor due to the decrease in tension. The calculation of sag of a conductor using parabolic equation found that sag increases from 7.60 to 7.73 m considering linear thermal expansion. The effects of elongation of the conductor can significantly reduce tension while the resistance of the conductor is assumed as similar to normal conditions. Fig. 21.14 shows that the increase in temperature during

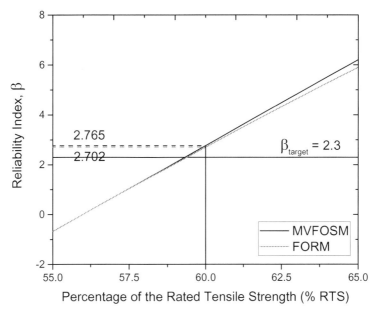

Fig. 21.13 Effect of the percentage of rated tensile strength (%RTS) on reliability index considering self-weight, ice, and wind load.

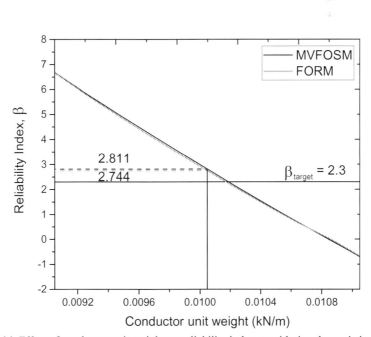

Fig. 21.14 Effect of conductor unit weight on reliability index considering thermal elongation.

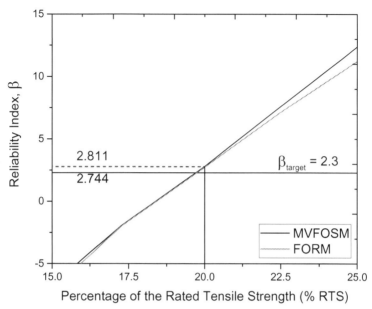

Fig. 21.15 Effect of the percentage of rated tensile strength (%RTS) on reliability index considering thermal elongation.

transmission can increase the reliability index compared to normal conditions. Also, 20% of rated tensile strength, which is a mean resistance, can ensure the safety and reliability of the conductor during transmission in Fig. 21.15. However, rate strength is very sensitive to the reliability index as this may cause breaking in the conductor when the rated strength is undermined.

21.5 Conclusions

It is important to ensure the safety and reliability of the conductor whereas the target reliability indices of overhead line conductor in tension have never been proposed and presented previously in any standards. This chapter aims to improve the safety criteria of the overhead line under normal serviceability and weather-related conditions with the consideration of uncertain variables that govern the problem in the design and construction. This chapter investigates the reliability safety indices of overhead line conductors using MVFOSM and FORM methods. The influence of self-weight of conductor, ice and wind load, and temperature increase during transmission on the performance of conductor is considered. It is found that both methods provide quite close results while MVFOSM tends to give slightly divergence results when considering the variables apart from the mean value. The obtained results show that the unit weight of the conductor influences the reliability index and the properties may not meet the target due to uncertainties while sag also plays a role in reducing the reliability index. Moreover, this study suggests the percentage of rate tensile strength in different conditions that the current use of resistance is slightly low and should be increased to reach the target reliability. Also, tension is more critical than torsion as the reliability index is generally less than those obtained from torsion. The findings of this study will quantify risks and therefore improve the reliability of overhead line conductors while taking into account the degree of uncertainties of arrangement and material properties of overhead line conductors during construction and installation. The results can be used as benchmark safety indices to improve the target reliability indices and standard design of overhead line conductors. The outcome of this study will help an engineer to improve the inspection regime and mitigate the risk of safety concerns of overhead line conductors.

Data availability

All data, models, and code generated or used during the study appear in the submitted article.

Acknowledgments

The authors are sincerely grateful to European Commission for the financial sponsorship of the H2020-MSCA-RISE Project No. 691135 "RISEN: Rail Infrastructure Systems Engineering Network," which enables a global research network that tackles the grand challenge of railway infrastructure resilience and advanced sensing in extreme environments (www.risen2rail.eu) [30].

References

[1] K. Sriram, Y. Prasad, Design of Electrical Transmission Lines, Taylor and Francis, London, 2016.
[2] A. Hatibovic, Derivation of equations for conductor and sag curves of an overhead line based on a given catenary constant, Period. Polytech. Electr. Eng. Comput. Sci. 58 (2014) 23–27.
[3] B. Hu, R.W.K. Chan, Time-dependent reliability analysis of railway overhead structures, SN Appl. Sci. 1 (2019) 1279.
[4] S.C. Yang, T.J. Liu, H.P. Hong, Reliability of tower and tower-line systems under spatiotemporally varying wind or earthquake loads, J. Struct. Eng. 143 (2017) 04017137.
[5] C. Ngamkhanong, S. Kaewunruen, The effect of ground borne vibrations from high speed train on overhead line equipment (OHLE) structure considering soil-structure interaction, Sci. Total Environ. 627 (2018) 934–941.
[6] C. Ngamkhanong, S. Kaewunruen, B.J. Afonso Costa, State-of-the-art review of railway track resilience monitoring, Infrastructures 3 (2018) 3.
[7] C. Ngamkhanong, S. Kaewunruen, C. Baniotopoulos, Far-field earthquake responses of overhead line equipment (OHLE) structure considering soil-structure interaction, Front. Built Environ. 4 (2018) 35.
[8] C. Ngamkhanong, S. Kaewunruen, C. Baniotopoulos, M. Papaelias, Crossing phenomena in overhead line equipment (OHLE) structure in 3D space considering soil-structure interaction, IOP Conf. Ser. Mater. Sci. Eng. 245 (2017).
[9] C. Ngamkhanong, S. Kaewunruen, R. Calçada, R. Martin, Failure of overhead line equipment (OHLE) structure under hurricane, in: Innovative Solutions for Soil Structure Interaction, Springer International Publishing, 2020.
[10] G. Saudi, Structural assessment of a guyed mast through measurement of natural frequencies, Eng. Struct. 59 (2014) 104–112.

[11] A. Ahmed, C. Ostowari, Longitudinally and transversely spaced cylinders in cross flow, J. Wind Eng. Ind. Aerodyn. 36 (1990) 1095–1104.
[12] J. Qi, D. Hongzhou, Comparison on wind load prediction of transmission line between Chinese new code and other standards, Procedia Eng. 14 (2011) 1799–1806.
[13] J.S. Barrett, Y. Motlis, Allowable tension levels for overhead-line conductors, IEE Proc. Gener. Transm. Distr. 148 (2001) 54–59.
[14] L.M. Tolbert, L.J. Degenhardt, J.T. Cleveland, Reliability of lightning-resistant overhead lines, IEEE Ind. Appl. Mag. 3 (1997) 17–21.
[15] S. Yang, W. Zhou, S. Zhu, L. Wang, L. Ye, X. Xia, H. Li, Failure probability estimation of overhead transmission lines considering the spatial and temporal variation in severe weather, J. Mod. Power Syst. Clean Energy 7 (2019) 131–138.
[16] D.O. Ehrenburg, Transmission line catenary calculations, Trans. Am. Inst. Electr. Eng. 54 (1935) 719–728.
[17] H. Alen, Advanced Application of the Catenary and the Parabola for Mathematical Modelling of the Conductor and Sag Curves in the Span of an Overhead Lin, Óbuda University, Budapest, 2019.
[18] L.L. Grigsby, Electric Power Generation, Transmission, and Distribution, CRC press, 2016.
[19] British Standards Institution, BS EN 50182:2001 Conductors for Overhead Lines. Round Wire Concentric Lay Stranded Conductors, British Standards Institution, London, United Kingdom, 2001.
[20] C. Huang, A.E. Hami, Overview of Structural Reliability Analysis, Research gate, 2017, pp. 1–3.
[21] S. Kaewunruen, A. Remennikov, Reliability Assessment of Railway Prestressed Concrete Sleepers, 2008.
[22] C. Huang, A. El Hami, B. Radi, Overview of structural reliability analysis methods—part I: local reliability methods, Incert. Fiabil. Syst. Multiphys. 17 (2017) 1–10.
[23] A.M. Hasofer, N. Lind, An exact and invarient first order reliability format, J. Eng. Mech. Div. Proc. ASCE 100 (1974) 111–121.
[24] D.A. Douglass, R. Thrash, Sag and tension of conductor, in: Electric Power Generation, Transmission, and Distribution, CRC Press, Florida, 2006. 2007.
[25] National Electric Energy Testing Research & Applications Center, 774–T53 ACCR Conductor Torsion and Resistance, National Electric Energy Testing, Research & Applications Center, 2006.
[26] National Electrical Safety Code, 2017 National Electrical Safety Code(R) (NESC(R)), 2017 National Electrical Safety Code(R) (NESC(R)), 2016, pp. 1–405.
[27] R. Van Coile, D. Hopkin, L. Bisby, R. Caspeele, The meaning of Beta: background and applicability of the target reliability index for normal conditions to structural fire engineering, Procedia Eng. 210 (2017) 528–536.
[28] E. Ghannoum, Probabilistic design of transmission lines Part II: design criteria corresponding to a target reliability, IEEE Trans. Power Syst. PAS-102 (1983) 3065–3079.
[29] P.L.I. Skelton, G. Poots, On the rotation of conductors of finite torsional stiffness by eccentric snow loading and control of rotation using counterweights, Int. J. Numer. Methods Eng. 28 (1989) 2829–2838.
[30] S. Kaewunruen, J.M. Sussman, A. Matsumoto, Grand challenges in transportation and transit systems, Front. Built Environ. 2 (2016).

Index

Note: Page numbers followed by *f* indicate figures, *t* indicate tables, and *b* indicate boxes.

A

ABA. *See* Axle box acceleration (ABA) measurement
Accelerometers, 361–363, 362–363*f*
Accessibility-based rail system vulnerability methodology, 10–15
  rail network accessibility
    land use characteristics, 13
    under line(s) failure, 13
    under link(s) failure, 12
    under station(s) failure, 12
  station-based rail network accessibility, 10
ACI code
  creep prediction on concrete railway sleeper, 287–288
  shrinkage prediction on concrete railway sleeper, 290–291
Adaptation Reporting Power (ARP), 49–50, 57–58
Adapting to Rising Tides project, 47
Adaptive management, 40–41*b*, 40*f*
Advanced condition monitoring, 3
Aggregate imaging system (AIMS), 300
AIMS. *See* Aggregate imaging system (AIMS)
American Railway Engineering Association (AREA) Specification, 248
Anchor-reinforced sleeper, 254–256, 255*f*
  results, 256, 256*f*
AREA. *See* American Railway Engineering Association (AREA) Specification
ARP. *See* Adaptation Reporting Power (ARP)
ATIP. *See* Autonomous Track Inspection Program (ATIP)
Australian Standard 3600-2009
  creep prediction on concrete railway sleeper, 288–289, 289*t*
  shrinkage prediction on concrete railway sleeper, 291
Autonomous Track Inspection Program (ATIP), 309
Axle box acceleration (ABA) measurement, 127–128, 128*f*

B

Ballast. *See also* Ballast specifications, effect on lateral resistance
  bed profile size, 296*f*
  condition assessment, 308–310
    data/signal processing, 308–309
    numerical simulations, 309–310
  conventional track, 296*f*
  definition of, 295
  drainage failure, 301
  degradation, 299–305
    bed hardening, 301
    compaction (bulk density), 302
    consequences of, 300–301
    deformation, 301
    loading type, 301–303
    low capacity, 301
    mechanism, 299–301
    parent rock material, 302
    particle size and shape, 302–303
    reasons for, 299–300
    types of, 300
  degradation mitigation, 303–305
    material, 304
    new geo-inclusion materials, 304
    sleeper innovations, 304–305, 305–306*f*
  function, 295–296
  idealization model, for train-track simulation, 105–107, 107*f*
  inspection, 306–308, 307*f*
  problems, solutions to, 298–299
    condition assessment, 298–299
    degradation, 298
    inspection, 298

Ballast *(Continued)*
　research on current problems, 296–297
　　bed degradation mechanism, 297
　　degradation mitigation, 297
　　maintenance, 297
　　performance assessment, 297
　　performance improvement on ballast bed, 297
Ballasted track, 243
Ballast specifications, effect on lateral resistance, 247–250
　material properties, 247–248, 247t
　particle size, 248
　profile influence, 248–250, 249f, 249–250t
　sleeper type and shape influence, 250–261
　　anchor-reinforced sleeper, 254–256, 255f
　　bi-block sleeper, 252–253, 253f
　　bottom-textured sleeper (frictional sleeper), 256, 257f
　　FFU synthetic sleeper (*see* FFU synthetic sleeper)
　　ladder sleeper, 253–254, 254f
　　mono-block sleeper, 252–253, 253f
　　nailed sleeper, 256–258, 257f
　　steel sleeper, 251, 251f
　　winged sleeper, 251–252, 252–253f
Bayes network (BN), 349–350
Bearers, 322–323, 322f
BEM. *See* Boundary element method (BEM)
Bi-block sleeper, 252–253, 253f
BN. *See* Bayes network (BN)
Bottom-textured sleeper (frictional sleeper, 256, 257f
Boundary element method (BEM), 111
Boundary lubrication, 143–144

C

CaDD (Capacity Diagnosis and Development) software, 57
Cancun Adaptation Framework, 37
Cantilever mast structure, 426f
Carregado test site, experimental characterization of, 210–229
　dynamic characterization of the structure, 220–225

　　dynamic stiffness of the footings, experimental assessment of, 222–225, 224–225f
　　general description, 220–221, 221t, 221–222f
　　modal identification tests, 221–222, 222–223f
　general description, 210–211, 210–211f
　ground characterization, 215–220
　　general description, 215–217, 216f, 216t
　　geophysical tests, 217–220, 217–220f
　mitigation measures based on seismic metamaterial (phononic crystal) concept, testing of, 225–227, 226–227f
　track characterization, 211–215
　　general description, 211–212, 212f
　　mechanical properties, assessment of, 213–215, 213–215f
　　unevenness measurement, 212, 212f
　vibrations induced by railway traffic, measurement of, 227–229, 228f, 229t
CCC. *See* Commission on Climate Change (CCC)
CCRA. *See* Climate Change Risk Assessment (CCRA)
China
　China Academy of Rainway Sciences, 302
　demand spikes, 73–74
　supply losses, 71
CIL. *See* Climate Impact Lab (CIL)
Climate change, 37–38
Climate change adaptation, 39–41, 44–47
　multinational examples, 44
　national examples, 45–46
　Rail Adapt Project (*see* Rail Adapt Framework, for climate change adaptation)
　regional and local examples, 46–47
Climate Change Risk Assessment (CCRA), 43, 45–46, 52–53, 57–58
Climate Impact Lab (CIL), 39
Climate Impacts Program, UK, 39
Coefficient of adhesion and the creep, relationship between, 139
Coefficient of friction (CoF), 130
CoF. *See* Consequence of failure (CoF); Coefficient of friction (CoF)
Commission on Climate Change (CCC), 45

Index

Concrete railway sleeper
  creep prediction on, 286–289, 287f
    ACI code, 287–288
    Australian Standard 3600-2009, 288–289, 289t
    Eurocode 2, 286–287
  shrinkage prediction on, 289–291
Conductor unit load actions, 449–451
  ice and wind load, 449–451, 449–450f
  thermal elongation, 451
Consequence of failure (CoF), 350
Contact contaminants, classification of, 140–141, 141f
Contact force measurement, 127, 127f
Contact-induced vibration and noise, 113–114
  friction induced vibration and squeal, 113–114
  impact vibration and noise, 113
Continuous Dynamic Measurement, 244–246
Continuous welded rail (CWR), 97, 243, 295
Conventional railway structure, 271, 272f
Convolutional neural network (CNN), 309
Copernicus Climate Change Service, 50–52
Corrugation, 113
COVID-19
  impact on rail demand and supply, 67–69, 68–69f, 74–77
Creep prediction on concrete railway sleeper, 286–289, 287f
  ACI code, 287–288
  Australian Standard 3600-2009, 288–289, 289t
  Eurocode 2, 286–287
CWR. See Continuously welded rail (CWR)

D

Damage detection, SHM for, 390–393, 391f
  using train induced dynamic responses, strategy for, 398–405
    feature discrimination (outlier analysis), 404–405, 405f
    feature extraction and modeling (double PCA), 400–403, 401f, 403f
    multisensor features fusion, 404, 404f
    overview, 398–399, 399f
Data/signal processing, 308–309
  autonomous maintenance projects, 309
  data process, 308
  deep learning, 308–309
  unsupervised learning methods, 309
DCDT. See Deflected Cantilever Displacement Transducer (DCDT)
DCPPT. See Discrete Cut Panel Pull Test (DCPPT)
Deep learning, 308–309
Deflected Cantilever Displacement Transducer (DCDT), 360, 361–362f
Demand losses, 74–75, 74f
  resilience, increasing, 75–77, 75t, 76–77f
Demand shocks, 66–67, 67t
Demand spikes, 72–74, 72f
DEM. See Discrete element method (DEM)
Differential subgrade settlement, vehicle-track interaction due to, 168–170
  effect on dynamic response, 169–170
  settlement amplitude, influence of, 170, 170f
  settlement wavelength, influence of, 169–170, 170f
  modelling of, 169, 169f
Discrete cohesive zone model, 172–173
Discrete Cut Panel Pull Test (DCPPT), 244–246
Discrete element method (DEM), 99, 298–299, 309–310
Dynamic amplification factor (DAF)
  value, calculation of, 364–374, 366f
    commentary on results, 373–374, 373f, 373–374t
    correlation with combined modulus of elasticity of mortar and masonry, 372, 372f
    correlation with natural frequency of the bridge, 370–371, 371t, 371–372f
    correlation with rise/span ratio, 370, 370f
    correlation with span length, 369, 369f
    correlation with structural parameters of the bridge, 366–368
    correlation with train formulation, 368–369, 368–369f
    statistical approach, 366, 367f, 367t

E

Earthquake, 102–104
  effect on vehicle-track interaction, 176–177
  OHLE under, 431–433, 432f
EHL. See Elastohydrodynamic lubrication (EHL)
Elastohydrodynamic lubrication (EHL), 144
Emergency and preincident management, 91–92
Engineering resilience, 41–42
Eurocode 2
  creep prediction on concrete railway sleeper, 286–287
  shrinkage prediction on concrete railway sleeper, 289–290, 290t
European Union
  Climate Change Adaptation Strategy, 44
Event tree analysis (ETA), 344, 345–348t, 349
EWENT (Extreme Weather impacts on European Networks of Transport), 39
Extensometers, 360–361, 361–362f
Extreme temperature, 97–102, 103t
  hot weather, 97–100, 98–100f, 103t
  low temperature, 100–102, 101f, 103t
Extreme weather conditions
  vehicle-track interaction under, 175–177
    earthquake, effects of, 176–177
    extreme high temperature, effects of, 175
    extreme low temperature, snow and ice, effects of, 175
    extreme wind, effects of, 176
    rainy weather and rail contamination from leaves, effects of, 176
    sand contamination, effect of, 177
  wheel-rail interface under, 141–155
    events and consequences, 141, 142t
    extremely high temperature, 151
    solid particles at wheel-rail contact, 152–155, 153–155f
    sub-zero temperatures, 151–152, 152f
    water at wheel-rail contact, 142–149, 143f

F

Failure mode and effect analysis (FMEA), 345–348t, 349–350
Fasteners, 324, 325f

Fastening/sleeper resistance, 246–247
Fatigue assessment for prestressed concrete sleeper, 278–284
  fatigue properties of materials, 280–283
    concrete, 280–282, 282f
    prestressed steel, 283, 283f, 283t
  structural performance over life cycle, 279–280, 281f
Fatigue life assessment method, 283–284, 284f
Fault tree analysis (FTA), 344, 345–348t, 349–350
FE. See Finite element (FE)
FEM. See Finite element method (FEM)
FFU synthetic sleeper, 258–261
  characteristics of, 258–260, 258–259t
  optimized type A, 258–260
  optimized type B, 260
  results, 260–261
    modified FFU sleepers, lateral resistance of, 261
    shoulder height and width, influence of, 260–261, 261t
  test method and plan, 260
Finite element (FE), 355–356, 361
Finite element method (FEM), 111, 114, 425–427
Finland
  Climate Policy Program, 46
  Finnish Transport Agency, 46
  Ministry of Transport and Communications, 46
First-Order Moment Reliability Method (FORM), 442, 445–446
Flanging
  and contact transition, 120–125, 122–124f
  noise, 113–114
Flooding, 104–107, 106–107f
Fluid-film lubrication, 144–146, 145t, 145f
Fluid viscous damper (FVD), 409–410
  laboratory testing of, 418, 418–419f, 420t
FMs. See Friction modifiers (FMs)
FORM. See First-Order Moment Reliability Method (FORM)
Frequency domain-based statistical analysis of acceleration records, 374–379, 375f
  bridge's natural frequency effect on transmitted accelerations to the bridge, 378–379, 383–384f, 385t

Index 467

span length effect on transmitted accelerations to the bridge, 377, 380–381f, 382t
train crossing speed effect on transmitted accelerations to the bridge, 377, 378f, 379t
train formation effect on transmitted accelerations to the bridge, 374–377, 376f
Frictional instability, at curves, 118–120, 119–121f, 119t
Friction induced vibration and squeal, 113–114
Friction modifiers (FMs), 130
FTA. *See* Fault tree analysis (FTA)
Fuzzy Bayesian network (FBN), 350, 351f
FVD. *See* Fluid viscous damper (FVD)

G

Geophysical tests, 217–220, 217–220f
Global warming, 38f
Grinding, 130

H

HA. *See* Hazard function (HA)
Hazard function (HA), 345–348t
Hazards, rail infrastructure systems and earthquakes, 102–104
extreme temperature, 97–102, 103t
hot weather, 97–100, 98–100f, 103t
low temperature, 100–102, 101f, 103t
flooding, 104–107, 106–107f
HCs. *See* Head cracks (HCs)
Head cracks (HCs), 130
Highways England, 107
Hong Kong
COVID-19 impact on rail demand and supply, 67–68
Hot weather, 97–100, 98–100f, 103t
Humidity, 149–151
Hurricanes
Hurricane Sandy, 47
OHLE under, 433–435, 433f, 434t, 435f

I

Impact vibration and noise, 113
India
COVID-19 impact on rail demand and supply, 67–68

Infrastructure-to-infrastructure communication, 3
Infrastructure-to-vehicle communication, 3
Interdependencies, 42–43b, 43t
Intergovernmental Panel on Climate Change (IPCC), 39
International Union of Railways (UIC) Rail Adapt Project, 41–42
Intervehicles, interactions of, 167–168, 167f
IPCC. *See* Intergovernmental Panel on Climate Change (IPCC)

J

Japan
supply losses, 71

K

Kyoto Protocol, 37

L

Ladder sleeper, 253–254, 254f
Land use characteristics, in rail network accessibility, 13
Lateral Load Evaluation Device (LLED), 246–247, 246f
Lateral resistance
ballast components contribution to, 262, 263–265f
ballast specifications, effect of, 247–250
anchor-reinforced sleeper, 254–256, 255f
ladder sleeper, 253–254, 254f
material properties, 247–248, 247t
particle size, 248
profile influence, 248–250, 249–250t, 249f
sleeper type and shape influence, 250–261
components of, 246f
factors related to, 245f
sleeper, numerical assessment of, 261–262
Leaves, rail contamination from
effect on vehicle-track interaction, 176
Line(s) failure, rail network accessibility under, 13
Link(s) failure, rail network accessibility under, 12
LLED. *See* Lateral Load Evaluation Device (LLED)

Longitudinal nonlinear interactions of slab track, modeling of, 172–173
Low temperature, 100–102, 101f, 103t

M

MA. *See* Markov analysis (MA)
Magnetorheological damper (MRD), 410
Markov analysis (MA), 345–348t, 349
Mean Value First-Order Second Moment Reliability Method (MVFOSM), 442, 445
Mono-block sleeper, 252–253, 253f
MOWE-IT (Management of Weather Events in the Transport System), 42
MPV. *See* Multi-Purpose Vehicle (MPV)
MRD. *See* Magnetorheological damper (MRD)
Multiple station failures
  Shenzhen rail network accessibility under, 31, 32t
Multi-Purpose Vehicle (MPV), 100–101
MVFOSM. *See* Mean Value First-Order Second Moment Reliability Method (MVFOSM)

N

Nailed sleeper, 256–258, 257f
NAP. *See* National Adaptation Program (NAP)
National Adaptation Program (NAP), 45, 57–58
Netherlands, the
  Dutch Delta Program, 46
Network Rail, 52
  Weather Resilience and Climate Change Adaptation 2014–2019 (WRCCA), 46–47
NOAA. *See* United State National Oceanic and Atmospheric Administration (NOAA)

O

OD-based importance-impedance network degradation measures under incidents, 16
OECD. *See* Organization for Economic Cooperation and Development (OECD)
OHLE. *See* Overhead line equipment (OHLE)

Operational resilience, 39
Organization for Economic Cooperation and Development (OECD), 37–38, 44
Overhead line conductor, reliability quantification of, 441–442
  First-Order Moment Reliability Method, 445–446
  load calculations, 447–453
    assumptions and statistical parameters, 452–453, 452t
    conductor unit load actions, 449–451
    tension critical equation in conductor, 447–449, 448f
    torsion critical equation in conductor, 451
  Mean Value First-Order Second Moment Reliability Method, 445
  reliability analysis, 443–447, 443–444f, 444t
    ice and wind load, 453, 454t
    self-weight, 453, 453t
    thermal elongation, 454, 454t
    torsion, 454, 455t
  reliability index, 453–454
  second-order reliability method, 446–447, 447f
  sensitivity analysis, 455–460
    ice and wind load, 456–458, 457–459f
    self-weight, 455–456, 455–457f
    thermal elongation, 458–460, 459–460f
Overhead line equipment (OHLE), 425–427
  under harsh environment, 431–437
    earthquakes, 431–433, 432f
    ground vibrations from trains, 435–437, 436–437f
    wind and hurricanes, 433–435, 433f, 434t, 435f
  maintenance criteria for, 427–428, 427–428f
  vibration characteristics of, 428–430, 429f, 430t, 431f
Oxidation, 149–151

P

Paris Climate Agreement (2015), 37–38
PIANC. *See* World Association for Waterborne Transport Infrastructure (PIANC)

Index

PIARC. *See* World Road Association (PIARC)
PoF. *See* Probability of failure (PoF)
Polygonal wheel under traction condition, vehicle-track interaction due to, 171–175
 longitudinal nonlinear interactions of slab track, modeling of, 172–173
  discrete cohesive zone model, 172–173
  longitudinal resistance model of the fastener, 172
 polygonal wheel wear, modeling of, 171–172, 171*f*
Portal frame mast structure, 426*f*
PRA. *See* Probabilictic risk assessment (PRA)
Prestressed concrete sleepers
 common damages of, 272*t*
 dynamic load, 273–278
  capacities, 274–278, 276–280*f*
  probabilistic dynamic loads on tracks, 273–274, 275*f*
 fatigue assessment for, 278–284
  fatigue life assessment method, 283–284, 284*f*
  properties of materials, 280–283
  structural performance over life cycle, 279–280, 281*f*
 time-dependent behavior of, 285–291
  creep prediction on concrete railway sleeper, 286–289, 287*f*
Prestressed steel, 283, 283*f*, 283*t*
Probabilictic risk assessment (PRA), 83–84
Probability of failure (PoF), 350

R

Rail Adapt Framework, for climate change adaptation, 41–42, 46–57, 48–49*f*, 51*f*, 53*f*
 consequences, 49–52
 hazards, 49–52
 implementation plan, 53–55, 53*f*
 managing and supporting, 55–57
 objectives, defining, 48–49
 option generation and analysis, 53–54
 organization hierarchy, 56*f*
 outcome review, 55
 risk appraisal, 52–53
 strategy, developing, 48–53
 vulnerabilities, 49–52

Rail Infrastructure Systems Engineering Network (RISEN), 1–4
Rail joint, wheel-rail impacts at, 115–117, 116–118*f*
Rail network accessibility
 land use characteristics, 13
 under line(s) failure, 13
 under link(s) failure, 12
 under station(s) failure, 12
Rail seat abrasion, 284–285
Rail system resilience, 70, 70*f*
Rail transport system, model of, 66*f*
Railway bridges using external dampers, improved dynamic resilience of
 bridge-damper systems, response of, 415–418, 416–417*f*
 damper performance, comparison of, 419–421, 420*t*, 421*f*
 equation of motion for a bridge with viscous damper, 411–413
 fluid viscous damper, laboratory testing of, 418, 418–419*f*, 420*t*
 real-time hybrid simulation and testing, 414–415, 414*f*
 structural vibration mitigation, 409–410, 411*f*
Railway bridge under increased traffic demands, 355–357
 aims and scope, 356–357
 analysis of results, 364–379
  DAF value, calculation of, 364–374, 366*f*
  frequency domain-based statistical analysis of acceleration records, 374–379, 375*f*
 geometrical bridges, 359*t*
 monitoring program, 357–363
  sensoring plan, 358*f*
  sensors and instrumentation technique, 360–363
  test train formation, 357–359, 360*f*
 processing data, 363–364, 365*f*
Railway ground-borne vibrations, 209–210
 Carregado test site, experimental characterization of, 210–229
  dynamic characterization of the structure, 220–225
  general description, 210–211, 210–211*f*
  ground characterization, 215–220

469

Railway ground-borne vibrations *(Continued)*
   mitigation measures based on seismic metamaterial (phononic crystal) concept, testing of, 225–227, 226–227*f*
   track characterization, 211–215
   vibrations induced by railway traffic, measurement of, 227–229, 228*f*, 229*t*
   experimental validation, 234–238
     dynamic response of the structure, 236–238, 237*f*
     general description, 234
     track-ground system, dynamic response of, 234–236, 235–236*f*
   numerical modeling of, 230–234
     overview, 230, 230*f*
     soil-structure interaction model, 232–234, 234*f*
     train model and train-track interaction, 231–232, 232*f*
     2.5D FEM-BEM model, 230–231, 231*f*
Railway information models (RIMs), 3
Railway infrastructure resilience, 2–3
Railway stations exposed to terrorist threat, management of, 81–82
   emergency and preincident management, 91–92
   percentage distribution of targets, 82*f*
   previous studies, 83–84, 83*f*
   security risk analysis, 84–87, 85*t*, 86–87*f*
   technology and terrorist threat, 89–90
   terrorism risk, management of, 87–89, 88*f*
Rainy weather, effect on vehicle-track interaction, 176
RBD. *See* Reliability block diagram (RBD)
RCPs. *See* Representative Concentration Pathways (RCPs)
Reliability analysis, 443–447, 443–444*f*, 444*t*
   ice and wind load, 453, 454*t*
   self-weight, 453, 453*t*
   thermal elongation, 454, 454*t*
   torsion, 454, 455*t*
Reliability block diagram (RBD), 349
Reliability index, 453–454
Representative Concentration Pathways (RCPs), 58
Resilience, 4, 44–47
   analysis, 7–8
     Shanghai Metro Network, 25, 25*f*, 27*f*
     shortcomings in, 8
   definition of, 41–42, 65
   engineering, 41–42
   measurement of, 18
   modeling, 16–19
   multinational examples, 44
   national examples, 45–46
   OD-based importance-impedance network degradation measures under incidents, 16
   operational, 39
   practices in Chinese cities, 19–32
   rail system, 70, 70*f*
   railway infrastructure, 2–3
   recovery, 4
   and vulnerability, relationship between, 8–10, 9*f*
RISEN. *See* Rail Infrastructure Systems Engineering Network (RISEN)
Risk appraisal, 52–53
Risk-based maintenance of turnout systems, 341–343
   analysis of, 344–349
   comparative evaluation of, 345–348*t*
   environmental impact consideration into a maintenance chain, 350–351, 351*f*
   establishment of, 349–350, 349*f*
   identification, 343–349
   risk and safety, 343–344
Risk management and vulnerability assessment, relationships between, 87*f*
Risk matrix (RM), 345–348*t*, 349
RM. *See* Risk matrix (RM)
Rockefeller Foundation, 44
Rolling contact fatigue, 113
Rolling-sliding contact, 138–139, 139*f*

S

Sado River railway bridge, 394–398
   description, 394, 394*f*
   monitoring system, 394–395, 395*f*
   numerical modeling and validation, 395–396, 396*f*
   structural conditions, simulation of, 396–398, 397*f*, 399*f*
Safe operating temperature limits, 244*f*
Salt environment, 150–151
Sand contamination
   effect on vehicle-track interaction, 177
Second-order reliability method (SORM), 446–447, 447*f*

Index

Security risk analysis, 84–87
  components of, 86f
  previous terrorist attacks, 85t
Sensitivity analysis, 455–460
  ice and wind load, 456–458, 457–459f
  self-weight, 455–456, 455–457f
Severe wear, 113
SFT. See Stress-free temperature (SFT)
Shanghai Metro Network, 19–26
  accessibility reduction under different disruptions, 24f
  operational incidents and duration time of, 20f
  resilience analysis, 25, 25f, 27f
  vulnerability analysis, 19, 21f, 22–23t
Shared Socio-economic Pathways, 58
Shenzhen rail network, 26–32
  accessibility under individual station failures, measurement of, 16, 30f, 31t
  accessibility under multiple station failures, 31, 32t
  passenger volume distribution, 28f
  urban rail transit, 28f
SHM. See Structural Health Monitoring (SHM)
Shrinkage prediction on concrete railway sleeper, 289–291
  ACI code, 290–291
  Australian Standard 3600-2009, 291
  Eurocode 2, 289–290, 290t
Signals Passes at Danger (SPADs), 139–140
Single Tie Push Test (STPT), 244–246, 248, 252–253, 253f, 256, 256f, 262–266
SITT. See Snow and Ice Treatment Train (SITT)
Sleeper design process, 272
Snow and Ice Treatment Train (SITT), 100–101
Soil-structure interaction model, 232–234, 234f
Solid particles, at wheel-rail contact, 152–155, 153–155f
SORM. See Second-order reliability method (SORM)
SPADs. See Signals Passes at Danger (SPADs)
Station-based rail network accessibility, 10
Station(s) failure, rail network accessibility under, 12
  Shenzhen rail network, 16, 30f, 31t

Steel sleeper, 251, 251f
Stock rail, 319–320
STPT. See Single Tie Push Test (STPT)
Stress-free temperature (SFT), 97, 100
Structural Health Monitoring (SHM), 389–390
  for novel detection, 390–393, 391f
  using train induced dynamic responses, strategy for, 398–405
  Sado River railway bridge, 394–398
    description, 394, 394f
    monitoring system, 394–395, 395f
    numerical modeling and validation, 395–396, 396f
    structural conditions, simulation of, 396–398, 397f, 399f
Supply losses, 70–72, 71f
Surface roughness
  effect on wheel-rail contact, 146–147, 148f
Sustainable Development Goals (2015), 37, 57
Sweden
  COVID-19 impact on rail demand and supply, 68–69, 68–69f
  Swedish Rail Administration, 46
Systems thinking approach, 2

T

Technology, and terrorist threat, 89–90
Tension critical equation, in conductor, 447–449, 448f
Terrorism risk, management of, 87–89, 88f
  consequences, 89, 90f
  threat, 88
  vulnerability, 89
Thames Estuary 2100 Project, 47
Thermal elongation, 451
  reliability analysis, 454, 454t
  sensitivity analysis, 458–460, 459–460f
Time-dependent behavior of prestressed concrete sleepers, 285–291
  creep prediction on concrete railway sleeper, 286–289, 287f
    ACI code, 287–288
    Australian Standard 3600-2009, 288–289, 289t
    Eurocode 2, 286–287
  shrinkage prediction on concrete railway sleeper, 289–291
    ACI code, 290–291

Time-dependent behavior of prestressed concrete sleepers *(Continued)*
  Australian Standard 3600-2009, 291
  Eurocode 2, 289–290, 290*t*
TLPT. *See* Track Lateral Pull Test (TLPT)
TOD. *See* Transit-oriented development (TOD)
Tomorrow's Railway and Climate Change Adaptation (TRaCCA) Project, 42–43, 46–47
Torsion critical equation, in conductor, 451
TRaCCA. *See* Tomorrow's Railway and Climate Change Adaptation (TRaCCA) Project
Track bed, 323–324
Track buckling, 98–99, 98–99*f*
Track Lateral Pull Test (TLPT), 244–246
Track numerical model, 190, 190*f*, 191*t*
Track slab, in-plane vibration of, 168, 168*f*
Track unevenness profile, 191–194, 194–195*f*
Track vibration measurement, 128–129
Train-track coupled dynamics model, 165–168
  intervehicles, interactions of, 167–168, 167*f*
  physical model, 165–167, 165–167*f*
  track slab, in-plane vibration of, 168, 168*f*
Train-track interactions, 161–162
  train-track coupled dynamics model, 165–168
    intervehicles, interactions of, 167–168, 167*f*
    physical model, 165–167, 165–167*f*
    track slab, in-plane vibration of, 168, 168*f*
  vehicle-track coupled dynamics model, 162–165
    vehicle-ballasted track coupled dynamics model, 163–164, 163–164*f*
  vehicle-track interaction due to differential subgrade settlement, 168–170
    effect on dynamic response, 169–170
    modelling of, 169, 169*f*
Transit-oriented development (TOD), 6
Turnout, 319–320
  component definition of, 343
  components of, 320–325, 342*f*
    bearers, 322–323, 322*f*
    driving mechanisms, 324–325
    fasteners, 324, 325*f*
    rails, 320–322, 321*f*
    track bed, 323–324
  detection methods, 330–333
    data-driven detection, 332–333
    nondestructive testing, 330–332, 331–332*f*
  inspection, 325–333, 326*f*
    modes of, 327–330, 328–329*f*
    period of, 327, 327*t*
  layout, 320*f*
  lifecycle cost, 338, 338*f*
  maintenance, 334–337
    crossing, 336–337, 337*f*
    installation/replacement, 334–335, 334*f*
    removal, 334
    repair activities, 336–337
    risk-based (*see* Risk-based maintenance of turnout systems)
    stock and closure rails, 336
    switch blades, 336
2.5D FEM-BEM model, 230–231, 231*f*

U

UIAIA. *See* University of Illinois aggregate image analyzer (UIAIA)
UIC. *See* International Union of Railways (UIC)
UKCP. *See* UK Climate Projections (UKCP)
UNFCCC. *See* United Nations Framework Convention on Climate Change (UNFCCC)
United Kingdom (UK)
  Adaptation Reporting Power, 55–56
  Climate Projections (UKCP), 50–52
  COVID-19 impact on rail demand and supply, 67–68
  Department for Threat (DfT), 84
  Met Office, 39, 50–52
United Nations Framework Convention on Climate Change (UNFCCC), 37
United State National Oceanic and Atmospheric Administration (NOAA), 39
United States
  COVID-19 impact on rail demand and supply, 67–68
  supply losses, 71

Index

United States National Climate Assessment, 46
University of Illinois aggregate image analyzer (UIAIA), 300
Unsupervised learning methods, 309

V

Vehicle-ballasted track coupled dynamics model, 163–164, 163–164f
Vehicle numerical model, 190–191, 192f, 192–193t
Vehicle-track coupled dynamics model, 162–165
   vehicle-ballasted track coupled dynamics model, 163–164, 163–164f
Vehicle-track coupling system, framework of, 190–196
   track numerical model, 190, 190f, 191t
   track unevenness profile, 191–194, 194–195f
   vehicle numerical model, 190–191, 192f, 192–193t
   vehicle-track interaction method, 195–196
   wheel flat geometry, 194–195
Vehicle-track interaction due to differential subgrade settlement, 168–170
   effect on dynamic response, 169–170
      settlement amplitude, influence of, 170, 170f
      settlement wavelength, influence of, 169–170, 170f
   modelling of, 169, 169f
Vehicle-track interaction due to polygonal wheel under traction condition, 171–175
   longitudinal nonlinear interactions of slab track, modeling of, 172–173
      discrete cohesive zone model, 172–173
      longitudinal resistance model of the fastener, 172
   polygonal wheel wear, modeling of, 171–172, 171f
   system dynamic responses, characteristics of, 173–175, 173–174f
Vehicle-track interaction method, 195–196
Vehicle-track interaction under extreme weather conditions, 175–177
   earthquake, effects of, 176–177
   extreme high temperature, effects of, 175
   extreme low temperature, snow and ice, effects of, 175
   extreme wind, effects of, 176
   rainy weather and rail contamination from leaves, effects of, 176
   sand contamination, effect of, 177
Vulnerability
   analysis, 5–7
      accessibility-based rail system vulnerability methodology, 10–15
      practices in Chinese cities, 19–32
      Shanghai Metro Network, 19–26, 22–23t
   and resilience, relationship between, 8–10, 9f
Vulnerability assessment and risk management, relationships between, 87f

W

Water
   as piezo-viscous fluid, 145–146
   at wheel-rail contact, 142–149, 143f
      boundary lubrication, 143–144
      effect of speed, 146, 146t, 147f
      fluid-film lubrication, 144–146, 145f, 145t
      surface roughness, effect of, 146–147, 148f
      temperature effect, 147–149, 149f
Weak spots on track, 111–112
   maintenance of, 129–130
      impact-inducing track sections, 129–130
      large-friction-inducing sections, 130
WEATHER2 (Weather Extremes: Impacts on Transport Systems and Hazards for European Regions), 42
Weather Resilience and Climate Change Adaptation 2014–2019 (WRCCA), 46–47
Weigh-in-motion (WIM) system
   numerical modeling, 187–196
      vehicle-track coupling system, framework of, 190–196

Weigh-in-motion (WIM) system *(Continued)*
    wayside system, 187–190, 189*f*
    static loads
        estimation of, 193*t*, 196–199, 198*f*
        obtaining, 184–187, 186*f*
        wheel loads, statistical evaluation of, 196–198, 197*t*, 197*f*
Wet Rail Phenomenon, 143–144
Wheel defect detection
    flat detection, 199–203
        with envelope spectrum analysis, 185–187, 188*f*
        geometry, 200–201, 200*f*
        noise of the signal, 202–203, 203*f*
        rail unevenness profile, 202, 202*f*
        random position of the defect with respect to the sensors, 201, 201*f*
    numerical modeling, 187–196
        vehicle-track coupling system, framework of, 190–196
        wayside system, 187–190, 189*f*
Wheel flat geometry, 194–195
Wheel loads
    statistical evaluation of, 196–198, 197*t*, 197*f*
Wheel-rail contact, 111, 112*f*
    basics of, 138–141
    contact positions in, 139–140, 140*f*
    solid particles at, 152–155, 153–155*f*
    water at, 142–149, 143*f*
        boundary lubrication, 143–144
        effect of speed, 146, 146*t*, 147*f*
        fluid-film lubrication, 144–146, 145*f*, 145*t*
        surface roughness, effect of, 146–147, 148*f*
        temperature effect, 147–149, 149*f*
Wheel-rail dynamic interaction, 111, 112*f*
    consequences of, 112–114
    contact-induced vibration and noise, 113–114
        friction induced vibration and squeal, 113–114

        impact vibration and noise, 113
    modeling of, 114–126
        case studies, 115–125
        methodology, 114, 115*t*
    -related problems, detection of, 126–129
        ABA measurement, 127–128, 128*f*
        contact force measurement, 127, 127*f*
        track vibration measurement, 128–129
    waves induced by, 125–126, 126*f*
    weak spots of track, maintenance of, 129–130
        impact-inducing track sections, 129–130
        large-friction-inducing sections, 130
Wheel-rail interface, 137
    under extreme conditions, 141–155
        events and consequences, 141, 142*t*
        extremely high temperature, 151
        sub-zero temperatures, 151–152, 152*f*
        water at wheel-rail contact, 142–149, 143*f*
    humidity and oxidation, 149–151
    issues related to, 138*f*
Wheel/rail structure degradation, 113
    corrugation, 113
    rolling contact fatigue, 113
    severe wear, 113
WIM. *See* Weigh-in-motion (WIM) system
Wind, extreme
    effect on vehicle-track interaction, 176
Winged sleeper, 251–252, 252–253*f*
WMO. *See* World Meteorological Organization (WMO)
World Association for Waterborne Transport Infrastructure (PIANC), 45
World Bank
    Adaptation Fund, 57
World Meteorological Organization (WMO), 39
World Road Association (PIARC), 45
WRCCA. *See* Weather Resilience and Climate Change Adaptation 2014–2019 (WRCCA)

Printed in the United States
by Baker & Taylor Publisher Services